TAMING THE RAYS

Second Edition

OTHER PITCHPOLE BOOKS BY THE AUTHOR

Winifred Brown: Britain's Adventure Girl No 1, 2013

Genes, Flies, Bombs and a Better Life: In the Footsteps of Hermann Muller, 2016

TAMING THE RAYS

A HISTORY OF RADIATION AND PROTECTION

Geoff Meggitt

Copyright © 2018 Geoff Meggitt

All rights reserved. No part of this book may be reproduced or transmitted in any form or by any means, electronic and mechanical, including photocopying, recording, or by any information storage and retrieval system, without permission in writing from the Publishers.

ISBN 978-0-9575549-8-6

For Joyce, Ann and Claire

Contents

Table of Contents

1 The Early Years of X-rays...1
2 Radioactivity...26
3 Fission, Weapons and Power..40
4 Natural, Medical and Other Sources..86
5 Interactions and Cell Death...101
6 Dose Concepts and Models...120
7 Genetic Effects..135
8 Somatic Effects...165
9 Measuring External Radiation..212
10 Measuring Internal Activity...249
11 The Evolution of Organisations...263
12 Changing Standards...300
13 Criticality Safety...316
14 Reactor Safety...349
15 Post Script...384
16 References...390
17 Index..436

Figures

Kassabian's laboratory	17
Kassabian's hands	17
Atmospheric and underground weapons tests	58
Dose contributions over time from nuclear testing	59
Collective dose	69
Per caput dose from different sources (UNSCEAR)	95
Exponential survival curves of bacteriophages	103
Survival of mammalian (HeLa) cells after x-ray exposure	104
Effect of dose rate on survival of human melanoma cells	105
Schematic counter tube performance	213
The Victoreen Model 70 [ORAU]	215
Walkie Talkie [ORAU]	220
The Chang and Eng neutron dosemeter [ORAU]	229
The Victoreen Minometer [ORAU]	231
The Lauritsen Dosimeter [ORAU]	231
Ernest Wollan's film badge [ORAU]	234
Components of an early film badge [ORAU]	234
Harwell Criticality Locket (adapted from Majborn 1980)	346
f-N curve for expected early fatalities (WASH-1400 Fig 5-3)	369
Probability consequence diagram Farmer 1967	374
Farmer release criterion	375

Tables

World-wide collective doses from nuclear power	73
Estimates of effects of doubling mutation rate	145
Mendelian disease risks from radiation	154
The background of multifactorial disorders	160
Radiation risk estimates for multifactorial diseases	162
Comparison of T65D and DS86	186
Excess lifetime mortality from all cancers	195
Risk estimates from various sources	204
PSA study results (from Hayns 1999)	377

Acknowledgements

The images of Mihran Kassabian's laboratory and his hands in Figures 1 and 2 are copyright Radiology Centennial Inc. Those of instruments in Chapter 9 are all reproduced with permission from the remarkable online and physical museum of radiation measurement maintained by Dr Paul Frame at Oak Ridge Associated Universities. Other images are credited, where appropriate, in their captions.

The images on the front cover are taken from those in the book or are in the public domain.

My thanks to the many people who helped by providing information or reading parts of the manuscript, notably David Sowby, Alan Jennings, Ron Loosemore, John Dunster, Monty Charles and Kathleen Gregory. Peter Thorne has helped with th chapter on criticality safety.

I am particularly grateful to Roger Jackson who read the entire manuscript and made many suggestions for improvements. Thanks also to the staff of the John Rylands Library at the University of Manchester. Without them all, this would be a poorer effort. However, any errors are entirely my fault.

Preface

Radiation protection has emerged from early years when gruesome injuries crippled or killed many of the early pioneers of x-rays, through attempts—often in secrecy—to understand the potential impact of nuclear weapons to a point where it is built around one of the most complex and transparent protection regimes on the planet. The story is one of science and engineering but also of organisation, politics and ethics. At one end it touches the benevolent possibilities of medicine and nuclear power; at the other the awesome and destructive ones of nuclear weapons (and, some would add, nuclear power too).

The main aim has been to trace, in some sort of synchronism, the development of sources of radiation and the attempts to understand and control the risks to people from them. One way to do this would have been to follow a straightforward chronological approach, tracing the developments right across the field year-by-year. However, this seemed (it was tried) to lose so much of the intellectual structure of the subject so that, with the exception of the early years, the evolution of different aspects (theory, measurement, standards and others) has been followed separately. While this risks fragmenting the subject (although the intent has been to bring the branches together in later chapters) it does, at least, give more flexibility to set some of the areas—such as genetics and the changing ideas of cancer—in a broader context.

Some readers will be disappointed that the work is not more anecdotal. Even if I had wanted to do this, it would have been difficult. There seems to have been little attempt to record the experiences of individuals, particularly in the UK (although those interested in the history of medicine have characteristically done better than most), that would have formed the basis for such an approach. The largest impact of this omission is probably on the coverage of practical health physics, where there would surely be fascinating stories to tell of the evolution of organisations, of working methods and of relationships between health physicists and operators. The availability of such material would have made it even harder to leave out areas like training, monitoring strategies, implementation of ALARA and emergency response than it was in the attempt to stay focussed on the evolution of the central ideas.

A preoccupation of part of the health physics community for several decades—the 1950s to the 1980s—was civil defence, with the aim of limiting the number of early radiation casualties. It was an important episode, with a strong emphasis on the acute effects of radiation, and it required an understanding of these to be won by the study of the bomb survivors, accidents and through experiments on animals. As a result the acute effects are classified, their symptoms recognised and some effective treatment regimes are available. However, the early effects are seldom encountered, occurring only after infrequent accidents, and the risks they pose are usually minimised by good engineering design and proper training. They have been deemed to lie outside the scope of this book.

Medical uses dominated for the first 50 years of the atomic age and remained the single largest man-made source of radiation exposure through the period. However the focus of this book is elsewhere. Partly this is an acknowledgement of the existence of a substantial body of work on the history of medicine—a profession which seems to take its past seriously. Here, after an account of the early years, the therapeutic side of radiation medicine, where the biggest developments have taken place, is ignored: the doses employed are far larger than usually experienced elsewhere and the cost-benefit equation is unique. In no other area are the potential benefits so great for the person exposed and the consequences of under or over exposure so potentially devastating. Diagnostic uses have (or should have) a similarly direct association of costs and benefits for an individual but the doses and perceived risks are lower. So much lower in fact that, after the first decades when diagnostic doses came under proper control, the risks associated with them were for a long time disregarded. This became a concern in perhaps the 1970s and even at the end of the century there were concerns that a large number of diagnostic examinations had a flimsy clinical basis.

So it would have been wrong to ignore the area. Diagnostic radiography has therefore been addressed—but without much clinical detail.

Even given this, medical applications have had a major impact on the developments described in this book, originally through the demands they made on clear definition of measurable quantities and on the measurement techniques themselves, and throughout as the motivation for much of the fundamental radiobiological research described here.

The evolution of standards is an important part of radological protection and it has probably received more attention than any other aspect. It has been approached here from an international perspective since the truly meaningful extension to national situations would need an account of their own interpretations in their own legal systems. These differ enough to make it a fascinating subject of study but a potentially tedious one for many readers so the account of standards here takes little account of national standards except where they contributed—as they did after the Second World War for example—to the formulation of international ones. Since all major nations follow, or aspire to follow however tardily, at least the general ideas of the ICRP, this is not too serious a omission.

Some might see as much more serious the absence of much reference to alternative views of the risks posed by radiation to those of the professional mainstream, the official view. The defence is that they have, right or wrong, had rather little impact on the direction of radiological protection (although they have succeeded in, at least temporarily undermining the credibility of nuclear power in the public mind). They will be found to crop up from time to time in the narrative.

Over the century a wide range of radiation units have been used and they have been applied to several related concepts: exposure, absorbed dose, dose equivalent are just some. While modern units and concepts have been used where an attempt has been made to put things in context over time, the original units and concepts have been used in many places. Translating these into modern terms is not always straightforward anyway. So the reader will find rads and rems scattered around the text without conversions into Grays and Sieverts. Where an activity has been quoted in Curies in an original document, this has generally been reproduced here—with a translation into Becquerels where this seems to help significantly in understanding. The simple conversion factors 100 rad=1 Gy, 100 rem=1 Sv and 1 Ci= 37 GBq will be enough to disentangle any confusion in most cases.

In this second edition the opportunity has been taken to make a number of corrections to the text and the citation system has been updated to fit more closely with modern practice.

Two chapters have been added. The first of these draws on and extends a review I wrote for the *Journal of Radiological Protection* on the history of nuclear criticality safety. The second is a short history of developments in the assessment of the safety of nuclear reactors. Between them they illustrate the breadth and sophistication of our subject.

Geoff Meggitt

June 2018

1 The Early Years of X-rays

Wilhelm Conrad Röntgen was a 50 year old professor at the University of Würzburg in 1895. Expelled from school for refusing to name a fellow student who drew a caricature of a teacher, he had entered academic life by taking an entrance examination to the Zurich Polytechnic School. He became assistant to the Professor of Physics there, August Kundt (best known for his resonant tube experiment from school physics), and went with Kundt to Würzburg. Kundt tried to get Röntgen an academic post there but failed. Röntgen then took a series of posts at Strasbourg, Hohenheim, back to Strasbourg (now as Professor of Theoretical Physics), Giessen University and then, in 1888, back to Würzburg.

He had worked on a variety of topics in physics: specific heat of gases, thermal conductivity of crystals, polarisation of light, electrical characteristics of quartz and compressibility of fluids. He was now Professor of Physics and Director of the Physical Institute at the University of Würzburg[1] and took an interest in the developing field of cathode rays.

Röntgen liked to repeat the experiments of others when he entered a new field and he was operating a Lenard tube, a discharge tube with a thin aluminium window that allowed cathode rays to escape to the outside. He covered the tube with a black card—to shield its fluorescent glow—so he could check some of Lenard's work and noticed that a barium platino-cyanide screen lying nearby became fluorescent. He soon realised that he was dealing not just with cathode rays but with a new phenomenon. This discovery was made on the evening of 8 November 1895 and he worked in secret for the next seven weeks investigating the properties of these new rays with a fluorescent screen and then photographic plates. He found that the rays originated from the fluorescing area of the tube (and later from an aluminium target if he introduced one) and that they travelled several metres through air.

But his most striking discovery was that these x-rays, as he called them, could pass through objects and affected photographic film. He made radiographs of a set of weights, a piece of metal and the bones of his wife's hand.

He handed his first paper to the president of the Physical Medical Society of Würzburg on 28 December 1895 and it was immediately printed and distributed. Röntgen contacted colleagues around the world—Lord Kelvin and Sir Arthur Schuster in the UK—sending them, on 1 January, the paper and radiographs (no copies went to the USA). An English translation was made by Schuster's assistant, Arthur Stanton, and it appeared in Nature on 23 January 1896 with an accompanying letter by A A Campbell Swinton (the Scottish engineer later to promote a television system based on cathode rays) and the radiograph he had taken of a hand on 13 January. On the 18 January the British Medical Journal carried a paper by Schuster announcing the discovery[2].

Most people got to know by another route [3]. The Austrian paper *Neue Frei Presse* published an account on 5 January and this was picked up by the British *Daily Chronicle* which published a short article the following day (spelling Röntgen's name as "Routgen"). The *Manchester Guardian* printed a longer account on 7 January with the London *Evening Standard* publishing on 7 and 8 January. The *New York Times* reported "Professor Routgen's experiment" on 16 January. Röntgen gave a public lecture on 23 January in Würzburg and it started a kind of fever.

When asked by a magazine reporter what he thought when he observed the x-rays, Röntgen gave his celebrated answer: "I did not think; I investigated". He obviously considered this an appropriate comment because he said it again to Sir James Mackenzie Davidson, who visited him soon after the discovery[4].

Others had, previously and unknowingly, generated x-rays. Lenard, in the investigations of cathode rays that led to his receiving the Nobel Prize in 1905, believed he had detected the waves (as he then thought) passing through his hand. This was probably an observation of x-rays and, when he became a leading Nazi, he claimed that he had discovered the rays before Röntgen[5]. Other experimenters had also produced the rays and even experienced some of their effects. For example, Arthur W Goodspeed of the University of Pennsylvania had been photographing sparks and discharges in 1890 and found strange images on the plates. These could later be attributed to the production of x-ray[6].

Since every laboratory of consequence already had the equipment necessary to study cathode rays they could produce x-rays and Röntgen's discovery was quickly confirmed. Campbell Swinton made the first radiograph in the UK on the evening of 7 January with an anode made from platinum sheet. It was poor and he made better ones the following day. The first in the USA was just days later. The first images of "clinical conditions" in the UK were made soon afterwards, probably by John Hall-Edwards in Birmingham[7] on 12 January when he stuck a needle in his associate Ratcliffe's hand and radiographed it. Schuster, in Manchester, radiographed a dancer's foot in February—she had a needle in it—and kept the image on his desk until his death[8].

The communication by Röntgen to Lord Kelvin in Glasgow found its way to John Macintyre, the physician recently appointed to the position of 'Medical Electrician" at Glasgow Royal Infirmary. He had worked with Kelvin, then in his 50th year in the chair of natural philosophy, and the amateur scientist and patron Lord Blythswood, to build the hospital's Electrical Room to administer the then- popular techniques of electrotherapy. He soon reproduced Röntgen's results and before the end of January was giving a demonstration to colleagues.

In 1896 alone some 50 books and pamphlets and nearly 1000 papers were published on x-rays. The term "skiagraph" (shadow picture) was used for a while but "radiograph" was soon widely adopted. The *Archives of Clinical Skiagraphy* were started in May 1896 and its title became *Archives of the Rontgen Ray* in 1897.

The first US x-ray was taken by A E Wright of Yale on 27 January 1896 and Dr. Henry Louis Smith, a Professor of Physics at Davidson College in North Carolina, performed one of the first x-ray experiments in the United States: in February 1896 *The Charlotte Observer* published his x-ray photograph of a bullet in the hand of a cadaver. The excitement provoked by x-rays is illustrated by the actions of three Davidson College students, Osmond L Barringer, Eben Hardie, and E Pender Porter. On the night of January 12, 1896, the three students bribed a janitor to let them into the medical laboratory on campus. After three hours of experimenting, they produced an x-ray photograph of two rifle cartridges, two rings and a pin inside a pillbox. They also radiographed a human finger they had sliced from a cadaver with a pocket-knife.

"We kept our picture and escapade a secret and it was not until later that we realized we were making history for the college instead of just breaking the rules", Barringer wrote some years later.

The extraordinary popular impact of the discovery has been well documented. The fascination with seeing into a living body was to some extent balanced by a concern for a loss of privacy. The latter was perhaps summed up in *Photography* soon after the discovery:

> Thro' cloak and gown—and even stays
>
> Those naughty, naughty Roentgen rays

We were, after all, still in the reign of Queen Victoria [9]

Röntgen published two more papers on x-rays. The first in March 1896 showed that x-rays could make air conduct electricity (although he could not explain this) and reported that x-rays could be produced by anodes of all the substances he tried but that platinum, inclined at 45 degrees, gave the most penetrating rays. In the second, published in 1897, he showed that any material exposed to x-rays would itself emit x-rays and studied the output of the rays from various tubes. He never gave another talk on x-rays and abandoned them in 1900 to return to his work on crystals when he moved to the University of Munich in 1900. He retired in 1920 but continued to work at the University until his death on 10 February 1923. He was awarded the first ever Nobel Prize for Physics in 1901 for his discovery and investigation of x-rays.

In England medical opinion on the potential value of x-rays, sceptical at first, was swayed by lectures, notably one by Silvanus Thompson at the Medical Society of London on 31 March. In February the *British Medical Journal* commissioned Sydney Rowland to investigate and he then produced 13 influential weekly reports between February and June[10].

1 The Early Years of X-rays

Röntgen, in his first paper, speculated that because these new rays shared some characteristics of light (they formed shadows, caused fluorescence and affected photographic plates) they might be related. He suggested that they might be longitudinal (rather than transverse) waves in the ether[11]. The serious possibility that x-rays were electromagnetic was considered early on and Lodge in July 1896[12] was able to draw on theoretical reasons to suggest that they were transverse waves of extremely short wavelength far beyond the ultra-violet. He also thought that "the Becquerel rays", which had subsequently been found, were "a less extreme extension in the same direction"—so he was not right about everything.

The conclusive demonstration that x-rays were electromagnetic waves was not made until 1913. Diffraction gratings were well-known for optical wavelengths. These were composed of very fine opaque lines drawn very close to one another in a regular way. If the repeat of the lines is close to the wavelength of light then light passing through it will show a diffraction pattern. Von Laue suggested, in 1913, that the regular atomic structure of crystals might make a sufficiently fine diffraction grating for x-rays. Friederich, Knipping and Laue placed crystals of various materials in a fine beam of x-rays falling on a photographic plate. They found that, as well as the spot caused by the main beam, there was a pattern of spots from the diffracted radiation that showed conclusively that they were dealing with a wave. The theory and practice of x-ray diffraction from crystals was rapidly developed by the Braggs and this led to a powerful tool for chemists and biochemists for the rest of the century.

By the end of 1896 x-ray tubes had been improved vastly. The 'Focus' tube was developed by Herbert Jackson, fro

m earlier work by Crookes. Its concave cathode directed the cathode rays into a small spot on an target anode set at 45 degrees, giving much sharper pictures. It was established that the best target materials were those of high atomic weight and that platinum was the best practical material (tungsten and uranium were better experimentally but were more difficult to work with). It was generally used in a thin layer on nickel because of cost. Initially those experimenting with x-rays had to construct their own equipment from scratch but the interest was so great that, within just a year of the discovery, the American General Electric company had produced a catalogue of x-ray equipment. Although it was still necessary to assemble the parts yourself.[13]

Both induction coils and Wimshurst machines (or similar electrostatic generators) were used to provide the high voltage for the tubes. The induction coils were better once reliability problems associated with the interrupter were solved. The power supply was generally an integral part of the package because of the unavailability or unreliability of mains power supplies and it was, anyway, two decades before good insulated cables became available[14]. As improved generators appeared, targets of osmium and tantalum were used, and this, with better arrangements for heat dissipation, allowed currents to rise ten fold to 50 mA[15].

The early x-ray tubes depended on the presence of some residual gas (they became called "gas tubes") for the electric discharge to take place: it was the positive ions colliding with the cathode that produced most of the electrons. In operation some of the gas became adsorbed onto the inside of the tube increasing the vacuum. This led to the tube becoming "harder": a higher voltage was required and the x-rays became more penetrating while their intensity decreased. A number of systems for keeping the pressure in the tubes nearly constant were devised. Some involved heating palladium membranes and allowing hydrogen to permeate into x-ray tubes; others used heat to release absorbed gases from mica and charcoal. By 1900 automatic regulation was possible using an auxiliary spark to heat mica and release absorbed gases. However, the popular description of a gas tube as "a glass bulb surrounded by profanity"[16] was appropriate for a while yet. But the industry grew quickly and when the British Röntgen Society held a competition for x-ray tubes in 1901 twenty-eight tubes were submitted; the winner came from Germany.

Improvements in resolution of radiographs came when it was realised that the scattered and secondary x-rays generated in the tissue of the body were a major cause of blurring. In Switzerland, Otto Pasche devised a system to address this as early as 1903[17] but the first practical system was that of the German Gustav Bucky (later after emigration to the USA to become Albert Einstein's doctor and be present at his death) in 1913. Bucky arranged two grids—one between tube and patient and the other between patient and photographic plate—in such a way that only un-scattered rays could reach the plate. Much better resolution resulted but the image was overlaid with an image of the grids. The system was later improved by the American Hollis E Potter who used a single slatted grid that moved around during the exposure, so it left no clear image. The Potter-Bucky diaphragm is still used. A side effect of the system was the possibility of getting larger images. Since large photographic plates were difficult to handle, this encouraged the search for a film-based system[18].

Early radiographs were made onto glass photographic plates coated with emulsion on one side. The emulsion had a habit of slipping off during developing and it was often the job of a junior to wax the edges of the plates to help to keep the emulsion in place. It was quickly established that a thicker emulsion was better than a thinner one and before the end of 1896 double-sided plates were being produced, although they were too expensive for general use. John Carbutt in the USA developed an improved emulsion in 1896 reducing the exposure required. Paper was tried instead of plates in 1897 but it found only limited use because of inferior image quality.

Eastman Kodak also improved emulsions and produced special x-ray film in 1913. This became commercially available in 1918 although it was not widely used, because the images were not of the high-quality produced on plates, until 1923. The film was coated with emulsion on both sides ("Dupli-Tized") and this improved its sensitivity. However, the film base was celluloid nitrate and highly inflammable: when the x-ray film store at Cleveland Clinic caught fire, 129 people died. In 1924 Eastman introduced the cellulose acetate base as safety film.

The intensifying screen was suggested by Campbell Swinton in January 1896 and in April he found that, with a photographic plate in contact with a fluorescent screen, it was possible to get an image of a hand in a few seconds rather than 1 to 2 minutes. The image quality was not very good and he discovered that a finely-powdered screen material gave good resolution but poor response; increasing the powder grain size improved the response but the resolution deteriorated[19]. However, priority should perhaps go to Michael Pupin in Chicago. He was trying to x-ray the hand of a wealthy New Yorker in February 1896. The man had been hit in the hand with shotgun pellets and could not bear to unclench it for long enough to get a good look with a fluorescent screen. Pupin put the screen on a photographic plate and obtained an acceptable image in just a few seconds. However, it was to be a long time before the intensifying screen was developed and used routinely.

TAMING THE RAYS

Röntgen's original discovery led directly to the fluoroscope. In this simple device a fluorescent screen covered the end of a light tight box and it could be observed through an eyepiece. This meant that the shadow pictures could be observed in daylight. It was one of the most used devices in the early days of x-rays with the advantage of simplicity and the immediacy of real-time observation. It was developed as the hand-held Cryptoscope by Professor Enrico Salvioni of Perugia using barium platino-cyanide for the screen, and by February 1896 several UK investigators had built copies[20]. It was improved by Edison, who selected calcium tungstate as the best screen material, and coined the name Fluoroscope. He later characteristically claimed the invention. It was common practice for a radiologist to check a set-up by putting his own hand between the x-ray tube and the screen and checking the image, a practice that was to be the cause of much pain and suffering later on.

There were some remarkable technical achievements early on. For example, Macintyre from Glasgow produced the first x-ray movie—a frog's leg in motion—and showed it to a meeting of the Royal Society in June 1897 [21].

The first major advance in x-ray tube design was the invention of the modern high-vacuum thermionic tube by William David Coolidge in 1913. Coolidge had found a way of making the metal tungsten suitable for use as a filament in electric light bulbs while working for General Electric in the USA. It was used by GE and they produced their first tungsten light bulb in 1910. Coolidge then turned his attention to x-ray tubes and introduced a tungsten filament as the cathode. This, when heated by passing an electric current through it, gave off a controllable supply of electrons. When the vacuum was improved (no ionisation of gas was necessary to create the electrons at the cathode) Coolidge had a reliable tube that allowed the voltage and current to be controlled independently. Radiologists could then accurately control the hardness of the x-rays produced (it depended on the voltage across the tube) and the intensity of the beam (it depended on the current heating the filament).

The Coolidge tube was so effective that the inventor's hair fell out after he tested it (he later used a severed human leg for experiments and lived to be 102). It was quickly taken up in the USA but its adoption in Europe was held up by the First World War so for a decade or more the gas tubes continued to be used.

1 The Early Years of X-rays

X-rays were a novelty: the Kaiser had his arm x-rayed, the British Prime Minister Lord Salisbury and his wife had their hands done in 1896 and the Portuguese queen, Emilia, required her ladies in waiting to have their chests radiographed to show the damage caused by tight corsets[22]. However, there were many serious uses in a remarkably short time as two examples from the USA show.

The first US medical radiograph was on 3 February 1896 by Edwin Frost an astronomer at Dartmouth College, New Hampshire. A boy who had injured his wrist was seen by Dr. Gilman Duboi Frost, Edwin's brother. Edwin was asked to make the radiograph and produced the first image of a Colles fracture. A photograph exists of the moment of exposure.

Over the Christmas period of 1895, Tolman Cummings was shot in the leg in a Montreal bar by one George Hodder. The local hospital was unable to find the bullet but an x-ray taken at McGill University (later Rutherford's home for a while) in early February 1896 clearly showed it and was used by surgeons to get it out. The bullet and x-ray were produced in court and Hodder was convicted and sentenced to 14 years. This was thus probably the first time x-rays were used as evidence in a court case anywhere. The first dental x-rays were taken in Germany in January 1896—needing a 25 minute exposure. C Edmund Kells took the first in the USA in New Orleans in April [23].

As 1896 progressed there were almost endless radiographs of bone fractures and of needles and bullets embedded in bodies with increasing clinical value. But there were also important, more experimental developments. Radiographs were obtained of tumours and, even in 1896, radio-opaque substances were being used to visualise organs. Within a very short time Cannon in the USA was using pearl buttons and bismuth mixed with food to show the gastro-intestinal tract and the Austrian Hashek had injected a metal-salt solution into an amputated hand and arm and produced the first angiogram. Soon a liquid contrast medium was introduced into the stomach as the bismuth meal (later the less toxic barium sulphate was used). The opaque meal to diagnose ulcers and cancers of the stomach and duodenum was developed in Vienna in 1904 by Reider and popularised in the UK by workers such as A E Barclay and Sebastian Gilbert Scott

Traumatic injuries of war were quick to receive the benefits of diagnostic x-rays. The first use of x-rays for examining military casualties was by the Italians in their Abyssinian War of 1896[24]. The British Army's first radiographs were made at the now-derelict Royal Victoria Hospital at Netley near Southampton in November 1896 and by mid-1898 sets were installed at Aldershot, Woolwich, Dublin and Gibraltar and portable sets had been ordered for field use.

The first British involvement in combat use was during the Graeco-Turkish war of 1897 when the British Red Cross, financially supported by an appeal by the *Daily Chronicle*, sent two hospital units to help the Greeks. They were accompanied by an "absolutely complete" x-ray outfit. The team was led by F C Abbott, a surgeon from St Thomas's Hospital in London and the man in charge of radiography was Robert Fox Symons (later Sir Robert). Over a period of six weeks, they treated 114 war casualties and took some 50 or 60 radiographs. It was a clear success in showing that good radiographs could be taken in forward hospitals. A number of problems were experienced in transporting the equipment safely but the major difficulty was in obtaining power; they depended on the warship *HMS Rodney* to recharge their accumulators[25]. A rival German Red Cross Society team took an x-ray set to support the Turks and this was similarly a success (except for problems with accumulators).

The first use with British troops on the battlefield was probably in 1897 when an "apparatus" was sent to the North West Frontier, then the border between India and Afghanistan and now part of Pakistan. An army of 100,000 was in the field to put down a rebellion of the local tribesmen and about 40,000 of these were in the Tirah Expeditionary force on the Tirah plateau. Here there were 23 field hospitals and, because of the appalling problems of transporting wounded, they were undertaking a great deal of surgery. Several hundred x-rays were taken by the regimental surgeon of the Coldstream Guards, W C Beevor using a prototype apparatus developed by A E Dean of Hatton Garden, London. This, with its three Cossor x-ray tubes, came in a collection of wooden cases carried, suspended on poles, over the rough terrain by Indian bearers. His problems were associated with batteries and the tendency for the emulsion of his Eastman x-ray papers to melt. A memorable image came from his work: that of the bullet fragment lodged in the leg of General Wodehouse. The general's wound had been probed under intense fire from the tribesmen and his stoicism had been praised in the British newspaper. When the wound had failed to heal after several weeks, one of Beevor's x-rays showed part of a bullet had been left in his calf muscles. On his return Beevor's report played a part in promoting x-rays as an important tool in military surgery [26].

The expedition of General (later Lord) Kitchener was sent to repossess the Sudan and avenge the death in 1885 of Gordon in Khartoum. The expedition was eventually supplied with x-ray equipment but only after cross words had been exchanged in the House of Commons. After this, x-rays were firmly established as a part of military medicine.

The effects of radiation on the skin—hair loss and burns—suggested that it might have some value in treating skin conditions and other ones which were resistant to then-current treatments[27]. After all, ultra-violet radiation was already in use for some of these. So, there were some early rather crude attempts to use the new x-rays as therapeutic tools. At the end of January 1896, Grubbé tried to treat an advanced breast cancer and lupus vulgaris, Voigt in Hamburg treated cancer of the nasopharynx in February and Despeignes tackled cancer of the stomach in July [28]. In December 1896 Leopold Freund made the first considered use of x-rays as a therapeutic tool when he irradiated a large hairy birthmark (hirsuites) covering the entire back of a 4 year-old girl. The treatments were spread over more than 10 irradiations each lasting 2 hours and the growth was removed but the patient suffered episodes of serious ulceration over many years. However, she was examined in 1956 (she walked in off the street) and found to be well but with some damage to the lower back and again 15 years later when she was 75.

In 1896 the physicist Michael Pupin in Chicago made what seems to have been first therapeutic application of x-rays in the USA for cancer in treating a woman with breast cancer[29]. Pupin had no medical background and is perhaps now best known for his work on long-distance telephony. By the early 1900s radiotherapy was being used quite extensively and successfully (and perhaps often unwisely) for non-malignant skin conditions conditions such as tinea capitis, acne vulgaris, eczema, lupus, skin tuberculosis[30] as well as for skin, breast and other cancers. Many who did used the technique saw it as a form of cauterisation rather than anything more sophisticated but during the 1910s it became rather clearer that it was indeed skin conditions that might benefit most with the equipment and techniques then available.

Within six months of Röntgen's announcement several of the leading hospitals in London had set up permanent x-ray units. Some were based on existing electrotherapy departments.

Possibly the first in Britain was set up by the surgeon Thomas Moore at the Miller Hospital in Greenwich after he made early (March 1896) radiographs with the scientist William Webster. Moore was Treasurer of the Röntgen Society and one of the first editors of *Archives of the Roentgen Ray* and when he died in 1900 his radiography was taken up by John Jewell Vezey, amateur scientist and another Röntgen Society man, who worked unpaid until he died at work in 1906. The apparatus installed in Vezey's time was based on a 12 inch coil with mercury interrupter and could be used on the wards with accumulators. Webster, a founder member of the Röntgen Society, had the first reported incidence in the UK of radiation "sunburn" on his right hand[31].

The London Hospital extended its Electrotherapeutic Department to include an x-ray service. The Department head, W S Hedley, took on Ernest Harnack as an assistant and they built up the x-ray service, constructing some of the equipment themselves. Their work covered a wide range of conditions but they were probably most useful, in the early years, with their radiographs of fractures and of needles stuck in hands—an occupational hazard in the tailoring industry centred around the hospital.

The Department grew over the next few years and in 1903 it split into two. One part was concerned with the treatment of skin lesions by electrical treatment and x-rays, with x-rays being used for rodent ulcers and ringworm. The other part dealt with the diagnosis and treatment of other conditions using x-rays. This second department was headed by E Reginald Morton, an electrotherapeutist who focussed on x-ray therapeutics as a natural extension of his electrical work. A further reorganisation in 1909 led to the appointment of S G Scott as the Department head and he became the hospital's first full-time radiologist with a primary interest in diagnosis.

In 1905 an article in *Archives of the Röntgen Ray* described two treatment rooms: one for skin tumours and diseases and one for other tumours. Some protection measures were in place: the x-ray tubes were enclosed in lead glass to protect the operator and the doses to the patient were measured by time and by pastille methods. Patients were protected from soft x-rays by lint soaked in calcium

1 The Early Years of X-rays

tungstate and then dried[32]. However, ringworm, a condition prevalent at the time and resistant to other treatment, was treated by irradiation of the scalp in a set of three barber chairs located side by side. Within 20-30 years many of the patients has developed radiation-induced rodent ulcers of the scalp. The method continued to be used until 1960.

The other major London hospitals developed along similar lines. Barts extended its Electrical Department to include x-ray diagnosis and skin condition treatment in 1896 and dealt with over 200 patients in its first year (rising to over 1000 in its sixth).

St Thomas's had the distinction of being the location of the first radiograph taken in a London hospital at a demonstration on 13 February 1896. The equipment was improved (a Jackson focus tube was quickly introduced giving much shorter exposure times) and used operationally from then on with a Department being set up in October 1896. In 1897 over 400 patients were examined. By 1914 St Thomas's was the leading hospital in radiology in the UK.

Guy's Hospital was slightly later. The first x-rays were attempted in April 1897 but it was probably a year or two before the x-ray service became a permanent feature with the appointment of two radiographers in 1899. By 1904 there was a clear division between electrical treatment and x-rays.

The explosion of interest in the rays meant that exploitation far outstripped understanding and when potential hazards were seen they were widely ignored. So, the absence of shielding around the early x-ray tubes resulted in considerable injury to the operators and the problem was compounded by the common practice of operators looking at their own hand with a fluorescent screen to test the apparatus.

Early warnings of the potential hazards came within a few weeks of Röntgen's discovery. In the UK, L R L Bowen, in a talk to the London Camera Club on 12 March 1896—reported in the *Lancet* – warned that x-rays might produce effects like sunburn. In April L G Stevens reported in the *British Medical Journal* that people exposed to x-rays suffered sunburn and dermatitis. Warnings also came from the USA: in March Thomas Edison reported sore eyes after extended exposure [33] and William J Morton saw burns[34]. Dermatitis and hair loss were reported by J Daniel of Vanderbilt University after he taken a one hour exposure of a man's skull.

An example from America illustrates some of the cavalier attitudes to x-rays. In the summer of 1896 Herbert Hawks was demonstrating x-rays in Bloomingdale Brothers' Store in New York. Hawks, an assistant to Dr. Pupin at Columbia University, experienced radiation burns and received an unusual diagnosis.

> Mr. Hawks, during the afternoon and evening of each day for four days, was working around his apparatus for from 2-3 hours at a time. At the end of the four days, he was compelled to cease active work, owing to the physical effects of the x-rays upon his body. The first thing Mr. Hawks noticed was a drying of the skin, to which he paid no attention, but after a while it became so painful it was necessary to stop all operations. The hands began to swell and assumed the appearance of a very deep sunburn. At the end of two weeks the skin all came off the hands. The knuckles were especially affected, they being the sorest part of the hand. Among other effects were the following: the growth of the fingernails was stopped and the hair on the skin that was exposed to the rays all dropped out, especially on the face and sides of the head. The chest was also burned through the clothing, the burn resembling sunburn. Mr. Hawks' disabilities were such that he was compelled to suspend work for two weeks. He consulted physicians, who treated the case as one of parboiling. [35]

The response to the revelations of these effects was varied, indeed Hawks thought his experience was probably largely due to electrical effects[36]. Others suggested that such effects came from the electric sparks in the high-voltage generator, from ultra-violet(uv) radiation, from chemicals used in developing plates, from ozone generation in the skin and from faulty technique[37]. The lingering doubt that radiation was the cause of injuries should have been eliminated when, in November, the American physicist Elihu Thomson purposely exposed the little finger of his left hand for half an hour close to an x-ray tube. Over a period of a week or two the finger became swollen, sensitive and painful. He was convinced that the effects were caused by the "chemical activity" of the rays and issued a caution. (One of his recommendations was "Do not expose more than one finger") [38].

Thomson's report prompted a number of others to publish their experiences of x-ray burns and the eminent UK physician, Sir Joseph

1 The Early Years of X-rays

Lister said in his presidential address to the British Association for the Advancement of Science in September 1896:

> It is found that if the skin is long exposed to their action it becomes very much irritated, affected with a sort of aggravated sunburn. This suggests the idea that a transmission of the rays through the human body may be not altogether a matter of indifference to internal organs... [39]

However, in January 1897 John Hall-Edwards, the man who probably made the first clinical x-ray in the UK and had exposed himself for hours each day for a year, could write:

> We have heard so much about the effect of the x-rays upon the skin; this I think must be due to some idiosyncrasy of the operators, for although I have myself been experimenting daily for the last eleven months I have failed to notice anything of the kind. [40]

Hall-Edwards had a long interest in photography and microphotography and was already an Honorary Fellow of the Royal Photographic Society before working with x-rays. By 1899 his opinions on x-rays were changing:

> Continued and protracted exposure to the rays at varying distances from the tubes has an effect upon the hands which although unpleasant is not dangerous. It interferes with the growth and nutrition of the nails. The skin round the roots of these become red, irritable and cracked, and the nails themselves thin and brittle. Most constant workers suffer in this way.[41]

By 1902 he was developing painful sores and warts and a photograph of "chronic dermatitis" printed in the *Archives of the Röntgen Ray* (of which he became editor in 1903) was probably of his own hands. At the annual meeting of the British Medical Association, held in Oxford in 1904, Hall-Edwards described his condition and this was subsequently published in the *British Medical Journal*. In this illustrated article about his own conditions he strongly urged young workers to take every possible precaution before it became too late. The pain was "as if bones were being gnawed away by rats". By 1906 his left arm was useless and carried in a sling. In 1908, when cancerous growths were found, his left arm was amputated just below the elbow and the fingers of his right hand were removed. He advised caution and went on to become a Birmingham City Councillor before dying in 1926.

Other cases developed. For example, at the pioneering London Hospital, Harnack had three assistants, Reginald Blackall, Ernest Wilson and Harold Suggars. By 1903 they all had radiation injuries. Wilson took a series of photographs of his hands showing progressive bony damage leading to malignancy and died in 1911. Harnack ultimately had both hands amputated. Suggars and Blackall worked for longer and helped to establish the College of Radiographers.

In the USA the cases of Clarence Dally and Mihran Kassabian received wide publicity and had great impact. Dally had joined Edison after he left the US Navy where he had been a Chief Gunner's Mate ("a little fellow, but a specimen of perfect manhood" according to his surgeon). He worked with Edison from 1896 and was responsible for testing tubes and assisting Edison with his x-ray development. For the tube testing he often placed his hand between the fluoroscope and the tube. His hair soon began to fall out, his face wrinkled and his hands developed dermatitis. The skin condition worsened over several years, leading to failure of the blood vessels in his left arm, and a cancerous condition developed. By 1901 it was necessary to amputate his left arm and in 1903 fingers were removed from his right hand. The right arm was later amputated and Edison supported him until Dally's death in October 1904. The experience caused Edison to give up work on radiation. He abandoned work on a fluorescent lamp based on radioactive material and said: "I could make the lamp all right, but when I did so I found that it would kill everybody who would use it continuously" [42].

Mihran Kassabian, through his work as a photographer and his interest in electrotherapy became the "skiagrapher" and instructor in electrotherapeutics at the Medico-Chirurgical College and Hospital inPhiladelphia. By combining his work as a skiagrapher with that of electrotherapist, he was able to confer the status of a clinical department on radiology. Over a two-year period he examined more than 3000 patients and exposed more than 800 radiographs and became director of the Roentgen Ray Laboratory at the Philadelphia Hospital in 1903. He developed x-ray "burns" by April 1900 and, by 1908, he had a malignancy on his left hand. Kassabian described the progress of the illness(and the amputations he underwent) at professional meetings and in his 1907 textbook *Roentgen Rays and Electrotherapeutics*.

1 The Early Years of X-rays

He died in 1910 of radiation-induced cancer. His exposure of the risks of x-rays was another important factor in making workers take protection more seriously.

Figure 1: Kassabian's laboratory

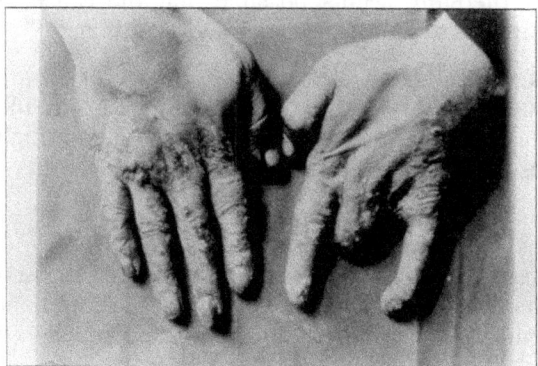

Figure 2: Kassabian's hands

These were just two of the many x-ray workers who suffered delayed but appalling injuries from x-rays. In 1936 a memorial was erected in Hamburg to the early pioneers of x-rays who suffered radiation injury or lost their lives due to their work.. Of the original 169 names from 15 nations, 14 are from Britain. The British names are Reginald Blackall, Barry Blacken, John Hall-Edwards, Cecil Lyster, Stanley Melville, Hugh Walsham, John Chisholm Williams and Ernest Wilson, William Ironside Bruce, William Hope Fowler, J W L Spence, Dawson Turner, James Riddell and G A Pirie. The last six were Scottish.

The citation reads:

> They were heroic pioneers for a safe and successful application of x-rays to medicine. The fame of their deeds is immortal.

The British Institute of Radiology compiled documents to support the inclusion of thirty-four additional British doctors and nurses. Most of these died of the complications of skin cancer, a few from aplastic anaemia or leukaemia[43].

Marie Curie died of a "pernicious aplastic anaemia" on 4 July 1934 and it was immediately, and for a long time, assumed that this was a result of exposure to radium. However, when her ashes were reburied, with those of her husband, in the Pantheon in Paris in 1995, the measurement made by French radiation protection experts showed that the levels of radiation associated with them were quite low. It therefore seems possible that the cause of her death was exposure to x-rays during the First World War [44].

For several years after Röntgen's discovery injuries were usually seen as temporary and superficial but, by about 1905 most workers were taking some precautions. With the increasingly powerful x-ray set-ups available, it became even more important.

Since the common practice of checking a set-up by placing the hand in front of the fluorescent screen was probably the single biggest cause of injury, the invention of the Chiroscope in 1903 must have made some difference. This was a skeleton hand with simulated flesh mounted behind a fluorescent screen. The Osteoscope was a similar device using a complete forearm [45].

The human arm, Chiroscope and Osteoscope were possibly adequate ways of checking apparatus for diagnostic use but there was obviously a need for a more quantitative measurement of tube output. One of the earliest methods for measuring this was based on the the comparison between the brightness of a fluorescent screen produced by x-rays and that of some standard source. Basing his system on a method where the standard source was an acetylene lamp, Guilleminot, in 1907, used as the standard a fluorescent screen irradiated with a radium source of known strength[46]. His unit was the "M". Butcher, in 1908, used a similar method and expressed his results in "radion" or "radio-lux" units in analogy with visible sources.

While the primitive methods of the arm and Chiroscope might have been adequate for diagnostic applications, a quantitative method was also needed for controlling the dose received by patients in therapy—where an over-exposure could have very serious consequences.

The dose delivered in therapy was measured by several means but pastilles that changed colour and film strips that blackened were developed from about 1902. The pastilles were calibrated against a standard epilation dose. One, the pastille of Sabouraud and Noiré using a barium platinocyanide compound, was available in booklets. It simply changed colour (or rather tint) at the epilation dose and this was called the B or pastille dose. Lower doses could be measured by placing the pastille closer to the tube than the organ being irradiated and applying the inverse-square law. Another system, prepared by Holzknecht in 1902 (and based on a "secret" recipe involving potassium chloride and sodium sulphate), allowed easy comparison between the pastille tint and a graduated standard and was based on the H unit chosen so that the epilation dose was 5H. The original calibrations were made with unfiltered radiation but adjustments for epilation doses were not too large with filtered radiation[47]. The pastilles survived as measurement systems for a long time; the Sabouraud and Noiré pastille devised in 1904 was still in use in the 1930's.

Strips of film were also used as the dose meter. Kienböck in 1905 used strips of silver bromide paper, later known as Kienböck strips, which were placed on the patient's skin during irradiation. After development in a standard way they were compared with a greyness scale in an instrument called a quantimeter[48] calibrated in x units, where 10x corresponded to the epilation dose. The strips had the disadvantage of requiring processing but they did provide a permanent record of exposure.

Ionisation as a measure of radiation was considered early on and both gold-leaf electroscopes and ionisation chambers were suggested as measurement devices. In fact, Paul Villard, the discoverer of gamma-rays, suggested a unit based on the charge liberated by radiation as early as 1908. However, the technology that allowed routine, reliable measurements of ionisation was not to become available for some years and most practitioners came to rely on a system like the pastilles. This led to a wide range of dose units and little scope for intercomparison. Mould has listed more than 50 units that were suggested or used up to 1937[49].

Measuring the penetration of the x-rays was important to diagnosis and therapy. Initially, using the Fluoroscope, the classification of radiation hardness was based directly on tissue and bone penetration but penetrameters soon became available. The earliest was actually made by Röntgen in 1897 but credit is generally given to Benoist in 1902 who used the same principle for his radiochromometer. This was a standard thin silver disc surrounded by aluminium foils of increasing thickness. It was placed behind a fluorescent screen and the brightness behind the silver disc was compared with that behind the aluminium foils. When the two brightnesses were the same, the thickness of that aluminium foil gave a measure of the penetrating power. Other instruments based on the same idea were developed elsewhere [50].

Shielding of the tube was unusual before about 1908 but some practitioners were careful throughout. Francis H Williams of Boston can be seen in the Mould article with a protective box around the tube in a 1902 photograph and he remarked later that he thought penetrating rays like x-rays must have "some effect upon the system" and took precautions accordingly. Williams's early caution came from his brother-in-law and collaborator, the remarkable William Rollins.

Rollins, a Boston dentist, stood for what we would now call a precautionary approach to x-ray safety. One of his experiments, reported in February 1901, resolved the debate addressed by Elihu Thomson's exposure of his fingertip. Instead of sacrificing a body part, Rollins put a guinea pig in a Faraday chamber—a set of electrically-earthed boxes that excluded any electric fields—and exposed it to an x-ray source outside the box. The exposure lasted two hours per day and, after 11 days, the guinea pig died. A second died, after similar treatment, after 8 days.

It led Rollins to propose three precautions:

- physicians should wear glasses that keep out x-rays when using fluoroscopes

1 The Early Years of X-rays

- x-ray tubes should be kept in shielded boxes with a small window to give a cone of radiation no larger than needed
- patients should be shielded except where necessary for examination or treatment.

Just two weeks later he reported another disturbing result. This time he placed a pregnant guinea pig in the chamber and irradiated it: the foetus it was carrying died. It was not unusual for physicians to examine pregnant women with x-rays to check the size of their pelvis and the condition of the foetus and Rollins cautioned against this. It should have been clear that these effects were a result of x-ray exposure and not some obscure electrical effect but not everyone was convinced and it was suggested that the guinea pigs had either suffocated or developed some infection. Rollins had in fact disposed of these objections by having retained a set of control animals, who showed no ill effects. One of Rollins's more alarming observations was—or rather should have been—that the guinea pigs that died showed no radiation burns—possibly because the softer x-rays were filtered out by the surrounding boxes. It was an early indication that there were hidden effects, much more serious than transient skin burns or cosmetic changes[51].

Rollins did not simply exhort radiologists to take more care but devised a number of practical ways for them to protect themselves and their patients. The enclosure of the x-ray tube already mentioned was improved in two ways over the next year or two. He invented a shutter that could be opened remotely and an adjustable rectangular collimator that could be used to achieve the smallest usable area of illumination of the patient. He suggested improvements to intensifying screens and tried to discourage the practice of "warming up" the tube while the patient was exposed to the beam. He also, probably, proposed the first numerical protection standard when he suggested that the test for adequate shielding of a tube should be to place a photographic plate on the outside. If the plate was not fogged after a 7 minute exposure, then the protection was sufficient.[52]

Rollins was a man ahead of his time. His impact was limited in the USA because the growing x-ray community was not disposed to accept that the astonishing new rays might have a serious downside that might limit their spread. He results and proposals may, anyway, not have been widely known outside the Boston area[53] and they certainly did not cross the Atlantic.

Protective wear, as an alternative or supplement to shielding, appeared as early as 1898 when Price in the USA proposed the use of lead-rubber gloves. By 1905 the Crusader-like Friedlander full protective suit (with apron, gloves, hoods and spectacles) was on the market in the USA for $30[54]. Other similar suits were available. Gloves and aprons became routine in Britain in about 1905. In 1908 Hall-Edwards published a list of 10 rules including: shielding of the tube with just a small aperture opposite the patient, shielding of the operator by a moveable panel, keeping your distance from the tube and using an opaque apron and lead glass spectacles when viewing a fluorescent screen. He emphasized that the effect of x-ray exposure was cumulative [55].

Even at the London Hospital, where the radiography team had already suffered the effects of radiation, protective measures were slow arriving. However, by 1908, the practice of checking with the hand had been stopped, tubes were enclosed in lead-lined boxes and, in 1909, shielded cubicles were installed for therapy.

The skin was clearly the organ most damaged by x-rays and some practitioners made use of the filter devised by George E Pfahler in 1905 [56] to protect themselves and patients. This simple disc of leather removed the less penetrating rays that damaged the skin but allowed the more penetrating ones, that produced the radiograph, through. This was not all good news: some therapeutic irradiations had been limited by skin reactions and when these were reduced much higher doses could be delivered to deeper tissues [57].

So, by perhaps 1910, the dangers of acute and disastrous tissue damage were widely recognised and there were some straightforward protection measures being adopted. The means of measuring larger doses were available and were used for control of patient exposures. Together these things could, if sensibly applied, reduce and perhaps eliminate the dreadful acute effects—and within a few years professional bodies would step in with recommendations on protection to do just this. However, many of the early workers were to die because of their injuries and even more were to suffer and die from unsuspected long-term effects that were a long way from being understood.

1 The Early Years of X-rays

For a more detailed account of the history of radiology see the the "History of Radiology" by Adrian Thomas and Arpan Banerjee.[58]

Notes Chapter 1

1 (Bowers B, 1970)
2 (Burrows E H, 1986)
3 (Bowers B, 1970)
4 (Kaye G W C, 1926)
5 (Bowers B, 1970)
6 (Mould R F, 1993)
7 ibid
8 (Burrows E H, 1986)
9 (Lentle B and Aldrich J, 1997)
10 (Burrows E H, 1986)
11 (Rontgen W C, 1896)
12 (Lodge O, 1896)
13 For a re-creation of an early x-ray set up and a study of its performance see (Kemerinck M et al, 2011)
14 (Kevles B H, 1997)
15 (Burrows E H, 1986)
16 (Kevles B H, 1997)
17 (Cardwell D S L, 1972)
18 ibid
19 (Anon, 1896b)
20 (Burrows E H, 1986)
21 ibid
22 (Kevles B H, 1997)
23 ibid
24 (Burrows E H, 1986)
25 ibid
26 ibid
27 (Thomas A M K, Isherwood I and Wells P N T(eds), 1995)
28 (Mould R F, 1993)
29 (Calder J F, 2001)
30 (Coppes-Zantinga A R and Coppes M J, 1998)
31 (Burrows E H, 1986)
32 (Hope-Stone H F, 1999)
33 (Anon, 1896a)
34 (Coppes-Zantinga A R and Coppes M J, 1998)
35 ibid
36 (Brecher R and Brecher E, 1969)
37 (Glasser O, 1933)
38 ibid

39 ibid
40 (Guy J, 1995)
41 ibid
42 (*New York World*, 1903)
43 (Guy J, 1995)
44 (Butler B, 1995)
45 (Mould R F, 1995)
46 (Mould R F, 1993)
47 (Schall W E, 1932)
48 (Mould R F, 1995)
49 ibid
50 ibid
51 (Brecher R and Brecher E, 1969)
52 ibid
53 His principal work "Notes on the X-Light" was privately published in Boston in 1904
54 (Kevles B H, 1997)
55 (Burrows E H, 1986)
56 (Pfahler G E, 1906)
57 (Coppes-Zantinga A R and Coppes M J, 1998)
58 (Thomas A M K and Banerjee A K, 2013)

---

2 Radioactivity

Henri Becquerel was third in a dynasty of four Becquerels who became professors at the Muséum d'Histoire Naturelle in the Jardin des Plantes in Paris. His grandfather, Antoine-Cesar, held the post from 1838 to 1878, when he died. Henri's father Edmond then took over the chair until his death in 1891 and was then followed by Henri.

One of the interests of Antoine-Cesar and Edmond had been luminescence. Antoine-Cesar had been prompted to study the subject, with his son, by observing the phosphorescence of the sea on a journey to Venice in 1835. Edmond became an established expert, publishing his authoritative *La Lumiere, ses causes et ses effets* in 1867/8, and the museum had an extensive collection of phosphorescent salts as a result.

This background played some part in Henri's momentous discovery.

One of the people Röntgen had communicated his results to in January 1896 was Henri Poincaré who presented them, with the radiographs, to a session of the Academie des Sciences on 20 January 1896. After the presentation Henri Becquerel asked Poincaré where in the tube the x-rays originated and, when told that they came from the area of the tube that fluoresced, immediately thought that the x-rays might have been caused by the luminescence. He began experimenting the next day.[1]

He soon retrieved some crystals of uranium salts he had lent to someone else: they were phosphorescent when exposed to sunlight and could therefore be expected, if he was right, to give off x-rays. To test this Becquerel wrapped a photographic plate in black paper, put

He repeated the experiment a few days later but there was not much sun so he put the experiment in a dark drawer. He left it there for a several days because it was cloudy and then decided, anyway, to process the plate, expecting to see very little because there could have been very little phosphorescence to cause x-rays. In fact, when he processed the plate on 1 March, he found a very clear image. He initially thought there was some kind of stored phosphorescence but over the next few weeks he established that the radiation that exposed the plate came equally from other uranium compounds that were not luminescent. He also found that the radiation discharged electrified bodies—just as x-rays did—and reported this on 9 March. Since uranium metal gave very strong images he concluded that the radiation ("une phosphorescence invisible") originated from the element itself. He reported this on 18 May 1896. He went on to conduct a series of experiments between November 1896 and May 1897 on the ability of the rays to discharge electrified bodies[2].

There was no public reaction to this discovery at all comparable to the response to Röntgen's work and there were rather few papers following it up (indeed by 1898 Becquerel himself had turned to other things). This was, as Badash[3] has pointed out, an heroic age of radiations and Becquerel rays, as they were initially known, were nothing special. Uranium was not widely available and the pictures produced had a poor quality compared with those made with x-rays.

At the end of 1897 a young Polish woman living in Paris with her husband and daughter was looking for a subject for her doctorate. Marie Curie had two degrees, a fellowship and a monograph on the magnetisation of tempered steel to show for her 30 years. She and her husband Pierre (already well-known for his discovery of piezo-electricity and his work on magnetism) became interested in the work published by Henri Becquerel. As far as they knew, no-one was following it up and it was therefore an ideal starting point for some doctoral work[4]. Her first step was to study the ionisation caused by the Becquerel rays using an experimental set-up devised by Pierre and his brother Jacques for their researches in piezoelectricity. It allowed her to measure the minute voltages involved with good accuracy by balancing them with the voltage from a weighted piezoelectric crystal[5]. She confirmed that the ionisation was proportional to the amount of uranium, was not dependent on the chemical state of the uranium and was independent of factors like temperature. She then began a systematic investigation of all the known elements to see if uranium was unique and discovered that there was one other element, thorium, with the same property. (The activity of thorium was found independently by Gerhardt Schmidt, a Professor at Munster University, in March 1898). She coined the term "radioactivity" to describe it.

She then embarked on the measurement of a large number of mineral samples from the collection at the Ecole Superieure de Chimie Industrielle (later the Ecole Superieure de Physique et de Chemie) and confirmed that all those that did not contain either uranium or thorium were inactive. However, she made the dramatic discovery that some of the unrefined mineral samples were more active than could be explained by their uranium or thorium content. The ore pitchblende was nearly four times more active than the uranium oxide extracted from it. Her earlier work and this discovery, all resulting from the most meticulous work, convinced her that she was dealing with a new element. This was announced, as a probability, on 12 April 1898. To convert the probability into a certainty she would need to isolate the element from the pitchblende and she knew that it could be there only in minute quantities [6].

Pierre dropped his own research to join her in this quest and they began a series of chemical separations to find the new element. Over the next few months they found that there were probably two elements but by July they were able to announce the discovery of "polonium", a radioactive metal with similar chemical properties to bismuth. After several months further work they (with their collaborator G Bemont) announced the second element, "radium", with chemical properties similar to barium but with unprecedented radioactivity[7] nearly 1000 times that of pure uranium[8].

Starting with a tonne of pitchblende residue given to them by the Austrian government they began a four-year toil in primitive conditions to separate the radium. The first stages involved precipitating out the active components until just barium and radium remained. The next stage was much more difficult because barium and radium are quite similar—and there was such a tiny proportion of radium. The method used involved dissolving the mixture in hot water, in 20 kg batches, and then stirring it before allowing it to cool. The radium and barium separated out somewhat into different crystals and by repeating the process—fractional crystallisation—many times purer and purer radium was obtained. In 1902 Marie, who had taken on the bulk of the separation task, finally had about a tenth of a gram of pure radium compound and could determine its atomic weight as 225

Other radioactive elements were soon found. André Debierne, working at the Sorbonne, but in close touch with the Curies, discovered actinium in 1900. Following an observation by R B Owens at McGill University in 1899 that the activity of thorium was affected by draughts and that activity was created nearby, F E Dorn discovered the new gaseous element radon (initially called radium emanation or niton) in 1900.

In 1902 Rutherford and Soddy realised that radioactivity was a random process that resulted in transmutation of atoms. They had been experimenting with the chemical separation of thorium nitrate. Instead of the complete separation they expected into an active precipitate and inactive solution they found that both components were active. Indeed the solution was initially more active than the precipitate and they concluded that they had found a new radioactive material they called thoriumX (ThX). They left the preparations over the Christmas period of 1901 and, on returning to them, were surprised to find that the solution had lost its activity and the solid was as active as it had been before separation. They concluded that Th produced ThX which gave rise to a gas, thoron (and this gas gave rise to induced activity). They found the characteristic exponential decay in activity and reached their conclusions that random transmutation was taking place in this series and in the production of UX from U (found by Crookes in 1900) and radon from radium.

Otto Hahn identified first RadioTh between ThX and Tn and then, once he had joined Rutherford, ThA, ThB and ThC.

In 1899 Ernest Rutherford identified two different kinds of rays and called them alpha and beta. The alpha rays were stopped by a thin aluminium foil; the beta rays were more penetrating. He found that the two rays could be separated by a magnetic field because they had opposite charges—although alpha-rays were difficult to deflect. In 1900 the Curies confirmed that beta particles carried negative charge.

In 1903, Rutherford managed to deflect alpha particles with a strong magnetic field and measured their velocity and e/m ratio. The velocity was 2.5×10^9 cm/s and e/m was close to a modern value. In 1904, Rutherford concluded that alpha particles were helium ions.

Becquerel demonstrated the deviation of beta particles by magnetic fields and Kaufmann measured velocity and e/m in 1902. He found the velocities involved to be very close to the speed of light and that e/m decreased as the velocity increased. The e/m values were very similar to those measured for cathode rays—establishing that they were probably the same particle—the electron.

The velocity distribution of beta particles was measured much later, in 1913, by Rutherford and Robinson.

Paul Villard, working in Paris in 1900 discovered a third kind of ray from radium. He was interested in the chemical action of cathode rays and was somewhat discredited by his conclusion that, because he found them to be a strong reducing agent, they were hydrogen. He studied the colouring of glass by x-rays and decided it was an oxidation process (we now know it is due to the creation of defects in the glass). It was while working with radium that he struck on the penetrating rays which could not be deflected by a magnetic field. In the second of his two papers to the Academie des Sciences in April 1900 he identified them as x-rays. Becquerel was not impressed: if the rays were there, he said, they would not have escaped him and the Curies. However, he was soon forced to admit their existence and made a grudging acknowledgement of Villard's role in *Nature* in 1901. Villard, perhaps disappointed by the reaction to his discovery, gave up work on the gamma rays (as they were named by Rutherford) and concentrated on cathode rays and x-rays [9]. Gamma rays were finally shown to be very high-energy electromagnetic radiation by Rutherford and Andrade.

The three types of rays were initially separated by their different abilities to penetrate matter and the details of this were investigated by several workers over the next decade.

For example, studies of alpha particles by Bragg and Kleeman in 1905 showed that they stopped ionising the air after they had travelled about 35 mm and that this was also the distance at which the particles lost their power to produce luminescence and affect photographic plates. They found that the density of the ionisation caused by the alpha particle gradually increased as it was slowed down and then suddenly dropped as the alpha reached the end of its range. Rutherford measured the velocity when ionisation stopped as

1.12x 10^9 cm/s. They also showed that the stopping power was proportional to the square root of the molecular weight. Later, in 1910, Geiger was to show that the range of alphas is proportional to the cube of the velocity of emission[10]. Rutherford and Richardson measured the mass absorption coefficients of gamma rays in aluminium in 1913.

The fluorescence of various substances when exposed to radiation was spotted quite early: in 1904 Marie Curie saw the fluorescence of diamond near to radium. Crookes exploited the fluorescence of zinc sulphide, caused when it is struck by alpha particles, in his spinthariscope in 1903. This was a small screen of zinc sulphide with a speck of radium behind it. The scintillations caused when the alpha particles struck the screen were seen through a magnifying glass. This was originally more of an amusement but still was a clear indicator of the particle nature of the alpha-particle. Subsequently the principle was used in serious research, first by observers counting the scintillations but later (as we will see in Chapter 9) it was automated.

Although there continued to be some scepticism about the existence of atoms, a number of workers speculated on how they might be constructed. The first person to suggest in print that they might contain the newly-discovered electron was Lord Kelvin. In 1903 he proposed that each atom was a sphere of positive charge with enough electrons (or "electrions" as he called them) embedded in it to make it neutral. This was taken up by J J Thomson in 1904[11] who attempted to analyse the stable arrangements of the electrons (which he was still calling "corpuscles") to obtain the observed spectra with no success. However, in 1906 he was able to conclude that the number of electrons was equal to the atomic number. Nagaoka suggested a planetary structure in 1904 with the interesting idea that radioactivity was the result of the flying apart of the atom.

One of the key instruments at the end of the century was the Wilson Cloud Chamber invented by C T R Wilson. In this, alpha particles from radium were allowed to pass through a chamber of air saturated with water. When the air was cooled by rapid expansion, the ions produced along the alpha particle tracks became centres for condensation, making the tracks visible. In 1897 Wilson found that, while the track was generally very nearly a straight line, just occasionally it would show a sharp deflection. Ernest Rutherford found similar behaviour in 1906 when alpha-particles passed through a thin metal foil and this was confirmed by his co-workers.

Rutherford proposed the nuclear model of the atom, based on these observations, in 1911 and this was confirmed by the alpha-particle scattering experiments of Geiger and Marsden in 1913[12]. It was possible to calculate the positive charge on the nucleus and this was found to be equal to about half the atomic weight, in agreement with the work of C G Barkla from his study of x-ray scattering[13].

A problem with this model was its stability: the electron should lose energy as radiation as it circles the atom and plunge into the nucleus. In 1913 Niels Bohr showed that, if the Rutherford model was adopted and combined with the quantum ideas developed by Max Planck and Albert Einstein, then the stability problem was removed. It was also possible to predict emission spectra.

There was an understanding that radioactive decay resulted in changed chemical behaviour. In 1913 van der Broek concluded that the nucleus did not contain just positive units of charge but both positive charge and negatively charged electrons. It was the net positive charge that gave rise to the chemical properties. It also meant that the atomic weight did not have to be twice the number of positive units of charge (as Rutherford had speculated in 1911).

Later in 1913, Soddy[14] considered the implications of this, given the detailed knowledge of the natural radioactive decay chains then available, and showed that there must be atoms with the same net nuclear charge but with different atomic weights. He called these "isotopes". A similar conclusion was reached by Kasimir Fajans at about the same time.

Also in 1913 Henry Moseley, working in Manchester, was measuring the spectra of wavelengths of x-rays emitted from solids when they were struck by cathode rays. He published his results in two papers in 1913 and 1914. The spectra were very simple and characteristic for each target element. The groups of spectral lines, known as the K,L,M ... series corresponded precisely in structure and the wavelength of corresponding lines was simply related between elements with an integer, characteristic of the element, that Moseley called the atomic number, Z. He identified it with the nuclear charge of van der Broek and Rutherford. The discovery of Z allowed chemists to finally understand the pattern of the periodic table (Moseley predicted three unknown elements with Z 43, 61 and 75 that were eventually found as Technetium in 1937, Promethium in 1941 and Rhenium in 1925). It strongly supported Rutherford and Bohr's nuclear atom[15].

So, by 1914 much of the basic structure of the atom had been mapped out.

The possibility of radium being used for therapy[16] is often said to have been suggested by radiation burns accidentally received by Becquerel in 1901 when he carried some radium in his shirt pocket. However, Mould [17] has unearthed evidence that, in fact, the burns were the result of planned experiments carried out by Becquerel and Pierre Curie. These experiments followed earlier work by Walkoff and Geisel in Germany in 1900. They involved first exposing Curie's left forearm to a radium source for 10 hours and then Becquerel placing a source in his jacket pocket for several hours. In both cases a skin burn was seen which repaired over a period of weeks with some scarring remaining[18].

The burns were recognised as similar to ones from x-rays by a dermatologist, Besnier, seen by Becquerel and the possibility of using radium in therapy was suggested. The Curies then loaned some radium to a colleague of Besnier, Henri Danlos, at the Hopital St Louis, Paris for experiments[19]. This was used by Danlos in 1901 to treat a lupus case and, later, by Louis Wickham and Paul Degrais to treat lupus and other skin conditions with considerable success, leading Becquerel to remark in his 1903 Nobel Laureate speech that radium was being used for therapy[20].

The availability of radium improved when industrial scale production began in Paris in 1902, moving to the Armet de Lisle factory at Nogent-sur-Marne in 1904. Also, the Austrian Government set up a factory at Jachymov. By 1914 around 12 gm of radium had been separated. Later, richer ores—with 10 times the yield of the Austrian pitchblende—were found in Colorado and by 1922 most of the world's radium was being produced in the USA. However by 1925 some 95% of the supply was being produced by Radium Belge, from ore from Katanga in the (then) Belgian Congo with separation at Oolen near Antwerp[21].

The success of the Paris work led to the setting up of the Laboratoire Biologique de Radium there in July 1906. It was funded and supplied with radium by de Lisle and by 1909 some 900 patients had been treated. Louis Wickham and Paul Degrais published their 1909 book on therapy *Radiumthérapie*... with an English translation *Radiumtherapy* following in 1910[22]. The book and the subsequent Paris work established radium therapy as an effective treatment and systematised the techniques involved. There had been other work outside France in the meantime.

At the London Hospital, a radium applicator had been developed by Dr Gustave Adler to treat tumours. To treat a tumour on the face, the base of the device was fixed to the forehead and a double ball-and-socket joint allowed the radium to be positioned over the tumour. It allowed the patient to move around and something similar was still being used in 1980 [23]. Dr James Sequeira, the first person to treat a rodent ulcer with x-rays at the Hospital, acquired some radium and undertook some treatments. He reported in a Brussels lecture in 1908 that it had not proved useful except for small malignant ulcers and tumours of the eyelid and inside facial cavities.

Early work in the USA was undertaken by William Rollins, the x-ray and radiation protection pioneer. He provided a capsule of radium salts to his collaborator F H Williams in 1900 for a trial on lupus—a very early and possibly the first one. By 1902 Rollins, a dedicated innovator, was suggesting the use of radium to treat lupus and cancer of the skin and he envisaged two ways of using it. It could either be used in capsules pressed against the skin or mixed with rubber or celluloid to make radioactive plasters. These plasters— coated on the inside with aluminium foil to remove soft rays and on the outside with lead foil for protection—could be made available on prescription for self-application. The plasters would, Rollins thought, save the poor the cost of frequent trips for treatment in hospitals[24].

Williams visited Europe in the summer of 1903 and was impressed by the radium therapy work in London and Vienna. In his report back he remarked on the convenience of the absence of complex machinery. He could see specific applications in the treatment of cancers in places inaccessible to x-rays, such a facial cavities.

Interest in radium therapy in the USA increased for one or two years, partly through the reports of European work and partly through the publicity received when the Curies were awarded the Nobel Prize in that year. Brecher and Brecher[25] list nearly 20 pioneers. However, over the next ten years radiotherapy continued to grow through the wider use of x-rays. A few (Robert Abbé, H A Kelly and Williams) continued to work with radium but as Brecher and Brecher comment: "At the end of the gas tube era, North America was lagging behind Europe in the use of radium for therapy".

By the outbreak of the First World War institutes had been set up—or were being set up—in France, the USA, the UK and Germany[26].

The diagnostic uses of radium were more limited. The radium rays were used by Marie Curie and others to create radiographs (she put one in her doctorate thesis in 1904[27]) but it was clear that for normal medical uses they were much inferior to those made with x-rays (a direct comparison between a radiumgraph and an x-ray radiograph can be seen in Mould's paper[28]. Radiumgraphs were an attractive idea because they could be produced without the elaborate equipment needed for x-rays but their poor quality meant that work on medical diagnostic applications ceased about 1904.

Marie Curie had discovered that rays from radium caused diamonds to fluoresce soon after its discovery. Compared with x-rays, radium gave rather few immediate dramatic effects and this effect was one of them. It seemed to offer some practical applications because while real diamonds fluoresced, imitation ones did not. Perhaps the most thorough examination was made by George F Kunz, the Tiffany and Company expert, and Charles Baskerville (who had earlier claimed two new elements that proved to be phantoms) in the USA in 1903. They looked at all 14,000 specimens in the collection of the American Museum of Natural History, for their response to uv, x-rays and radium rays. Their work attracted wide public and scientific interest but led to no better understanding of radioactivity, radiation or diamonds. Baskerville went on in 1904 to claim that thorium was actually a combination of two further elements: carolinium and berzelium[29].

The unit of measurement for radium activity that was established in 1904 continued to be used into the 1920's. It expressed the strength as the ratio between the activity of the radium and that of the same mass of uranium. Strengths of sources used in medical applications varied between 10,000 and 2,000,000. As an example of how it was used, Kassabian, in 1907, said that "the most powerful and pure radium manufactured is 10 mg of radium bromide with a strength of 1,800,000" [30].

The use of pastilles for measuring radiation was discussed in Chapter 1. Other methods, based on measuring ionisation currents (one of the basic scientific techniques used to investigate radioactivity since Mme Curie's initial work) were suggested for both x-ray and radium radiation measurement. In 1906 Belot proposed that the ionising power of the rays might be used in radium measurement with better precision and in 1907 C E S Phillips made a similar suggestion for measuring x-rays and radiation from radioactive elements. However, credit is usually given to Villard, the discoverer of the gamma-ray, who, in 1908, defined the unit as "the quantity of X-radiation which liberates by ionisation one electrostatic unit per cubic centimetre of air under normal conditions of temperature and pressure". The unit was not much used for many years—probably because ionisation measurement was only suitable for laboratory conditions—but was adopted as the roentgen (r) in 1928. By then instruments had been developed that could make the ionisation measurements in a practical environment. Mould[31] has explained some of the other ionisation-based units defined between 1908 and 1928.

In expressing doses to patients in brachytherapy the milligram-hour (the mass of pure radium salt in milligrams multiplied by the time of contact with the tissue in hours) was widely used after 1909 and gave a straightforward and fairly reliable way of measuring radium dose long before this could be done with x-rays. By 1912 it was modified to refer to the mass of pure radium.

Although Curie and Becquerel had experienced burns when they kept samples of radium about their person as described earlier, there are few accounts of permanent damage caused by radium in the first two decades after its discovery. The material was quickly recognised to be valuable so that it was stored carefully and the inverse square law must have given a high degree of protection. It was also not as widely available as x-rays. Certainly Marie Curie suffered unpleasant consequences to her hands (she always wore gloves to hide them) but there is some evidence that her time operating an x-ray set during the First World War was much more damaging physically and may have caused her death.

Colwell and Russ[32] in 1933 give a few examples of early radium injuries. Ordway in 1915 gave some examples of effects suffered by workers handling radium applicators. They were unpleasant but not, compared with what had been seen for x-rays, very serious. The first effect was a blunting of sensitivity of the fingers to touch but an increased sensitivity to heat "especially by some who also handled freshly boiled eggs at table". Thickening of the skin, a loss of elasticity and cracking of the nails followed. The skin also became more easily damaged: when one worker scorched the skin of his hand with a match, a large area of the skin peeled away.

G S Willis worked with x-rays from 1905 to 1917 and from 1912 he also handled radium salts in glass tubes "freely and with no attempt at protection" several times a day : the amounts were significant fractions of a gram. In 1918 he noticed a numbness in his fingertips and a weakness in his left arm. A tenderness and then a soreness developed in the fingers and then fissures, one so painful that it had to be treated—in a bizarre hair-of-the-dog antidote—with radium. The fissures worsened and ulcerated and when one of them was excised malignant tissue was found.

A few years later Thomas and Wakeley reported another case of a man who had worked with radium in increasing amounts since 1904. The skin on his hands were discoloured and by 1920 he had developed numerous warts: one of them was found to be a squamous cell carcinoma.

There was plenty of evidence of how dangerous radium could be from patients: while there were remarkable cures or remissions there were also some terrible injuries from ignorance, carelessness and misjudgement. They could be regarded as normal risks associated with the necessary large doses while for workers the effects for a long time seemed to be limited to the hands. However, by the early 1930s evidence of the more sinister consequences began to gather from radium as well as x-rays—which were now recognised as being similar in their effects.

In 1933 Colwell and Russ could point to 16 cases of leukaemia among x-ray and radium workers, with just two of them in radium workers. They recognised that this total was probably higher than it should have been but concluded that the question of whether radiation actually caused them was "sub judice".

TAMING THE RAYS

The protection measures used in those early years of radium are not as well documented as those for x-rays. That they were basic we can surmise from the recommendations on protection against radium published in 1921 by the X-ray and Radium Protection Committee[33]:

> The following protective measures are recommended for the handling of quantities of radium up to one gram:
>
> In order to avoid injury to the fingers the radium, whether in the form of applicators of radium salt or in the form of emanation tubes, should be always manipulated with forceps or similar instruments and it should be carried from place to place in long-handled boxes lined on all sides with 1 cm. of lead.
>
> In order to avoid penetrating rays of radium all manipulations should be carried out as rapidly as possible and the operator should not remain in the vicinity of radium for longer than is necessary.
>
> The radium which is not in use should be stored in an enclosure the wall thickness of which should be equivalent to not less than 8 cm. of lead.
>
> In the handling of emanation all manipulations should, as far as possible, be carried out during its relatively inactive state. In manipulations where emanation is likely to come into direct contact with the fingers thin rubber gloves should be worn. The escape of emanation should be very carefully guarded against and the room in which it is prepared should be provided with an exhaust electric fan.

Notes Chapter 2

1. (Allisy A, 1996)
2. (Allisy A, 1996)
3. (Badash L, 1978)
4. (Curie E, 1938)
5. (Brown G I, 2002)
6. (Curie E, 1938)
7. ibid
8. (Brown G I, 2002)
9. (Gerward L and Rassat A V, 2000)
10. (Starling S G and Woodall A J, 1953)
11. (Thomson J J, 1904)
12. (Geiger H and Marsden E, 1913)
13. (Findlay A and Williams T, 1965)
14. (Soddy F, 1913)
15. (Findlay A and Williams T, 1965), (Moseley H G J, 1913)
16. Mould has discussed priorities in R F Mould, Curr Oncology 2007, 14 (3), 118-122, Priority for radium therapy of benign conditions and cancer
17. (Mould R F, 1998)
18. (Becquerel H and Curie P, 1901). A translation into English is given in (Mould R F, 1998)
19. (Thomas A M K, Isherwood I and Wells P N T(eds), 1995)
20. (Brown G I, 2002)
21. (*Radium, its production and therapeutic applications*, 1925)
22. (Mould R F, 1993)
23. (Hope-Stone H F, 1999)
24. (Brecher R and Brecher E, 1969)
25. ibid
26. (Brown G I, 2002)
27. (Gerward L and Rassat A V, 2000)
28. (Mould R F, 1998)
29. (Badash L, 1979)
30. (Mould R F, 1993)
31. (Mould R F, 1993)
32. (Colwell H A and Russ S, 1934)
33. (X-ray and Radium Protection Committee, 1915)

3 Fission, Weapons and Power

In 1932 Cockcroft and Walton split the beryllium nucleus with artificially accelerated protons and James Chadwick discovered the neutron. In 1934 the Joliot-Curies showed that stable elements could be made radioactive by bombarding them with alpha- particles. Their success led Enrico Fermi, also in 1934, to test the results of bombarding elements with neutrons. In a series of experiments that covered most of the periodic table he found by chemical analysis that when most elements were bombarded they were changed into elements with atomic numbers one or two lower (or in a few cases they stayed the same). However, when uranium was irradiated with neutrons, a different effect emerged and the elements produced could not be identified with any of those with atomic numbers slightly less than uranium. Instead, Fermi speculated that the products might in fact be heavier elements. In his Nobel Lecture on 12 December 1938 he gave the proposed names for the two transuranic elements as ausenium (Z=93) and hesperium (Z=94) but, almost as he was speaking, an alternative explanation was emerging that would lead to weapons of almost unimaginable destructive power and to a promise of unlimited energy supplies.

The explanation was prompted by the work on uranium bombarded by neutrons by Otto Hahn and Fritz Strassmann working in Germany in 1938. Hahn had been encouraged to follow-up on Fermi's investigation of the subject by his long-time collaborator in Berlin, the physicist Lise Meitner. Hahn and Strassmann showed that neutrons striking the uranium produced an element that followed

3 Fission, Weapons and Power

barium in a chemical separation just as radium did (it was part of the Curies separation procedure). They were working with extremely small quantities and initially assumed that it was radium. Meitner, an Austrian Jew, had emigrated to Sweden in 1938 following the German annexation of Austria, but she and Hahn had corresponded on his results and they met secretly to discuss them in Copenhagen in November 1938. After the discussion, Hahn returned to Germany and after much inspired and delicate chemical separation work concluded, by the end of the year, that the mystery element that behaved like barium probably was barium. The tentative results were published in January 1939 and confirmed a few weeks later with the addition of strontium and yttrium as other products. So, instead of Fermi's transuranics with atomic weights greater than uranium, atoms with atomic weight around half that of uranium were produced.

Hahn was a chemist and was unable to underpin his results with a physical explanation. Meitner had been the physicist in the collaborations and she, with her nephew and fellow refugee Otto Frisch, interpreted the results: the uranium atom had split (or fissioned) into two roughly equal halves. Such an event, they estimated, would lead to the release of large quantities of energy, many millions of times larger than that produced in chemical reactions. Meitner and Frisch are usually credited with the fission explanation but it had been suggested to Fermi's team as a possibility by the German chemist Ida Noddack, the co-discoverer of the element rhenium, in 1932. Her suggestion was ignored.

The news from Hahn and Meitner was carried to the USA by Niels Bohr. He arrived on 16 January 1939 with a plan to keep the news secret, to protect priority, but it leaked out through his travelling companion Rosenfeld (because of a misunderstanding) and was soon being discussed by physicists there. Several confirmatory experiments were started and Bohr was obliged to make an announcement at a conference in Washington on 27 January. By the middle of February, fission was confirmed by experiments in Copenhagen, France (Frédéric Joliot-Curie) and four US laboratories. There were sensational articles in the press and, by the end of the year, more than 100 papers on fission appeared in the scientific literature. However, even before this, another possibility was being discussed: Fermi's suggestion that neutrons might be emitted when fission took place leading to the possibility of a chain reaction.

The possibility of such a chain reaction had been conceived by Leo Szilard in 1934 and he had filed a patent for it (the Patent was transferred to the British Admiralty in 1936 to prevent disclosure). When he heard about fission he quickly recognised the possibility for a fission chain reaction and that this depended on whether, as seemed likely, neutrons were emitted on fission and if there were enough of them to sustain the chain reaction. By early March Szilard and Zinn had shown that neutrons were indeed released. This was encouraging but there was at the same time one problem. Bohr had suggested on theoretical grounds that U-235 was the likely candidate for fission and by March this was confirmed experimentally in the USA. The wonder isotope formed only 0.7% of uranium and this scarcity seemed a serious obstacle to achieving criticality.

Another key element in reaching criticality was moderation. This characteristic of hydrogenous materials to quickly lower the energy of neutrons to thermal energies (where they were more likely to cause fission) had been discovered by Fermi in 1934. Halban and Frisch, in 1937, found that heavy water had a much lower neutron absorption cross-section than "normal" water, so that neutrons were not lost from the system as in normal water and in 1939 the Paris team realised that heavy water could therefore be used as an efficient moderator. In 1939 Szilard and Fermi showed that it was possible to achieve neutron multiplication—evidence of the chain reaction taking place—with a uranium lattice in very pure graphite.

Szilard and others, recognising the potential for a weapon, attempted to keep secret the information that neutrons were emitted in fission. Both he and Fermi had submitted papers on the topic to *Physical Review* but asked that they should not be published until it was clear that the consequences would not be harmful. They contacted British and French scientists working on the topic proposing they did the same. The British were prepared to go along with this but the team at the Collège de France in Paris (Frédéric Joliot-Curie, Hans von Halban and Lew Kowarski) responded differently. They were probably leading the European field and were focussed on the idea of a reactor rather than a bomb. They decided to publish, perhaps influenced by the appearance of a report in the press that fission neutrons had been found by a team at the Carnegie Institution of Washington. By April the US and French teams had all

3 Fission, Weapons and Power

published papers, the French one in *Nature* on 22 April reported the average number of neutrons per fission, rather optimistically, as between 3 and 4.

The first estimates of the critical mass necessary before a chain reaction could be sustained came from the Paris group where, in the summer of 1939, Francis Perrin estimated that this corresponded for fast neutrons to a ball of uranium oxide about 3 m in diameter with a mass of about forty tonnes[1]. Rudolf Peierls in Birmingham later extended Perrin's calculations. Otto Frisch, in *Annual Reports on Chemistry* in early 1940, thought that even with uranium enriched by ten times a chain reaction would not be fast enough to cause an explosion. He concluded: "our progressing knowledge of the fission process has tended to dissipate these fears (of fission bombs) and there are now a number of strong arguments to the effect that the construction of such a super-bomb would be, if not impossible, then at least prohibitively expensive...". James Chadwick, the discoverer of the neutron, had just commissioned the first British cyclotron and he was consulted on the likelihood of a bomb. He thought that an explosion was likely if enough uranium were available; between one and forty tons would be required. The uncertainty was down to the lack of good data on cross-sections.

In February 1940 the prospect of nuclear weapons changed dramatically[2]. Frisch and Peierls made some computations, with the available data, of the critical mass of pure U-235. They guessed the fission cross-section (rather too large a value in fact) and obtained the astonishing result that the critical mass was about 600 gm. Peierls calculated the staggering power of the explosion that might result. In March they put their thinking in a Memorandum to Sir Henry Tizard. He, although sceptical, set up a committee under George Thomson that first met in April 1940 (without Peierls or Frisch who were regarded as aliens) and they quickly began to commission work to establish the feasibility of separating the uranium isotopes although there seems to have been a view that the possible weapon's power would be around 100 tons of TNT. Although impressive, this would not change the course of the war.

The committee became known as the MAUD committee through a celebrated confusion. When Meitner returned to Stockholm after a visit to Bohr in Copenhagen in April she sent a telegram at his request to Bohr's friend Owen Richardson, a British physicist. It read "Met Niels and Margrethe recently both well but unhappy about events please inform Cockroft and Maud Ray Kent". Intended to reassure friends in Britain that Bohr and his wife were well but troubled by the German occupation (which had taken place just after Meitner arrived in Denmark), the telegram caused consternation. The three words "Maud Ray Kent" were taken to be some sort of code: Chadwick thought they meant "uranium taken" and others pointed out they were an anagram of the exhortation"Make Ur Day Nt". It turned out that Maud Ray had been the governess of the Bohr children, that she lived in Kent and that Bohr wanted to let her know that the family was safe. However, when a codename was sought for the committee, MAUD seemed to offer a particularly opaque one—although people partly in the know speculated that it stood for "Military Applications of Uranium Disintegration".

One of the MAUD committee's concerns was the fate of the heavy water, the world's entire stock of 180 kg, held in the cellars of the College de France in Paris. This had been recognised by the French team as a possible moderator for their "boiler" reactor and spirited out of Norway in March 1940 by plane to Scotland and then by train and boat to France. It now seemed vulnerable as the Germans advanced through France and in June it was brought back to Britain to be stored first at Wormwood Scrubs prison and then in the cellars of Windsor Castle. It was accompanied by Halban and Kowarski.

Frisch and Peierls had devoted a section of their memorandum to health effects. They estimated that, one day after the explosion, the radiation level would correspond to that from 100 tons of uranium and would decay about inversely with time. It would be spread by the wind and kill everyone within a strip several miles long. Rain would make the situation worse. Houses would offer only marginal protection. Tunnels and cellars would provide protection against radiation provided uncontaminated air could be supplied. Since the effects of radiation would be delayed by some hours, it would be necessary to have an organisation which determined where the danger area was. In May 1940 Cockcroft was warned of the dangers of radioactive dust by Mayneord at the Institute of Cancer Research.

However, much more important concerns were establishing the theoretical basis for a weapon and developing an isotope separation process. Frisch and Peierls (who both remained involved in the project and eventually joined a MAUD technical committee) and

3 Fission, Weapons and Power

Chadwick (at Liverpool) tackled the theoretical work while Francis Simon at the Clarendon Laboratory in Oxford began work on the gaseous diffusion method for separation.

There was another turn in the story at the middle of the year when it was reported from experiments in the cyclotron at Berkeley that bombarding U-238 produced a new element with mass number 239. A suggestion that the element might be produced in a uranium chain reaction had been made slightly earlier by Louis Turner at Princeton and at about the same time a similar suggestion emerged in England. The new element was named plutonium, was expected to be fissile and therefore might provide an alternative to U-235. Halban and Kowarski, working in Cambridge, obtained some encouraging results from an experiments in the autumn with a sphere of uranium oxide powder in heavy water—in fact they believed they had produced a chain reaction. Whether they had done this or not, by the end of the year it was accepted that a self-sustaining chain reaction could be produced with a few tons each of uranium oxide and heavy water. In the light of the plutonium discovery, this would not only allow the French to build their "boiler" reactor but would provide another route to a weapon. The work of the French on slow neutrons is described later.

By March 1941 there was finally a reliable estimate for the fission cross-section for U-235. From the work it had sponsored the MAUD committee was able to conclude in Part 1 of its July 1941 report that "the scheme for a uranium bomb is practicable and likely to lead to decisive results in the war". They estimated that a bomb containing 25lb of U-235 would have the destructive effects of 1800 tons of TNT and would release large quantities of radioactive material making bombed areas dangerous to life for a long period. They recommended collaboration on experimental work with the USA and thought that the U-235 separation plant might be located there. In Part 2 of the report they examined the use of nuclear energy for power and thought it promising but not worth considering until after the war. However, because of the weapon potential of the Pu-239 that could be produced in reactors, they considered that some work should continue on them.

The MAUD report was passed for consideration to a special panel of eminent scientists but, before they had time to comment, Professor Lindemann (later Lord Cherwell) briefed Winston Churchill on 27 August. Although he thought the odds of success in making a weapon during the war was only evens, he recommended the work went ahead. Churchill supported this and made Sir John Anderson, a Cabinet Minister (and a physical chemist who had worked on uranium) responsible. The special panel made a similar recommendation on the bomb at the end of September 1941 and coupled it with another that the Government should maintain close control over the development of nuclear power. (They also recommended that there should be study of the effects of radiation). There was some expectation that Imperial Chemical Industries (ICI) might be made responsible for the development of nuclear power but in fact "Tube Alloys" was set up. This was a division of the Department of Scientific and Industrial Research and was to take over all the work on nuclear energy. Overnight all the scientists involved became civil servants.

The USA stayed aware of progress through representatives at the MAUD meetings and through the exchange of documents leading to the handover of the MAUD report in early October 1941. The clear conviction of this that a practical weapon was possible soon enough to affect the course of the war was striking and Roosevelt quickly authorised information exchange. The USA was probably more interested in the energy possibilities than uranium bombs until 7 December 1941, when Japanese planes attacked Pearl Harbour.

Work was already progressing in the USA on a number of areas relevant to the bomb but this was "in the spirit of scientific curiosity". The first step towards a more coordinated and directed approach came with the formation of the National Defense Research Council under Vannevar Bush in 1940. This became the Office of Scientific Research in June 1941 and, soon after, the British MAUD reports arrived bringing a "sense of urgency".

By then it was understood that there were two routes to a bomb: U-235 and plutonium. For the U-235 route isotope separation was necessary and three methods initially seemed promising: gaseous diffusion, centrifuge, and thermal diffusion. By the end of the year electromagnetic separation was to make a fourth. Small quantities of plutonium were made in the Berkeley cyclotron in early 1941 and by May it was clear that it was fissile. However, it was obvious that the only way of making the new element in the quantities needed for a weapon would be in a reactor. By the end of the year the

3 Fission, Weapons and Power

responsibilities for isotope separation had been assigned by Bush to Harold Urey and Ernest Lawrence and for developing the plutonium route to Arthur Compton.

Compton centralised research on plutonium in Chicago as the cryptically-named Metallurgical Laboratory (Met Lab for short) and by the early 1942 had two possible reactor models: uranium/graphite and uranium/heavy water.

The programme was reviewed in May 1942 with James B Conant, Bush's advisor, and it was decided to pursue three separation methods (gaseous diffusion, centrifuge and electromagnetic separation) and both the reactor models. The thermal diffusion methodology was not considered. The decision to follow so many expensive alternatives was mainly driven by the perception that the Germans had a two-year lead in the work but by the end of 1942 two of these routes had been dropped: centrifuge separation because it involved so much precision engineering and the heavy water pile, because the graphite work was progressing well and it was, anyway, near-impossible to produce the necessary heavy water soon enough. In fact, a heavy water reactor was to be built at Argonne (and to operate successfully) but it was to have no part in the bomb programme. However, well before the end of the year there were dramatic changes in emphasis, organisation and pace.

The time had come to focus on production rather than research and the project was handed over to the US Army Corps of Engineers in the summer of 1942 and a new organisation, the Manhattan Engineer District, was created in August to oversee the entire project. (The District was—and is—the basic organisation unit of the Corps of Engineers and the project headquarters was originally in Manhattan. It later moved to Tennessee.) On 17 September, Col Leslie R Groves was told he would lead the project (he was about to go overseas). On 18 September Groves bought 1250 tons of uranium ore, on 18 September he bought 52,000 acres of land in Tennessee and by 23 September he was a Brigadier General. In October Groves appointed Du Pont to drive the plutonium production project and J Robert Oppenheimer to lead Project Y, the weapons team.

Fermi had moved to Chicago in the summer and by September was receiving the uranium and graphite for the first pile. In November construction of CP-1 was started at Staggs Field and critical operation was achieved on 2 December. Also in November Los Alamos was chosen as the site for Project Y and was codenamed site Y. Construction began the following month.

The plans for the Tennessee site, Clinton (code named site X) later to become Oak Ridge, were drawn up, starting in November, to include the isotope separation facilities, an experimental reactor and a pilot plutonium separation plant. Construction of the electromagnetic separation plant Y-12 began in February 1943 and the first stage was completed in October 1943 but it was to be over a year before U-235 started to flow in production quantities. The planning for K-25, the gaseous diffusion plant, started in May 1943 and construction began in October 1943. There were severe delays because of problems with the diffusion barrier and first feed to this was made in January 1945.

In the meantime a third isotope separation process had re-emerged. Thermal diffusion had been ignored in Conant's programme review but had been taken up by the US Navy because of their interest in submarine power. A pilot plant was built at the Naval Research Laboratory and when this was seen by Manhatttan Project experts in June 1944 they recommended that such a plant be built at Oak Ridge. In July Groves contracted to have S-50, the liquid thermal diffusion plant, built on a three-month time-scale. By early 1945 all three plants were operating and in March they were hooked together in tandem, S-50 feeding K-25 and this then supplying partly enriched material to Y-12. Between them, in this configuration and separately, the three plants produced 90 kg of enriched uranium by the summer.

After the success with CP-1 in Chicago (the pile was to be dismantled, rebuilt at Argonne and renamed CP-2) it was decided to build a larger pilot reactor, X-10, at Oak Ridge. This started operation in November 1943 and began to produce the first significant quantities on plutonium—until then there was no more than a few milligrams of the element in existence. In December the plutonium separation plant at Oak Ridge came into operation using X-10 fuel and by April 1944 the first sample was sent to Los Alamos. However, the Oak Ridge plutonium facilities were pilot plants (the "semiworks") and plans were well advanced for production facilities at Hanford in Washington state. This 800 square mile remote site (code named site W) had been acquired by Groves in December 1942 and the work on the first plutonium production reactor, B-100, was

3 Fission, Weapons and Power

started there in October 1943. Just one year later this reactor, containing 200 tons of uranium metal, went critical but quickly shut itself down. This was soon understood to be the result of the growth of the fission product xenon-135, a neutron poison. The problem was overcome but it was December 1944 before large quantities of fuel from the pile were sent to the Hanford separation plant.

Of course all this engineering activity was directed towards providing material for the bomb: getting the weapon to work was the responsibility of Project Y and Oppenheimer at Los Alamos. Staff had begun to arrive there in March 1943 and after the initial April conferences in April (where Serber made the presentations summarised in the *Los Alamos Primer* [3]) the initial organisation and programme were put in place. Broadly it was necessary to: confirm and extend the measurements of nuclear properties already made, start the research on the gun-type weapon and begin to look at the implosion mechanism. The development of the gun-type weapon was, by Los Alamos standards, relative straightforward and it formed the basis, without the need for testing, of the U-235 Hiroshima bomb in 1945. The plutonium route proved much more difficult. It was appreciated early on that a high spontaneous fission rate would make the gun method difficult for plutonium; the fissile material might prematurely blow apart giving just a fizzle. Implosion—in which the plutonium was quickly crushed from a non-critical to a critical configuration by a conventional explosion—might work. better.
Research on implosion showed that that it would produce an efficient weapon requiring less plutonium than a gun. However, it was not until a reasonable amount of plutonium became available from Oak Ridge in April 1944 that it was confirmed that implosion was the only possibility. The work was boosted when the idea of explosive lenses was introduced by a British physicist, James Tuck, in May 1944. In July Los Alamos was reorganised to give the highest priority to the implosion research and by December one of the major problems was dented with the first successful test of an explosive lens. While the design of the uranium gun was finalised in January 1945—all that was needed was enough enriched uranium to make it work—the work on the implosion weapon continued at a high intensity. The explosive lens design could be frozen in March and the first indications that the compression of the fissile mass could be achieved came soon after. The uranium weapon was ready—minus uranium— in May but it was only then that critical mass experiments could begin on plutonium. The implosion design was confirmed in July

1945 and the preparations for a test (code-named Trinity) of the implosion "gadget" could begin. This took place on 16 July with a yield of about 20 ktons. Little Boy, the uranium weapon, was dropped on Hiroshima on 6 August and Fat Man, the plutonium implosion device, was dropped on Nagasaki three days later.

Once the USA committed to the bomb project, with all their determination and resources, progress astonished the British mission. As Margaret Gowing describes in rich detail[4], they fought to make a contribution through periods when the Americans just wished to get on with the project. Some work was done in the UK and in the Montreal laboratory set up jointly with Canada but the major British contribution (and it was thought by some that it might have advanced the availability of the bombs by a few months) was in providing some 50 talented individuals for the US project. On the arguably generous arrangements with the US, the contribution did give them access to information that was invaluable to both the UK bomb and civil nuclear programmes after the war. The Montreal laboratory was Britain and Canada's stake in the slow neutron work and allowed then to work on piles and also on plutonium separation chemistry—initially with samples provided by the Americans.

The French have made various appearances in the narrative in their attempts to demonstrate a chain reaction with natural uranium and a heavy water moderator. Halban and Kowarski, when they escaped France, joined the slow neutron team working at the Cavendish Laboratory in Cambridge but there was soon pressure for some of them to move to the USA and become involved in the Manhattan Project. The possibility of their moving to the USA was rejected by the Americans for security reasons—they may have thought there were enough aliens already. There were also questions about the patents that the French held on some of the early work.

As an alternative to going to the USA Halban moved to the Montreal laboratory as Director with some of his colleagues in late 1942 (Kowarski did not go until two years later—he and Halban had some professional and personal differences). This might be a suitable base for collaboration with the US and they expected too that it would give them access to uranium and the heavy water then being produced in Canada (as well as their own much-travelled supply). They could then build the heavy water pile and produce plutonium.

3 Fission, Weapons and Power

However this coincided with the Chicago reactor going critical and a temporary cooling in relations between the USA and Britain. Access to heavy water was very restricted and little progress with the pile idea could be made until after the Quebec Agreement in the autumn of 1943. In fact, full agreement on the role of the French at Montreal was not reached until the following spring and by then the Montreal lab staff (now around 100) were dispirited. While the news was not what was hoped—agreement to go ahead with a full-scale heavy water pile—it was enough to bring the laboratory to life. They were to build a large-scale prototype pile.

John Cockcroft became Director of the Laboratory in April 1944, an appointment enthusiastically welcomed by Halban who suspected that he had become something of an embarrassment with the Americans. Chalk River was chosen as the site for reactor development and it was here that the zero energy pile ZEEP (designed by Kowarski who had been tempted to Canada by Cockcroft) and the pilot plant reactor NRX were built. ZEEP began operation in September 1945 and NRX in July 1947. By the time ZEEP started plans were well advanced for setting up a laboratory at Harwell in southern England as the spearhead of British post-war nuclear research and development. Halban left the project in 1945— after a visit to Joliot-Curie in Paris that enraged both the Americans and the Canadians— but stayed in North America until 1946 when he took up a post in Oxford; Kowarski and the other French scientists returned to France in 1946.

A major benefit of the wartime Montreal project was probably in reprocessing rather than pile construction. The USA had provided, around the time of the freeze in relations, some samples of fission products. In experiments with them and in enforced ignorance of the US approach to separation of plutonium (a precipitation process), by the end of the war the Montreal team were well ahead with the development of a continuous solvent extraction process that was to be the preferred process for the rest of the century.

The decision was taken in 1945 that there should be experimental establishment devoted to the development of atomic technology in Britain and the selection of Harwell near Oxford was announced in October 1945; by December a decision was also made that a pile should be built for plutonium production for a weapon. The plans developed quickly and the three leading players were picked and named by January 1946. John Cockcroft was to lead Harwell,

William Penney was to lead the weapon development and Christopher Hinton was to be responsible for producing the plutonium. Hinton, who had previously had no atomic experience but had made his name by streamlining munitions production during the war, was to be based at Risley in the north-west of England where the production factories were to be sited. The work on the weapon itself was initially centred at Fort Halstead and Woolwich Arsenal near London but this was transferred in 1950 to Aldermaston about twenty miles from Harwell. The date for a test explosion was set as 1952.

The expectation when the work began was that the pile would be based on heavy water but constraints soon dictated otherwise. There was very little heavy water available to the UK and supplies for a pile could not be secured; there was no enriched uranium available either. The pile therefore had to use natural uranium and this meant that the only two practicable moderators were heavy water and very pure graphite. Since there was no heavy water supply, graphite was the choice. The possibility of using normal water as the coolant—as the Americans had done—was considered but there were real safety concerns about this since it could, should the cooling fail, lead to a serious accident. While this risk could be accepted in a sparsely populated region like Hanford, there was nowhere similar in the British Isles. The choice was thus the intrinsically safer gas cooling. First designs assumed that the gas would have to be pressurised but it was soon realised that air at atmospheric pressure would provide enough cooling for a plutonium pile. It was recognised that it would not be adequate for a power-producing pile—but such issues were still well in the future. However, the path that British nuclear power would take was defined in those first few months as the design and construction of a natural uranium-fuelled, graphite-moderated and gas-cooled reactor got under way.

The first experimental pile GLEEP (Graphite Low Energy Experimental Pile) was built at Harwell and went critical in August 1947; the bigger BEPO (British Experimental Pile Zero), to be used for isotope production, general irradiation and as a neutron source, went critical a year later. Meanwhile, in the northwest Hinton was building the production facilities. Construction of Springfields as the fuel production facility started in 1946 and by 1947 Windscale on the Cumbrian coast had been picked as the location for the production piles (two were needed) and the chemical processing required to

3 Fission, Weapons and Power

separate out the plutonium and produce the metal for the bomb. The gaseous diffusion enrichment plant at Capenhurst was announced in 1949 with its initial role as re-enriching the uranium depleted in the Windscale piles (it was later to be extended to produce pure U-235). The first Windscale pile went critical in October 1950 and the second one in June of the following year. The chemical separation plant was completed in early 1951 and by March 1952 the piles were both in production and the plutonium finishing plant prepared the first metal billets from their output. In fact, some metal had been produced earlier at Harwell from Chalk River material.

In the meantime Penney's team had designed and made the first weapon, a one-off not unlike the Nagasaki bomb, for a test at the Monte Bello Islands (there had been discussions with the Americans and Canadians about testing in North America but they were inconclusive). The weapon minus the core was shipped out on the target vessel *HMS Plym* (which undertook a rather bizarre goodwill tour as it made its way round the Cape of Good Hope) and the core was flown out later. The test, arranged partly to show the effects of an atomic explosion in shallow water (a scenario of a weapon exploding on a ship in harbour was a prime one at the time), took place successfully on 3 October 1952. It was Britain's first step in an activity that was to have political world-wide consequences and draw the public's attention to the awesome potential of nuclear weapons and the hazards of nuclear radiation.

Since 1945 there have been something over 2500 tests of nuclear weapons with yields ranging from around a kilogram (generally termed "fizzles" and regarded as failures) to many megatons. The USA's 1039 tests were conducted to improve fission weapons and for the development of thermonuclear weapons and took place mainly at the Nevada Test Site and on small Pacific Islands (although there was one in Mississippi in the 1966/7 series and and another on the Aleutian Islands in 1969). After the August 1963 Partial Test Ban Treaty, which prohibited atmospheric and underwater tests, all the tests were conducted underground and, even given the examples where the blast was not fully contained, releases to the atmosphere were insignificant. The Comprehensive Test Ban Treaty came into force in 1996, prohibiting all nuclear tests by signatory nations.

During some of the above-ground tests at Nevada between 1951 and 1957, servicemen were deployed in a series of seven exercises known as Desert Rock. These were intended to accustom troops to conditions on the nuclear battlefield and to see how they reacted. Most involved some kind of manoeuvres and some of them resulted in troops receiving radiation doses. There were subsequently concerns, particularly for the Smoky shot of the Plumbbob series in 1957, about how large these had been and whether there had been any consequences for veterans. The epidemiological studies that resulted are touched upon in Chapter 8. Some of the tests also caused civilian exposures from fallout, notably the Harry shot of the Upshot-Knothole series in 1953.

The first post-war US tests, Crossroads, were at Bikini atoll in the Marshall Islands in 1946 with Fat Man type weapons and they then moved to Enewatak, some hundreds of kilometres east of Bikini, and later to Johnston Island (for high altitude explosions) in 1958 and 1962 and Christmas Island for tests in 1962.

These were generally much larger explosions than those at Nevada and normally above ground level. The first thermonuclear explosion Ivy Mike in 1952 at 10 Mt erased the island of Elugelab but it was the Bravo shot of the Castle series in 1954 that was to be most significant radiologically. This was expected to have a yield of 6 Mt but it was actually 15 Mt (making it the biggest explosion caused by the US). This combined with poor judgement of the weather caused fallout to drop on Marshall Islanders and engulf the *Lucky Dragon*, a Japanese fishing vessel.

Bravo was conducted at Bikini, no more than 100 miles east of the inhabited islands of Rongelap, Ailinginae and Utirik. The planning had all been based on the fallout drifting west but the test went ahead in unpredictable weather conditions and it was in fact carried to the east. This contaminated many of the test facilities but more importantly it irradiated the islanders, who it had not been judged necessary to evacuate, and US servicemen on Rongerik nearby. It was several days before the (several hundred) islanders could be evacuated and by then they had received doses up to 1.75 Gy on Rongelap, 0.69 Gy on Alinginae and 0.14 Gy on Utirik.

The *Lucky Dragon* (Daigo Kotoshiro Maru) was fishing outside the previously declared exclusion zone about 85 miles from Bikini. Impressed by the glow of the explosion which they saw "like the sun rising in the west" they collected samples of the white ash that rained on the boat as souvenirs. It was, of course, fallout and the twenty-three crewmen each received about 3 Gy, one of them dying of associated kidney failure some months later. The incident caused outrage in Japan not just because of the fishermen but because of the contamination of the ocean and fish stocks. The world became much more aware of the dangers of fallout and the underestimation of the size of the explosion dented confidence in the scientists: Oliphant and Frisch in the UK were called upon to deny publicly that H-bombs could ignite the earth's atmosphere. In the USA, Lewis Strauss, the Atomic Energy Commission chief, admitted that a single H-bomb could destroy New York city, perhaps the first public statement of the devastation such a weapon could cause.

Publications in the *Bulletin of the Atomic Scientists* by Dr Ralph Lapp in late 1954 and early 1955 revealed that the fallout from fifty 15 megaton bombs would be enough to devastate the US: the fallout from the Bravo test had been large enough to cover the state of Maryland. Lapp's conclusions were subsequently supported by US official reports.

In the UK the Strath report in 1955 confirmed the devastating extent of fallout in a large-scale war and, although about civil defence, it contributed to the concern about global fallout from testing. It was followed in June of 1956 by the report[5] from the UK Medical Research Council committee chaired by Sir Harold Himsworth which did address bomb testing. The conclusion of this was that, if testing continued at the expected rate, the effects would be insignificant: it would add 0.02-0.04% to the population dose from background. In the US the report of Detlev Bronk's Biological Effects of Atomic Radiation (BEAR) committee[6] appeared on the same day with much the same conclusions—not entirely surprising given that the two committees appear to have compared notes. As testing continued world-wide concern grew and the need for independent assessments was recognised by the UK and USA. The result was the setting up of the UNSCEAR committee in December 1955[7]. The UK conducted 21 tests between 1952 and 1958 and there were then a further 24 jointly with the Americans at Nevada, all underground, primarily to develop the Chevaline and Trident warheads.

The first British test, Hurricane (45 kt), took place in the Monte Bello Islands off the coast of northwest Australia on 30 October 1952 and there were two more tests there in 1956. Testing moved in 1953 to Emu Fields in the west of South Australia for the two Totem explosions but this was found too remote and the 1956/7 activity moved to Maralinga, 190 km south, where the seven tests in the Buffalo and Antler series were conducted. Up to 1962 Emu Fields and Maralinga were used for component trials; some of the Maralinga trials included dispersal of plutonium to simulate weapon accidents. This was not the place to test very large weapons (by far the largest at Maralinga was the final one at 27kt) and the Grapple series was conducted in the Pacific first on Malden Island (3 tests) and Christmas Island (Kiritimata), 700km to the north, where there were 6 tests. The Grapple series took place in 1957 and 1958 and included five weapons over 1Mt in the Christmas Island series. All the British tests were conducted in the atmosphere. The early and smaller ones were detonated near the ground or ocean; the larger explosions were air bursts from balloons or aeroplanes.

The British programme did not experience anything like the disaster of the US Castle Bravo shot but British and Australian servicemen were involved in a number of the series and there have subsequently been epidemiological studies of them, briefly described in Chapter 8. The contaminated Maralinga site was finally cleaned up by the end of the century although there were claims that the plutonium contaminated soil had merely been buried in shallow pits.

The French testing programme, to create General de Gaulle's "Force de Frappe", began in the Algerian desert at Reggan with three atmospheric tests in 1960 and 1961. Following that there were 14 further detonations, all underground at In Ecker, 600 km to the east. After this, from 1966, all the tests took place in the Tuamoto archipelago on the islands of Mururoa and Fangataufa, 800 km southeast of Tahiti. The 181 tests there were in the atmosphere until 1974, in a series that included weapons in the megaton range. France then, under pressure to comply with the Partial Test Ban Treaty (which it had not signed), switched to underground testing. The final test was conducted in 1996.

There seem to have been no proper studies of the impact of the French tests either on the local populations or the servicemen who observed them and exercised after them. Polynesian islands, including Tahiti, were subject to fallout on at least one occasion and there were reports of high levels of radiation both in North Africa and in the Pacific after atmospheric tests. The French Government insisted that there had been no hazard to health and refused to release any supporting information. They did however begin a health register in the four islands most likely to have been affected for 1984 onwards. A study by Florent de Vathaire in 2000 found increased thyroid cancers in Polynesia but concluded that this was unlikely to be due to radiation. However, further work unpublished at the time of writing was reported to have found a significant increase of 10-20 in the 239 cancers studied. In the meantime, in 2001, both the test veterans and Pacific islanders formed associations to demand more information and their cases were supported by several awards of compensation to servicemen involved in the tests. The French Government announced medical surveillance for 2000 islanders but did not appear ready to release any more information or sponsor epidemiological studies. A 1998 IAEA report commissioned by them concluded that the radiological hazards posed by residual activity on the test islands was negligible.

Russian testing began in 1949 at the test site to the west of Semipalatinsk in Kazakhstan on 29 August 1949, when Russia's first atomic weapon was tested; their first H-homb test was in August 1953. The tests there were low level or surface explosions, some of them approaching the megaton range, until the early 1960s when they were conducted underground. From 1957 to 1990 larger weapons were tested at the Novaya Zemlya test site of the north coast of the country. Here weapons of many megatons were exploded including the largest nuclear detonation by any nation, 50 Mt, in October 1961. The last atmospheric test was in December 1962. There were also about 120 underground explosions at other locations in the USSR (principally in what is now the Russian Federation and Kazakhstan) for peaceful purposes. Altogether, it has been estimated that 969 devices were detonated in 715 tests.

There were reports that local people at Semipalatinsk were exposed to high levels of radiation and this was confirmed in the 1997 study by Gusev and others[8] who found that nearby villagers had received effective dose equivalents of 0.7 to 4.47 Sv from fallout in the years since the tests. The exposures came predominantly from a few tests conducted between 1949 and 1956. The area and local population near Semipalatinsk became a subject of multinational studies but very little information has emerged about other locations.

China's testing has all been done at the Lop Nor test site in Sinkiang province in the west of the country. The first test was in

TAMING THE RAYS

October 1964. By the time of the final test in 1996 the Chinese government had conducted 45 tests, 23 in the atmosphere. The first thermonuclear weapon (3.3 Mt) was exploded in the atmosphere in 1967 and later in the year the largest atmospheric test at around 4 Mt took place. Atmospheric testing continued until 1980 (China was not a signatory to the 1963 Partial Test Ban Treaty) and the last underground test took place in 1996. The Chinese claim to have undertaken medical reviews of the local population since the early days showing that there had been no harm but groups in neighbouring states of the former USSR alleged serious health consequences. However, no reliable information seemed to be available.

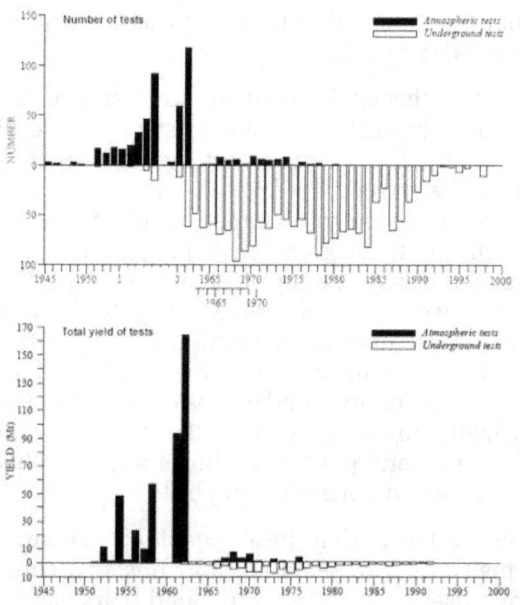

Figure 3: Atmospheric and underground weapons tests

Three other nations are known to have tested weapons: India in 1974 (1 test) and 1998 (2 tests), Pakistan two tests in 1998 and North Korea one in October 2006. All tests have been underground with, as far as is known, no significant radiological consequences. Figure 3,

3 Fission, Weapons and Power

taken from UNSCEAR (2000), summarises the history of tests and yields.

The atmospheric tests not only affected the local and regional populations but they injected large amounts of radioactive debris into the upper atmosphere and this spread around the world. Quite what proportion of the activity produced by a bomb followed this route depended upon how high above the ground it was detonated and how large the explosion was. The model used by UNSCEAR in 2000 had the atmosphere composed of four levels in four geographic zones (equatorial and temperate in each hemisphere) and the amount injected into each level by lower level explosions depended upon how high the radioactive cloud rose. Using this, calibrated with world-wide measurements of Sr-90 deposition since 1958, they were able to calculate the hemisphere-average effective dose equivalents for each year since 1945. The predominant pathways, after the early 1960s, are external irradiation and ingestion. The external pathway, after an early contribution from Zr-95+, has been dominated by Cs-137. The ingestion pathway component was largely down to Sr-90 and Cs-137 in the early years but after about 1970 C-14 became dominant.

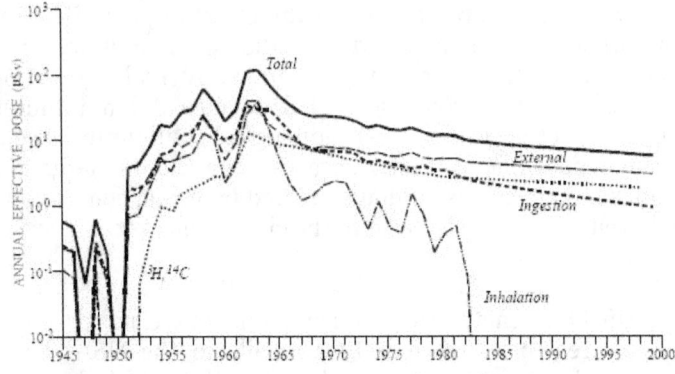

Figure 4: Dose contributions over time from nuclear testing

Northern hemisphere doses peaked in 1963 at 0.125 mSv/y (with all three pathways contributing about equally); by 1999 they had fallen to 0.006 mSv/y. In the southern hemisphere they peaked at 0.059 mSv/y and by 1999 were down to 0.003 mSv/y. UNSCEAR 2000's calculations (Figure 4) were much more detailed than the ones of 1982 but the effective dose commitments arrived at were similar: The average for someone in the northern hemisphere by all pathways was close to 1 mSv between 1945 and 1999. For an inhabitant of the southern hemisphere it was one third of that. The conclusions are, not surprisingly, very similar to those of 1982 but it is difficult to compare them with those of the 1958 report when only bone marrow and gonad doses were estimated. The earlier report made much play of genetic defects (with enormous uncertainties) but the overall impact as assessed now is probably rather higher than the equivalent estimates made then. One other difference is that it should all be behind us now; then it was part of an uncertain future.

Protection of the population against nuclear attacks was a preoccupation from the mid-1950s onwards under the title of "civil defence", a term dating from the Second World War. Much of it was concerned with organisation of warnings and advice on protection against blast and initial radiation effects. The radiological consequences of this and of fallout were concentrated on the early effects rather than on late somatic effects—which for much of the time were considered to have a threshold anyway. UK Government support for civil defence faded in the 1970s as it was realised that there was little chance of providing any real protection, a view reflected in other countries. Civil defence did not therefore have much impact on the general radiological community except perhaps through instrumentation, where the demand for large numbers of simple and cheap instruments created by it (particularly in the USA) may well have spilled over into the civilian market.

The US first reactor to produce significant electrical power was a breeder reactor, EBR-1 at Idaho Falls in the USA. The apparent scarcity of uranium encouraged an early interest in breeder reactors and EBR-1 construction was started in 1949. It went critical in August 1951 and on 20 December produced enough electricity to light four bulbs. The following day it produced enough to power the reactor building. The real purpose of EBR-1 was to demonstrate that breeding could be achieved and this it did. It ran until 1964 when it was decommissioned and replaced with EBR-2. EBR-1 is now a National Historic Landmark.

Fermi's CP-1 and the Hanford plutonium production reactors had to use natural uranium and the only moderator available was the high-purity graphite already being produced for electric furnace electrodes. This and many other reactor types were tested in the USA in the 1950s but it was systems based on light (ordinary) water as a moderator and coolant that came to dominate power production. These took advantage of the enrichment technologies developed during the war that could produce slightly enriched uranium fuel. With this, the neutron balance was favourable enough to counteract the losses in the hydrogen.

The first BORAX reactor was built at Idaho Falls in 1953 and used for the boiling reactor experiments that laid the foundations for the safety of modern light water power reactors; it was destroyed unexpectedly in a final test. A second BORAX reactor was fitted with a turbine and in 17 July 1955 was briefly connected to the local power grid to light the nearby town of Arco. The total electrical output, including power supplied to the facility itself, was about 2 MW.

At about the same time the STR—the land-based prototype of the PWR intended for nuclear submarines—was constructed at Idaho Falls and this design was used to power the *USS Nautilus*, launched in 1955. A parallel project using a sodium-cooled reactor was abandoned after the reactor of this type, fitted to the *USS Seawolf*, proved unreliable and it was realised that the consequences of an accident at sea involving sodium would be severe. *Seawolf* was then fitted with the second PWR and this reactor type—compact and reliable—became the standard. By 1962 there were 26 nuclear submarines in operation with many more under construction. The success of the Westinghouse PWR as a civil reactor system is at least partly down to its selection for powering submarines.

The first PWR designed for commercial power production was the prototype PWR built at Shippingport. This 60 MWe demonstration reactor started operation in 1957 and operated to 1982 but the first truly commercial PWR was the 167 MWe Yankee Rowe PWR in Massachusetts which started up in 1960 and was shut down in 1991. Seventy nine PWRs were subsequently built in the USA and around 230 world-wide. A small BWR prototype (5 MWe) was built at Vallecitos, CA and operated from 1957 to 1963. The Dresden-1 commercial 250 MWe BWR began operations in 1960 and was the first of 44 built in the USA and 110 world-wide. PWRs and BWRs with generating capacity up to about 1500MWe were built.

British and French reactor designers did not, in the 1950s, have enrichment facilities available to them and both Governments were unwilling to rely on the USA for supplies. They were thus forced to use natural uranium and the best moderating material available to them: graphite. They therefore developed the similar metal-fuel gas-cooled Magnox (UK) and UNGG (France) reactors as evolutions of their plutonium producing reactors. The first Magnox station opened at Calder Hall in 1956 (the world's first fully-commercial reactor to go into operation) and the UNGG G2 at Marcoule in 1959. As enrichment facilities became available the British moved on to the more efficient Advanced Gas-cooled Reactor (AGR) using oxide fuel and running at higher temperatures. The Magnox and AGR designs were the backbone of the UK programme and over 30 were still in operation at the end of the century but, by then, the UK had abandoned the AGR to build its first PWR—based on an American design—at Sizewell. The French meanwhile, after building eight gas cooled reactors, had switched to their own version (designed by Framatome) of the Westinghouse PWR and had built 60 of these by the end of the century.

A not-dissimilar programme pattern developed in the USSR. The original Obninsk reactor which was developed into the RBMK reactor series. These are fuelled with low enriched uranium in a graphite moderator but cooled by boiling water circulating through pressure tubes. This was a relatively accessible technology that did not require a large pressure vessel and meant that the reactors, like the early British and French ones, could be used to generate weapons-grade plutonium. Seventeen large RBMKs were built in the USSR, starting with the Sosnovy Bor reactor in 1973, but after the accident at Chernobyl several were shut-down permanently and no more were ordered. The USSR also experimented with a BWR but went on to develop the VVER series of PWRs starting with a small prototype in 1964. Around 30 were built in the USSR and its satellites (the RBMK was retained within the USSR because of its potential to produce plutonium).

3 Fission, Weapons and Power

Other reactor types have been the subject of experiment and some, notably Fast Breeder Reactors, have produced electricity in commercial amounts but the only other reactor type of real significance for power generation has been the Canadian CANDU. Canada had been the source of heavy water for the Manhattan Project and built plants to provide enough of this efficient moderator for a series of reactors based upon it. The heavy water is both the coolant and the moderator and the CANDU system can use natural uranium efficiently as a fuel. The use of pressure tubes rather than a giant pressure vessel means that refuelling can be done on-line. It is therefore a potential weapons-grade plutonium producer. Forty units were in operation around the world (18 in Canada) at the end of the century with another 10 or so under construction or ordered.

The peak in orders and construction in the mid-1970s brought a peak in grid connections a decade later. The number of units operating rose from under 100 in 1970 to over 400 by 1990 and the net capacity rose even more steeply because the capacity of individual units grew. By 1990 nuclear power capacity was 328 GW (GW electricity) but it rose by only 30 GW in the next decade as confidence in nuclear power dissipated in the face of environmental concerns and the Chernobyl accident. At the end of the century LWRs accounted for 85% of the nuclear electricity generated worldwide; around 65% came from PWRs if we include VVERs. Nearly 80% of French electricity came from nuclear power but in the UK and the USA the share was no more than 20%.

As a revival in nuclear power seemed possible at the beginning of the 21st century in the face of the threat of global warming, novel and advanced reactor concepts began to be reviewed more systematically. Most of the ideas that seemed likely to have practical potential revolved around LWR technology.

The early reprocessing at Hanford used a process (the bismuth phosphate process) that recovered the plutonium to be used in weapons. The process, developed and trialled at Oak Ridge, did not however recover the valuable uranium and was replaced by the PUREX (Plutonium and Uranium Extraction) process developed, again at Oak Ridge, in the late 1940s. This was first used at the Savannah River site in 1954 but quickly became, and remains, the preferred extraction process. The Hanford (now using the PUREX process) and Savannah River plants were devoted to reprocessing of materials generated on-site for military purposes. A reprocessing facility at Idaho Falls dealt mainly with Navy surface ship and submarine reactor fuel.

A plant intended for commercial fuel reprocessing was constructed at West Valley NY. It operated from 1966 to 1972 but, in fact, handled only spent fuel from the weapons programme. It shut down in 1972 for upgrades but the expense of meeting seismic protection requirements meant that it never re-opened. A plant constructed at Morris, IL failed to work properly and was declared inoperable in 1974. A third plant, planned for Barnwell, SC was ruled out when President Jimmy Carter announced in 1977 that, for non-proliferation reasons, the USA would not undertake commercial reprocessing. As a result, there was more than 50,000 tonnes of spent fuel in store in the US at the end of the century with around another 2000 tonnes being generated each year.

In the UK the B204 reprocessing plant was built at the Windscale (now Sellafield) works in Cumbria to reprocess fuel to extract weapons plutonium and operated from 1951 until replaced by the B205 plant for reprocessing Magnox fuel in 1964. The B204 plant was then converted into a pre-handling plant for B205 (so that it could be used to reprocess oxide AGR fuel) but an accident during maintenance work in 1973 led to its permanent closure. The B205 plant continued to operate and by the end of the century had reprocessed over 35,000 tonnes of spent fuel; it is expected to operate until all Magnox fuel had been reprocessed in about 2012. The THORP reprocessing plant for oxide fuels was the subject of a public enquiry in 1977-8. Approval to go ahead was granted in 1978 but it was not until 1997 that it finally went into full-scale operation.

Most of the spent fuel from the UK's fast reactor project at Dounreay in Scotland was reprocessed on-site. After the reactor was shut down in 1994, the reprocessing plant, which had started up in 1979, continued until a problem with the dissolver caused it to close in 1996.

French reprocessing started in 1958 at Marcoule in the UP1 PUREX plutonium recovery plant and the APM pilot plant opened soon after for experiments on reprocessing and waste management. French civil GCR fuel began to be handled at UP1 in 1965. The plant ran until 1998 but a new plant, UP2, was commissioned for GCR fuel at Cap de la Hague on the Cherbourg peninsula in 1966. It was extended with a new head-end plant in 1976 so that it could process the oxide fuel from the new generation of PWR reactors. It was upgraded in 1994 to increase capacity to 800 tHM/y (tonnes of heavy metal per year) and was used predominantly for domestic fuel. The UP3 plant was commissioned at Cap de la Hague in 1989, with a similar capacity, specifically for commercial reprocessing of foreign spent fuel.

The USSR undertook military reprocessing from the late 1940s and started to process civil fuel in 1977 when the RT-1 plant at the Mayak site at Ozersk north of Chelyabinsk started up. With a capacity of 400 tHM/y it continued to process spent fuel, including that from the VVER-440 reactors up to the end of the century and beyond. Since it could not handle the VVER-1000 fuel a new facility, RT-2, was proposed near the Siberian city of Krasnoyarsk and construction was started in the 1970s. However, money and conviction ran out and the collapse of the USSR meant the plant was left half built. As the century closed there was some suggestion that construction might restart or that the RT-1 plant might be modified to accommodate the VVER-1000 fuel.

A small plant (90tHM/y) was operated at Tokai-mura but most of the Japanese spent fuel was reprocessed in Europe. A large plant (800tHM/y) was being built at Rokkasho. By the end of the century around 80,000 tonnes of spent fuel had been reprocessed worldwide and most of this (around three-quarters) was fuel from the British and French gas-cooled reactors. The remainder (apart from relatively miniscule quantities) was LWR fuel processed in the French La Hague plants and, to a much lesser extent, in the Russian Federation[9].

India built its first reprocessing plant to separate plutonium at Trombay. Commissioned in 1964 it continued to operate, with major refurbishment, to the end of the century. A plant for oxide fuels was built at Tarapur and commissioned in 1974 and a third was constructed at Kalpakkam and cold commissioned in 1996. A pilot PUREX reprocessing plant was under construction at the end of the century at Lanzhou Nuclear Fuel Complex in China.

Spent nuclear fuel has either been put directly into storage or reprocessed and the waste products stored temporarily as very active liquids. The intent in most nuclear nations who have adopted reprocessing is to convert the liquids into a more-or-less inert solid, by vitrification for example, and to store or dispose of this in geological formations. The plutonium and recovered uranium are not regarded as wastes since they can potentially be recycled.

In the USA, where virtually all spent fuel from the commercial programme remains in store at 72 power stations, the intent, at the end of the century, was to dispose of it in a repository at Yucca Mountain in Nevada. The site was first identified in the late 1970s as suitable for a deep repository but the very long-term environmental standards that would have to be met were still being debated when the site was formally designated in House Joint Resolution 87 in 2002. The repository faced several regulatory hurdles and was not expected to open until, at the earliest, 2017.

In most other countries there were plans to dispose of high level wastes deep underground in either hard rocks or salt caverns and a number of countries have set up underground laboratories. However, by the end of the century, none of them had a clear plan for disposal. Sweden was perhaps closest with a repository for spent fuel to be constructed in granite rock at Oskarshamm, close to their central storage centre for irradiated fuel. In the UK, a deep repository experiment at Sellafield was undertaken but abandoned in 1997. However underground disposal was still the preferred option. In several countries interim storage solutions involving near-surface facilities were being considered. The Netherlands constructed such a facility (HABOG) at the Borsele site and this was expected to accept spent fuel elements and other wastes in the early years of the 21st century. The Dutch were looking for a long term solution in a retrievable deep repository.

Until the early 1990s many countries disposed of radioactive waste by dumping it at sea. Fourteen countries disposed of around 85 PBq at more than 80 sites in the Atlantic, Arctic seas and the Pacific. By far the largest dumpers were the UK in the NE Atlantic and the USSR in Arctic areas. Each of them accounted for more than 40% of the total activity dumped. Much of the UK waste was packaged in drums but these were expected to rupture as they sank or leak once they hit the bottom. The expectation was that the radioactivity would either decay or disperse.

3 Fission, Weapons and Power

The UK dumping grounds extended from 100 km off the west coast of Ireland as far south as the Canaries. They included some smaller disposals in the Irish Sea and the southern North Sea in the 1970s and around 16,000 tons (57 TBq) was dumped into the Hurd Deep in the English Channel (about 20 km north of the island of Alderney) between 1950 and 1963. Altogether the UK dumped 74,000 tons at 15 sites in the North Atlantic and English Channel. This contained 631 TBq alpha and 34 PBq beta/gamma (including 10 PBq of tritium). The last UK disposal was made in 1982 before the international moratorium on sea dumping was signed the following year[10]. The UK was not quite alone: Belgium disposed of 2 PBq of beta/gamma waste(including 0.8 PBq of tritium) in the NE Atlantic and Switzerland dumped nearly 4 PBq of tritium between 1969 and 1982. The USA disposed of nearly 3 PBq off its east coast between 1949 and 1967.

The USSR dumped six reactors with spent nuclear fuel and the shielding assembly from the nuclear icebreaker *Lenin* in the Kara Sea along the coast of Novaya Zemlya between 1965 and 1981. The activity of the *Lenin* assembly, dumped in 1967 in Tsivolka Fjord in 49 m of water, has been estimated at 19.5 PBq and it accounted for about half the total activity dumped[11]. A further 10 defuelled reactors were disposed of in the same area, the last in 1988, with (relatively) minimal additional activity. There were also disposals of solid and liquid wastes in the same period.

The USSR also disposed of defuelled reactors in the Sea of Japan in the 1970s and the last disposal, of liquid wastes, took place in 1993. In all the USSR is estimated to have disposed of about 0.9 PBq beta/gamma in the waters off its south eastern coast in solid and liquid form. The USA, between 1946 and 1970 disposed of around 0.6 PBq of waste in the Northern Pacific.

When they ceased sea dumping countries, were forced to either find some other disposal option or store the wastes. Since there was no legitimate alternative immediately available for any country, all sea dumping material was stored. Most remained stored at the end of the century, including in the case of the UK, drums that had already been prepared for the abandoned sea dump of 1983. Some countries did find partial solutions for wastes that might have been dumped. Sweden has operated a disposal facility in granite rock under the Baltic at Forsmark for short-lived ILW since 1988 and in 1999 the USA opened the Waste Isolation Pilot Plant (WIPP) in the deep salt beds of the Chihuahuan Desert in New Mexico but this was restricted by Congress to transuranic defence wastes. Facilities also operated between 1981 and 1998 at Morsleben, Germany and from 1998 at Himdalen, Norway.

While high and intermediate level wastes both remained a preoccupation—perhaps both solved in principle with underground disposal—low level wastes continued to be generated and generally buried in shallow pits.

As well as the solid wastes generated at all stages of the fuel cycle, all facilities discharge liquid and gaseous wastes to the environment, usually after some kind of treatment. The liquid discharges were usually made to the sea—although there were some examples of discharge to rivers and lakes. The reprocessing plant at Sellafield, discharging to the Irish Sea, showed rising annual discharges of both alpha and beta activity up to the mid-1970s reaching peaks of 10 PBq beta and 0.2 PBq alpha. These arose from both the fuel storage ponds and the reprocessing operations and, until the mid-1970s, they reflected the rising throughput at the site. Increased storage and corrosion of Magnox fuel in the ponds in the mid-1970s led to increased discharges. These were reduced dramatically from then on first by treating pond water with zeolite clay and then by a series of clean-up plant: SIXEP in 1985 and SETP in 1993 reduced fission product discharges and EARP in 1995 began to reduce actinide discharges. As a result of these measures beta discharges fell from their peaks of close to 10,000 TBq in the mid-1970s to 160 TBq in 1986 and (in spite of THORP coming online) 64 TBq in 2000. Alpha discharges similarly fell from 200 TBq at their peak to 3.9 TBq in 1986 and 0.15 TBq in 2000. Tritium, excluded from the above figures remained more or less constant after a sharp peak in the early 1970s.[12]

The Cap de la Hague discharges into the English Channel showed a broadly similar pattern at a lower level. High annual discharges from the mid-1970s to the late 1980s had betas at around 1000 TBq and alphas fluctuating around 0.5 TBq. Both were reduced by factors of 30-40 by the end of the century.

The collective dose to the European Union population[13], Figure 5, from all nuclear industry liquid discharges in the year peaked at 280 manSv for 1978 but by 2000 it fell to 14 manSv. Sellafield and La

3 Fission, Weapons and Power

Hague between them completely dominated the picture (accounting for 98% of the industry contribution) and Sellafield, apart from a few years in the mid-1980s, made the bigger contribution of the two. Liquid discharges from power stations were responsible for just a few percent of the total collective dose from liquid discharges[14].

The atmospheric contribution to collective dose to the EU population from the nuclear industry in 1996 was 140 manSv—more than ten times that from liquid discharges. Half of this came from power stations and most of this from UK gas-cooled reactors which contributed around 50 manSv. The other half was from the two reprocessing plants: La Hague

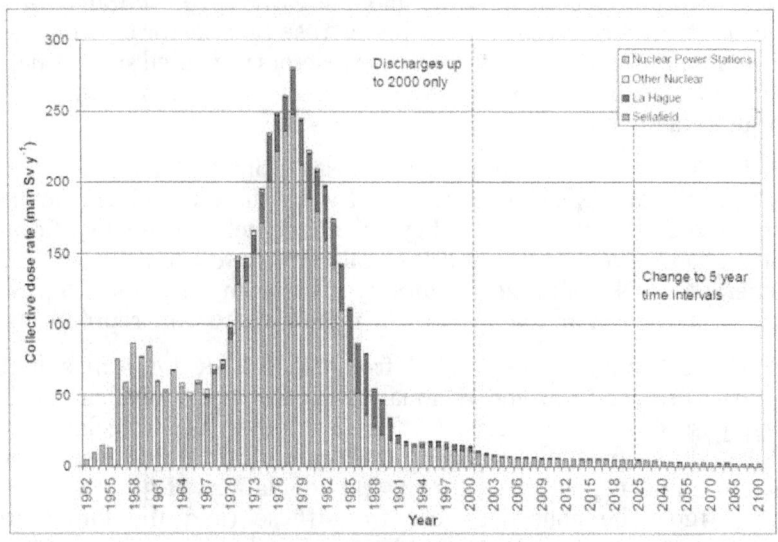

Figure 5: Collective dose

contributed 52 manSv and Sellafield 16 manSv. The calculations took account of the first pass of the radioactive plume over Europe and its subsequent dispersion and both of these made significant contributions to the collective dose[15].

The world picture and the time trend of collective dose from reactors was summarised by UNSCEAR in 2000[16]. The releases per GW year (/GWy) of electricity generated (what UNSCEAR called the "normalised releases") showed decreases in noble gas emissions, I-131 and liquids other than tritium by factors of near 100 between the early 1970s and the end of the century. Tritium (both air and liquid) and C-14 remained roughly constant. Normalised particulate releases dropped until the mid 1980s and then rose again but this was solely because of the (poor) performance of the Swedish Ringhals 1 BWR. These changes were due partly (maybe largely) to the increasing dominance of the PWR; there are very marked differences in normalised release performance with the PWR coming out, for most emissions, on top. There were also effects of improved emission control technology with all reactor types. The effect of this was that, while between the early 1970s and the early 1990s the amount of electricity generated by nuclear power rose by a factor of about ten, the annual collective dose from power reactors remained substantially constant at about 100 manSv. Carbon-14 and tritium came to be large components.

UNSCEAR also considered the doses from the front and back ends of the fuel cycle. While in recent years these have been smaller than the dose from reactors they have nonetheless been significant. In the late 1990s for example the collective dose from the 250 GW generated was about 200 manSv/y: 110 from reactors, 60 from mining and fuel manufacture and the remaining 30 from reprocessing.

The collective doses quoted from UNSCEAR for atmospheric releases are the local and regional totals and are equivalent to the first pass doses of reference [17]. However, they also calculated the global doses from the isotopes that enter into global circulation patterns and expose the entire world population: H-3, C-14, Kr-85 and I-129. The collective dose from these (and the long term component from the release of radon from mill tailings) over the next 10,000 years is something over 50 times the local and regional component. However, it is delivered over a very long time to a very large number of people. Around 10% of this dose is delivered in the first hundred years and, if nuclear power continued at about the same level as now, the collective dose delivered from the globally-dispersed nuclides in the year 2050 would be about 1000 manSv. The average dose to the world's population of 10^{10} would be around

3 Fission, Weapons and Power

0.1 µSv: 0.1% of the natural background and 6 or 7 orders of magnitude below levels at which health effects have ever been seen.

Collective dose is by no means the whole story of radiological impact: the maximum doses received by people from the discharges are at least as important. The critical groups for the Sellafield liquid discharges have almost certainly been studied more closely than any others and one striking fact to emerge is that different groups of people have been most highly at risk at different times. Up to about 1970 the most exposed group was consumers of laverbread, a dish made from the seaweed *Porphyra umbilicalis*. This was harvested on the Cumbrian coast and transported to South Wales, where it is regarded as a delicacy. The dose arose from Ru-106 and doses as high as 1 mSv/y were assessed between 1952 and 1970. However, the pathway declined in importance because key rail links were closed making the trade uneconomic. The Ru-106 levels in the discharges fell anyway and, for a few years, the most exposed group were those people exposed to external radiation from the mud of the nearby Ravenglass estuary. The doses were from Zr-95/Nb-95 and Ce-144. However, by the mid-1970s radiocaesium discharges were rising and the highest doses were received by people eating fish and shellfish: the highest critical group doses at any time were assessed in 1975-6 at 1.9 mSv/y. Shellfish concentrations fell with the decline in caesium emissions over the next decade but concentration in mud fell less quickly so, for a brief period, external exposure became more important. From 1988 the assessed dose from the seafood and external exposure pathways both hovered around the 0.1 mSv/y level. The critical groups for these pathways were distinct: local fishermen formed the seafood group while houseboat dwellers on the river Ribble, 60 km to the south, were the group most exposed to external radiation[18].

In a broad study of the impact of EU discharges[19] estimates were made of critical group doses: while not always the result of detailed studies, the estimates are useful in putting individual doses in perspective. The individual doses from atmospheric discharges in 1996 at 0.5 km from the plant (whether there was actually someone there or not) show annual doses from 0.01 to 1 µSv/y after 50 years of discharge (to allow for the residual effects of earlier discharges) from reactors. The Ringhals 1 reactor showed up at 28 µSv/y and the old British GCRs were up to 120 µSv/y. The atmospheric discharges at Sellafield led to a maximum dose of 39 μSv while those at La Hague were rather higher at 69 μSv. There had been no dramatic changes in the reprocessing plant doses over the previous decade.

Liquid discharges from Sellafield led, as indicated above, to annual doses around 100 μSv while La Hague were rather lower at 19 μSv. Reactors whether coastal or inland delivered around 1 μSv by liquid discharges—although some of the British AGRs were slight worse than that.

Until the 1990s the occupational exposures in the fuel cycle were dominated[20] by those in the mining sector where radon in underground mines was a major problem. In the 1990s not only was there a large reduction in the amount of uranium mined (a response to reduced demand and stockpiling) but a switch to opencast and leach mining, with reduced worker doses. The annual collective dose dropped from levels hovering around 1200 manSv to something closer to 300 manSv. Since the number employed probably fell too (UNSCEAR had some trouble believing some of the figures submitted to them) the individual doses to monitored workers showed little change.

The other large collective dose came from reactor operation; with the fall in mining doses it became the largest component in the 1990s. It climbed from 600 manSv/y in the 1970s to around 1000 manSv/y in the 1980s before falling slightly. This was a reflection partly of the increase in nuclear power generation but it was also the result of decreases in manSv/GWy for all reactor types except RBMKs. In LWRs the principal source of occupational dose is maintenance activity during shutdown. Better control of work and reduced reactor circuit activity brought the normalised dose in PWRs down from 8 manSv/GWy in the late 1970s to 2.8 in the early 1990s; BWR doses fell from the much higher initial level of 18 manSv/GWy to around 5. Similar proportional falls were seen in GCRs and HWRs. The fall in collective dose was matched by the decrease in individual dose. The average annual dose received by monitored workers fell from 4.1 mSv to 1.4 mSv.

The occupational exposures in the reprocessing sector also fell. The collective doses in the UK and France from reprocessing both fell by more than 50% between the late 1970s and early 1990s as did that in the USA (although this was defence work rather than civil

3 Fission, Weapons and Power

reprocessing). There was little change in India although data were sparse. If the Russian Federation is excluded, the world collective dose fell from around 50 manSv in the late 1970s to 33 manSv in the early 1990s. The only data available to UNSCEAR2000 from the Russian Federation was for the period 1990-1994 and this showed an annual collective dose of 33.9 manSv, doubling the world total. The annual dose to monitored workers fell from 2.94 to 0.36 mSv in France and from 8.31 to 2.03 mSv in the UK. The number of monitored workers tripled in France over the period—so this tended to reduce the average—but the dose to measurably exposed workers also fell, from 4.31 mSv to 1.43 mSv.

For the nuclear fuel cycle as a whole, the collective dose per GWy has fallen: from 20 manSv in the late 1970s to 9.8 manSv in the early 1990s. The annual collective dose to workers and the public (Table 1) has also fallen, in spite of massive rise in the amount of power generated.

Annual collective dose (manSv)	1975-9	1980-4	1985-9	1990-5	1995-7
Workers	2300	3000	2500	1400	
Public (local+ regional)	589	314	152	184	200

Worker doses from UNSCEAR 2000 Annex E Table 12 and Public doses from Annex C Table 46

Table 1: World-wide collective doses from nuclear power

There have of course been many minor mishaps, incidents and some more serious accidents from the use of fission. Most of these events have caused insignificant harm but some have been recognised for the potential for harm they revealed and a few have had serious consequences. Most affected just a few people (criticality accidents were particularly deadly), some affected a larger number of people but in a way which meant that victims (because of the statistical nature of cancer induction by radiation) could not be easily identified. One at least, Chernobyl, had effects that were not just disastrous locally but affected distant parts of the region and had major repercussions. But the first to really reach public attention to place in Cumbria, England in 1957.

The two Windscale Piles[21] were the first large reactors built in the UK; they became operational in 1950 and 1951. Each was a large stack of graphite blocks with an array of horizontal channels running right through in a design not too different from the earlier experimental pile BEPO. Natural (and later slightly enriched) uranium fuel in aluminium cans was inserted into channels at the front of the reactor; they were discharged by pushing them out through the back of the pile where they fell into trolleys in a water channel for recovery and reprocessing to separate plutonium. A variety of materials could also be loaded in "cartridges" into the channels for isotope production; Po-210, an alpha-emitter used in a weapon initiator, was produced in the pile and this (eventually) proved very significant in the accident consequences. Each pile was surrounded by a massive concrete shield—with access to the channels through holes in what was called the "charge face"—and was cooled by air. The air was blown through the pile channels by blowers and was then discharged up a 125 m tall stack through filters. The filters were added as an afterthought at the insistence of Sir John Cockcroft and were intended, in the event of an accidental release, to remove large particles that might contaminate the site. Because construction was already under way, they were mounted at the top of the stack.

One of the problems recognised with the piles was the build up of Wigner energy: the intense neutron flux caused displacement of some of the carbon atoms in the graphite and when they returned to their original positions there was a release of energy. It was understood that, if this happened in an uncontrolled way, there might be a large and potentially dangerous release of energy (a mysterious rise in temperature in Pile 2 had been put down to a Wigner energy release) so controlled releases were performed. These involved allowing the temperature of the graphite to rise to about 250C; the increased thermal vibration of the graphite atoms would then allow those which had been displaced to slip back into place. A number of these annealing cycles had already been performed (although, it seems, with increasing difficulty) when one was begun on 7 October 1957. The flow from the blowers was reduced and the reactor power increased from a low level. The energy release started but petered out before the whole core had warmed up. A second heating was then started. The first indication that all was not going to plan came on the 10 October when an air sampling point 0.8 km from the stack gave high readings and these were confirmed by further measurements. It was clear that a release was taking place but what had happened within the pile was not appreciated. The temperature monitoring system was never designed for annealing operations and it gave ambiguous results but it was clear that something unusual was happening; it was only when someone looked through one of the charge face holes that the graphite could be seen glowing cherry red. In retrospect it seems that there were hot spots in the graphite as a result of the annealing process and that either a fuel can or an isotope cartridge had ruptured and started the fire. The priority at the reactor was extinguishing the fire which had spread to involve over 10 tonnes of graphite. Attempts were made to push the burning cans through the channels but they were stuck. The best that could be done was to eject the cans in the region surrounding the fire creating a fire-break. An attempt to smother the fire with carbon dioxide failed and it was not until large quantities of water were pumped into the pile (a risky process) during the 11th and 12th October that the fire was put out. By then large amounts of radioactivity had been released from the stack.

Monitoring on site and in the surrounding area showed that the levels of gamma radiation from the plume and from deposition were relatively low and that inhalation levels, while briefly "worrying, but not dangerous" quickly declined. However, levels of iodine-131 in milk were considered a potential hazard and a milk ban was imposed over the next few days, extending eventually over 500 km^2 of the coastal region to the south of the site. The criterion used was based on keeping the thyroid dose to children below 0.2 Gy (it seems to have been devised in the heat of the incident by H J Dunster and F R Farmer of UKAEA) and the ban remained in place for several weeks[22].

Initial assessments of the radiological impact were published in
1958 and 1959 and suggested that the main hazard came from the iodine-131 released. The presence of Po-210 was mentioned but not quantified and proper estimates of the quantities released were not made available until the early 1980s when its significance began to be recognised as approaching that of the iodine-131. A 1990 reassessment, using the latest risk factors, concluded that the accident would be responsible for around 100 fatal cancers with

under 10 being thyroid cancers due to I-131 and about 70 lung cancers caused by Po-210 [23].

The Windscale accident had remarkably little impact at the time outside the technical community given the potential it revealed and the disruption it caused. Seen by most as part of an essential national commitment to nuclear weapons, with the trust then invested in scientists, plutonium piles were not optional—they were almost a fact of cold war life. The risks that went with them had not been discussed in public and—since there was only the shakiest technical framework for doing it—within the technical community either. Windscale changed some of that. After the fire the UKAEA was reorganised to set up a (somewhat) independent safety assessment group within it: the Authority Health and Safety Branch.(AHSB). Part of this, the Safeguards Division, was responsible for assessing the safety of UKAEA nuclear plant and giving an independent view of the advisability of letting it run. This resulted in perhaps the first step towards formal risk assessment in the UK. The main consequence for radiological protection was probably a closer focus on accident consequences and emergency planning (much had been improvised on the day) and the technical issues that went with it. The other AHSB Division, Radiological Protection, went on to help found the UK National Radiological Protection Board (NRPB).

The Sodium Reactor Experiment was a sodium-cooled graphite-moderated 20 MW reactor at the Rocketdyne/Boeing site in the Simi Hills about 50 km north of Los Angeles. It had been running as a commercial supplier of power to about 1000 local homes since July 1957 (slightly pre-dating Shippingport) when it suffered an unexplained power excursion on 13 July 1959. It was shutdown but then started again and ran until 26 July when it was found that there had been damage to around one-third of the core because of sealing materials contaminating the sodium and clogging cooling channels[24]. The hold-up tanks containing activity from the reactor were discharged to the atmosphere through filters over the next few weeks releasing, according to the company, a total of 700 GBq of Xe-133 and 330 GBq of Kr-85. All other fission products were claimed to have been retained within the reactor or on filters so that the maximum dose offsite was no more than 1 µGy.

The estimates of the amount released were vigorously disputed by local objectors to Rocketdyne's activities (which included rocket testing as well as nuclear work) who claimed that large quantities of solid fission products had also leaked out and were responsible for as many as 1800 extra cancers. None of the estimates and conclusions appear to have been subjected to peer review.

The Stationary Low Power Reactor Number 1 (SL-1)was a small BWR built by the US Army at Idaho Falls. At just 3 MW thermal it was a prototype for a power source that could be used at remote facilities. After shut-down in late December 1960 for maintenance and the holidays, it was being worked on when, at 2101 on the 3rd January 1961, it went prompt critical. The sudden surge of power (peaking at 20,000 MW according to McLaughlin[25]) vaporised the water in the reactor blowing out the control rod and throwing the reactor vessel upwards until it struck the roof. The operator who had been standing on the top of the reactor was killed instantly and his body was pinned against the ceiling. Two other technicians were also killed. When other staff were alerted to the accident they found that the radiation levels were so high that individuals could enter for no more than a minute. However, one person was recovered at about 2230 that night. He was alive but died soon after. A second body was recovered the following day but the operator who had been on the top of the reactor was not found for some days because of all the debris that had been created. His body was recovered on the 9th January. All the victims were so radioactive that they had to be buried in lead coffins.

It seemed clear that the excursion had happened because the main control rod (there was really only one control rod) had been withdrawn manually far too far and probably very quickly. The maintenance procedures called for it to be pulled out 10 cm to set up the automatic control system but it had in fact been pulled out something like 50 cm. Why this had been done has never been established: theories range from it being a simple error to sabotage, suicide and homicide.

There was a small release to the atmosphere of mainly I-131 and 22 people in the rescue team received doses up to 0.27 Gy while three more people received doses in excess of this. The reactor type was abandoned and no-one ever built another reactor that could go so catastrophically critical if just one control rod was withdrawn.

The accident at Three Mile Island Unit 2 PWR began at 0400 on 28 March 1979 when the pump circulating water around the secondary cooling circuit failed. The secondary cooling circuit is non-nuclear: it circulates water through the reactor's steam generator where it is converted to steam and this powers the turbine turbine. If the secondary cooling stops, the steam generator no longer removes heat from the reactor and it automatically shuts down. Since the pressure in the reactor then starts to rise, a pressure-operated release valve (PORV) on the pressuriser (a large pressure vessel used to control the pressure in the reactor) opens to vent steam from the primary circuit. All this was supposed to happen and did happen at TMI. Unfortunately, the valve did not close once the pressure had fallen and continued to vent steam and water from the reactor primary circuit without the operators' knowledge. Since the level of water in the pressuriser was indicated in the control room and since this showed as high, the operators assumed that the level of water in the reactor was also high (in fact they thought it too high and stopped water being pumped in automatically by the emergency core cooling system). The pressuriser water level was, in fact, being shown as high because its contents were now a mixture of water and steam as a result of the continuing loss from the PORV. Just after 0500 the main circulating pumps were switched off because they were cavitating, in the expectation that natural convection cooling would take over. In fact there was a steam-lock in the system so circulation had stopped. The venting of primary water was stopped after a shift change at 0600.

The level of water in the reactor fell and the reactor core began to be exposed. Around half of it was eventually exposed and a large part of it melted. It was the afternoon before the real situation was understood and then new water was pumped into the reactor and circulating pumps were restarted. The core temperature began to fall and control was restored.

The contaminated water that escaped from the PORV (there was nearly 1000 cubic metres of it) leaked into the containment building and some had been automatically pumped into an auxiliary building. There were renewed concerns on 31 March about a hydrogen bubble in the reactor pressure vessel but the event of radiological significance was the decision on the previous day to vent the auxiliary building, releasing contamination into the atmosphere.

These factors prompted widespread concern and, given the uncertainty about the state of the reactor, evacuation of vulnerable people within a 5-mile radius was advised by the state Governor. The crisis ended on 1 April when experts decided that the hydrogen bubble could not explode because there was no oxygen present. The plant was closed down, the core removed and by 1993 the large quantities of contaminated water had been processed.

The activity released at the time of the accident has had to be estimated because the monitoring systems were overwhelmed. A range of values has been arrived at but all indications are that the release was a few hundred PBq of noble gases, mainly Xe-133, and perhaps 500 GBq of I-131.

The NRC estimates of the radiological impact[26] were an average dose to the 2 million people in the region of 1 µSv (and so a collective dose of 2 manSv). A person at the site perimeter throughout the incident would have received less than 1 mSv.

While TMI had trivial radiological consequences it was the most significant event in the development of nuclear power in the USA. It revealed that simple malfunctions of plants could lead to serious consequences if mismanaged. The accident occurred as the film *The China Syndrome*, showing the serious potential consequences of a core meltdown, was on release in the USA. There were a number of organisational and technical changes after TMI (operator training was much improved) but its greatest impact was on public and therefore industry confidence in nuclear power. While critics played up the horrifying potential of the accident, the industry emphasised the tiny actual consequences. Nonetheless, no new nuclear plant was ordered in the USA for the remainder of the century.

In the early hours of 26 April 1986, the operators of Chernobyl Unit 4, an RBMK-1000 reactor, were running tests to see how long the turbines would spin after reactor shutdown and so produce power to drive the cooling systems before a backup diesel generator started. This had been done before at Chernobyl and elsewhere but meant that the reactor had to be operated at very low power, even though this was a region in which RBMKs were known to be unstable because they had a positive void coefficient: at low power. This meant that any increase in steam in the cooling system would lead to an increase in power and thus more steam—so that the power ramped up further in a potentially runaway power surge. The emergency core cooling system had to be switched off to perform the test and the operators made the fatal mistake of disabling other protection systems.

As the operators juggled the reactor power to prepare for the test there was, in spite of a last-minute attempt at an emergency shutdown, a sudden and enormous increase in power (to something like 100 times the reactor's rated value) which disrupted the cooling system and caused some of the fuel to blow apart. This caused an explosive generation of steam and this was followed quickly by a second explosion when exposed fuel vaporised. The top of the reactor was blown off, much of the core was ejected, fires were started all around as graphite burned and radioactive material began to be discharged to the atmosphere. The initial fires outside the core were brought under control quite quickly (at the cost of the lives of five fire-fighters) but the core remains continued to burn for nearly two weeks. The fire was finally extinguished after the molten core broke through the bottom of the biological shield partly through a decrease in reactivity as it became mixed with debris and partly, perhaps, because of the 5000 tonnes of neutron absorbent and fire control materials that had been dumped from helicopters.

After the expulsion of massive amounts of activity in the initial explosions the release rate fell. Most of the release occurred in the first 10 days but it did continue at lower levels for three or four weeks. By that time about 1760 PBq of I-131 and 85 PBq of Cs-137 (respectively about 50 and 30% of what was in the core) and large quantities of other isotopes had been released [27].

Several hundred emergency workers were quickly on site and 132 of them were subsequently found to have symptoms of radiation sickness: not surprising because even at 2 km from the reactor the dose rate were as high as 1 Gy/h. Ninety three workers were found to have had whole body doses greater than 2.1 Gy and 43 of these more than 4.3 Gy. Twenty eight people died within a few months. One of the fifty people with doses between 2.2 and 4.1 Gy died and only one of the 21 people with doses greater than 6.5 Gy survived. The main causes were intestinal damage and skin lesions. There were 11 further deaths from other conditions, some possibly related to the accident, between 1987 and 1998.

Some 600,000 people (the "liquidators") were involved in the cleanup and recovery after the accident; around a quarter of a million were servicemen. Around 220,000 of them were employed within the 30 km zone. They worked on decontamination of the reactor and site, the building of the sarcophagus that was made to enclose the reactor and general civil engineering work. The doses received in the early days are uncertain but the dose control standards used put some upper limits. Civilian workers were restricted to a 50 mSv annual limit with, exceptionally in 1986, voluntary exposures up to 250 mSv. This exceptional limit was reduced year by year and was soon eliminated for other than sarcophagus workers. For them it was 100 mSv. Military workers were allowed up to 500 mSv (an existing battlefield limit) in 1986 but after that the civilian limits were adopted. In total these recovery workers received around 40,000 manSv and some 40 individuals received doses over 250 mSv.

The plume from the release drifted first to the north and northwest and then to the east. The highest deposition of Cs-137 and I-131 took place around the reactor site and north of Gomel some 200km to the north-north-east. Contamination was generally in the north with some patches in the west and as far away as 500 km to the east where contamination had been deposited by rain. Altogether some 200,000 km² of the former USSR were contaminated at levels of greater than 37 kBq/m² (equivalent to 1 Ci/km²), an area that was home to around 5 million people. Contamination was found farther afield, notably in Scandinavia, Austria and Bulgaria.

In total about 116,000 people were evacuated from areas where exposures were thought to be high. Of these around 50,000 were evacuated from the town of Pripyat on the 27th of April and another 40,000 were moved elsewhere in the Ukraine. Some 25,000 were evacuated in Belarus. Individual external doses to the evacuees were as high as 380 mSv and the collective dose was just over 2200 manSv. Internal doses were mainly due to I-131 and were highest in those moved from Pripyat where some children received a thyroid dose of 1.4 Gy. As well as the people evacuated at the time of the accident, a further 220,000 people were relocated later.

The average effective dose equivalent to the 5 million living in the contaminated areas who were not evacuated was between 10 and 30 mSv in the 20 years following the accident according to the Chernobyl Forum [28]. By the end of the century most were receiving less than 1mSv/y but around 100,000 people exceeded this annual level. Thyroid doses largely arose from contaminated milk and ranged from 0.03 to a few Gy.

Since the accident 4000 cases of thyroid cancer have been identified and it is, according to the Chernobyl Forum, most likely that a large fraction can be attributed to the accident. No increase has been seen in other cancers so estimates of what it might be must be made from the doses received using dose models. As the Forum report points out, small differences in assumptions on the dose-effect relationship lead to large differences in the health consequences. The Forum suggested that, among the most-exposed 600,000 people there might be several thousand fatal cancers on top of the 100, 000 expected otherwise—a difference unlikely to be observable. The fatal cancer increase predicted among the 5 million living in contaminated areas is even more uncertain but might be 1% of the normal rate.

Contamination was significant in some other European countries. In the UK for example, although nearly 3000 km from Chernobyl, deposition of Cs-137 in upland areas meant that sheep in Cumbria, Wales, Northern Ireland and Scotland were placed under restrictions to prevent the isotope entering the human food chain. With a limit on sheep meat of 1 kBq/kg set in 1986 by an Article 13 group of experts, some 9000 farms and four million sheep were located in Restricted Areas. The animals here had to be tested to ensure that their meat was below the limit before it was released for human consumption. The chemistry of the peaty soils in upland areas were such that the transfer of radiocaesium from soil to grass persisted for much longer than anticipated and twenty years after the accident restrictions were still in place. By then 374 farms (355 of them in Scotland) and 200,000 sheep remained under restriction.

The Mayak reprocessing site near the town of Kyshtym and northwest of the city of Chelyabinsk started operating at the end of the 1940s to produce plutonium for the USSR's nuclear weapons. In the first few years it discharged large quantities of radioactive waste into the Techa river and then into the small nearby Lake Karachay. The large subsequent exposures of people who lived near the river and lake (including from the airborne contamination when the lake partly dried in the 1960s) were a subject of investigation up to the end of the century. One particular episode achieved notoriety as the Kyshtym explosion when it was revealed to the outside world by the Soviet dissident Zhores Medvedev in 1976.

From about 1953, after discharge to the Techa river ceased, radioactive wastes from the Mayak site (also known as Chelyabinsk-40 and then Chelyabinsk-65 from its postcode) were held as solutions in stainless steel tanks cooled by water and located in a concrete-lined canyon. The process then used for plutonium separation meant that the solutions were very rich in sodium nitrate and sodium acetate. The cooling system developed faults so that the water became contaminated and itself required treatment. As a result the cooling for the waste tanks was reduced and on one of the tanks it then failed completely. This tank then began to dry out, the nitrates and acetates precipitated and began to heat up. On 29 September 1957 the tank exploded ejecting around over 500 PBq of waste into the environment. While much of it fell to earth quickly, some 70 PBq drifted to the northeast and contaminated an area of around 23,000 km² to a level greater than 0.1Ci/km². There were 217 towns in the area, later known as the East Urals Radioactive Trace, with a total population of 270,000 [29]. The main contaminants of radiological significance were Ce-144, Zr-95 and Sr-90.

Water supplies were contaminated and 1000 or so people were evacuated after the first week from three villages contaminated at levels of greater than 500 Ci/km². Further evacuations were ordered eight months after the accident and in total 23 villages were cleared and nearly 11,000 people moved. Large areas were deep ploughed to bury contamination and several hundred square kilometres of land were withdrawn from agricultural use. Individual doses up to 500 mSv have been assessed and Bennett[30] has estimated the collective dose from the accident as 2500 manSv.

The accidents associated with unplanned criticality excursions are summarised and analysed in McLaughlin et al[31]. Up to 1999 there had been 60 accidents worldwide, mainly in the 1960s. Twenty two of these involved process operations with nine deaths. Thirteen of them had occurred in the USSR, seven in the USA and one each in the UK (a relatively minor incident at Windscale in 1970) and Japan (Tokai-mura in 1999). As well as the nine fatal exposures there were a further 13 of more than 1 Gy. From the 1960s it was much more common to install criticality alarm detectors to limit exposure. These devices required careful design to be effective and high reliability was demanded of them but they were effective in reducing exposures: in at least 10 of the accidents after 1960 and must have saved lives.

Up to 1999 there had been 38 accidents, including that at the SL-1 reactor, involving reactors and critical experiment systems with 12 deaths (including the three who died as a result of the steam explosion at SL-1). All but five of these accidents were before 1970. The last was at Sarov in the Russian Federation in 1997 with one fatality. As well as the fatal exposures perhaps 12 people were exposed as a result of the accidents. Generally, criticality alarms appear not to have been present at many of the accidents, perhaps because neutron detectors for experimental studies were thought enough. While they did function when present and give some warning they were sometimes swamped by the large signal generated in an excursion. By the nature of the experimental set-ups, staff were often behind shielding when an unexpected excursion took place perhaps explaining why in at least half of the incidents doses were very small.

A feature of many of the accidents in both categories was that they happened because of procedural violations. These seem to have happened for a number of reasons: ignorance, time pressure and overconfidence all played a part and some very experienced criticality technicians and scientists were involved. Some actions remain unexplained. What seems to be true is that, in spite of the complexity of criticality calculations and assessments, no accidents were caused by errors in them.

The 1999 Tokai-mura criticality accident was the only one in which there seems to have been any exposure of the public. An evacuation was ordered of residents within 350m of the plant and the highest doses measured subsequently were about 20 mSv among the 200 people moved. A release of fission products into the atmosphere occurred but the resulting doses were negligible.

Notes Chapter 3

1. (Perrin F, 1939)
2. Some of what follows is considered in more detail in Chapter 13.
3. (Serber R, 1992)
4. (Gowing M, 1964)
5. (MRC, 1956)
6. (BEAR Committee, 1956)
7. (Arnold L and Smith M, 2006)
8. (Gusev B, Abylkassimova Z N and Apsalikov K N, 1997)
9. (IAEA, 1999b)
10. (IAEA, 1999a)
11. ibid
12. (Hunt G J, 1997)
13. ibid
14. (CEC, 2002)
15. ibid
16. (UNSCEAR, 2000)
17. (UNSCEAR, 2000)
18. (Hunt G J, 1997)
19. (CEC, 2002)
20. (UNSCEAR, 2000)
21. (Arnold L, 1995)
22. (Dunster H J, Howells H and Templeton W L, 1958)
23. (Clarke R H, 1990)
24. (Daniel J, 2005)
25. (McLaughlin P M, 2000)
26. (USNRC, no date)
27. (UNSCEAR, 2000)
28. (The Chernobyl Forum, 2006)
29. (Cochran T B, Norris R S and Suokko K L, 1993)
30. (Bennett B, 1995)
31. (McLaughlin P M, 2000)

---

4 Natural, Medical and Other Sources

Since all the radioactive material available before the 1940's came from natural sources, it would have been accepted from the start, had the question been put, that man lived against a background of radiation. An understanding of just how significant that was came only some 80 years after Mme Curie's discovery.

The earliest studies of radioactivity in the air date back to 1900 when C T R Wilson and Hans Geitel independently found that the air inside an electroscope contributed to its steady discharge. Geitel and his colleague Elster, following this up in 1901, found that a charged wire strung up above the ground collected radioactivity which showed a half-life of about 30 minutes; Rutherford and Allen Canada showed that the activity was both alpha and beta. Studies soon revealed the same radioactivity in rain and snow.

By about the same time it had become clear that radioactivity was widespread and, over the rest of the first decade of the century, measurements were made on a wide range of rocks and soils. Water from the seas, oceans and springs was tested and found to contain radium and other radioactive materials. The link between radioactivity in the air and that in the soil began to be apparent with Geisel and Elster's work in 1902 on activity in air in caves and in air drawn from the ground, which both showed high levels of activity. They concluded that the radioactivity came from the ground itself and Ebert and Ewers showed that it was radium emanation (radon) [1].

Much more was learned about the radioactivity of the earth and how it varied from place to place. Even by 1905 Rutherford was able to conclude that the radioactivity in the air was closely tied to the radium content of the ground. The first calculations of dose rates above ground containing different concentrations of radioactivity seem to have been done by Victor Hess (who was interested in cosmic rays as we will see) in 1934. Hultqvist, in the mid-1950s found rather simple relationships that allowed UNSCEAR in 1958 to calculate dose rates above different types of rocks in their estimates of the contribution of terrestrial gamma-rays to per caput dose. They estimated the dose to bone and gonads from them at around 50 mrad/y and these estimates, even when UNSCEAR adopted the effective dose equivalent in 1982, remained remarkably constant for the rest of the century. It was found that doses indoors were rather higher than those outside and that some building materials caused elevated levels (a concern uncovered first in Sweden). Igneous rocks were found generally to give higher doses than sedimentary ones and some areas of very high natural ground activity were found in Brazil and India where, in Kerala, the monazite sands gave dose rates of 12 mGy/y and more. Investigations of people living in these high background areas seemed to show no elevated rates of cancers.

There is however another important source of external radiation and the first significant step in finding it was taken by Theodore Wulf who, in 1910, measured the ionisation rates at the top and bottom of the Eiffel Tower. He expected to get lower values at the top (because he would be further away from the radioactive earth) but found just the reverse. Wulf (who had invented the electrometer we will encounter later) also made measurements in caves and tunnels but his result interested Victor Hess who sought both to eliminate any effects from the Tower itself and to get higher. He took to a balloon. In 1912 he found that ionisation decreased as the balloon rose to about 1000 metres but then started to increase so that, by the time he was 5 km up, the rate was several times that at ground level [2]. He eliminated the sun as a direct source by making similar measurements during the partial solar eclipse of April 1912. Evidence pointed to an origin in deep space and the daily variations seen were put down to atmospheric (we would now say magnetic) effects but it took some time to make progress. While Millikan coined the catchy term "cosmic rays" in the 1920s, Hess could still point to how much needed to be done when he made his Nobel prize acceptance speech in 1936.

Cosmic rays were an important source of high energy particles for research and there were consequently dramatic developments in understanding their source and behaviour. The variation of their intensity across the earth and with height and time has been documented. However since 1958 the estimates of their contribution to doses have remained at around 0.3 mSv/y at ground level with perhaps three times that at the highest altitudes with significant permanent populations. They are of concern for astronauts and do noticeably increase the radiation exposures of aircrew and frequent fliers in private jets but for most of us they are a minor component over which we have little control.

The radium emanation was of no great significance as an external radiation source in the open air and its effects once taken into the body do not seem to have been given serious consideration until Hultqvist in 1956 published[3] model that accounted for the ventilation rate in homes in estimating doses to the lung. It also allowed for retention of the small aerosol particles to which the radon daughters became attached. The model was picked up by UNSCEAR in 1958[4] and, when combined with measurements of radon and thoron in Swedish buildings, suggested that the combined lung dose from the two isotopes lay between about 0.12 and 0.3 mGy/y (1.2 and 3.0 mSv/y when absorbed dose was multiplied by an RBE of 10) in wooden and brick houses but could be as high as 0.44 mGy/y (4.4 mSv/y) in buildings constructed of a lightweight alum shale concrete. While UNSCEAR recognised that the lung would be the critical organ, they devoted only one paragraph to the calculations and made no comment on the results. They were focused on the gonads and bone. In the subsequent report in 1962[5] radon inhalation was again covered in a paragraph but a few more authors [6] could be referenced on the magnitude of lung doses. For 0.5 pCi/l (19 Bq/m^3 and considered a "world average" value) of radon in equilibrium with its daughters, the estimated lung dose ranged from 0.1 (Hultqvist) to 0.90 mGy/y (1-9 mSv/y dose equivalent using RBE of 10) for continuous exposure. The Committee commented: "It is apparent that the lungs receive a higher irradiation from natural sources than any other body tissue."

By 1977 there had been much more work on lung models (although they still gave predictions of dose that varied by a factor of ten), the variation of radon in the atmosphere and the effects of ventilation. There was also a better understanding of the difficulty in making global estimates of average doses given the variety of environments and housing types. When all the factors were taken into account, the average lung dose they arrived at by UNSCEAR[7], 0.3 mGy/y, was close to the centre of the ranges published as far back as 1958. It was however now elevated to the main summary table and the value was remarked on in the main text. Its new prominence was emphasised by a claim that it was the first time it had been estimated.

The first indoor measurements in the UK had been made by Haque, Collinson and Blyth-Brook [8] at London's Borough Polytechnic in 1965 and there were important surveys in Austria, Norway, the USA, Hungary, Poland and the Soviet Union but the first events that stimulated a sharper interest in residential radon were probably the surveys of houses built in North America on (or using) uranium mine tailings. These tailings had been used in Grand Junction, Colorado, USA and Port Hope, Ontario, Canada and large surveys were undertaken in both places (and several similar others) during the 1970s and early 1980s. They resulted in remediation action being taken to reduce radon levels.

The growing general appreciation of radon was boosted dramatically when the new methodology of ICRP Publication 26 was applied to the problem. Instead of the lung being a separate critical organ, the dose to it was now incorporated as part of the effective dose equivalent. For the first time the full significance of the dose from radon could be seen. Not only did radon and thoron daughter inhalation between them contribute half of the entire mean effective dose equivalent from natural background (0.97 out of 2.0 mSv/y), but the high levels found in some areas pointed to annual doses to some members of the public that were well above the occupational dose limits.

In December 1984 Stan Watras of Boyertown Pennsylvania triggered the portal radiation monitor as he left work at the Limerick Nuclear Power Station nearby. This was unexpected: the plant was not yet operational. This was repeated several times over the next two weeks; when Watras, frustrated, tested himself as he arrived at work, the alarm went off again. With some difficulty he persuaded the health physics staff at the plant to check his home and they then found around 2600 pCi/l of radon daughters in the air of his basement; his effective dose equivalent over the preceding year could have been several Sieverts. This was not the result of uranium mine tailings but the result of sitting on a local region of igneous rocks with high uranium content. With help from his Congressman Watras managed to get remediation action on the house and the levels were reduced to close to normal. However the Watras house has become part of the folk-lore of radon—at least in the USA.

These various events led to reviews in many countries. In 1986 the World Health Organisation recommended action levels and in 1987 an action level for radon of 400 Bq/m³ was set for the UK. High levels that required action were found in Devon and Cornwall (some effective doses as high as 100 mSv/y were assessed) and the Building Regulations were modified to protect people from excessive radon levels. Three years later the UK action level was halved. All the US states were making radon measurements in homes.

In its 2000 report UNSCEAR estimated the worldwide average annual effective dose from inhalation as 1.2 mSv, with most of it coming from radon daughters. The range was given as 0.2 to 10 mSv. The risks to people from radon in homes were not seen in epidemiological studies (indeed what effects there were were claimed as support for radiation hormesis [9]) until 2005 when a combination of thirteen separate studies[10] showed residents of higher radon houses to be more likely than others to develop lung cancer, particularly if they were current or recently-reformed smokers. As it is, radon is the single most significant source of exposure to radiation for almost everyone on the planet, unrecognised for the first eighty years of the century.

On the other hand, lung cancer caused by radon has been recognised as a risk for uranium miners for a long time. Since it has formed one of the bases for radiation protection policy development, miners are dealt with in the chapter on somatic effects.

Like most natural materials our bodies contain radioactivity and they irradiate us. Many nuclides have been considered over the years (there was a particular enthusiasm to establish a background level against which fallout intake could be compared) but really only K-40 and the long-lived nuclides in the uranium/thorium series have been assessed to makes much of a contribution to global averages. Before the appreciation of the significance of radon, K-40 was thought responsible for around 20% of the dose to gonads and bone. After radon there was no change in the dose estimates but its contribution dropped to less than 10%. Once the effective dose equivalent began to be calculated the uranium/thorium series had more impact: with better understanding of the metabolic processes, its assessed contribution(and this mainly from Pb-210 in the uranium series) had increased to be similar to that of K-40 by 2000 [11].

Once the new science of diagnostic radiology began to settle down, with the modest standards of care for the radiologists established, the routine doses to the patients undergoing procedures became of no great concern. Tubes became better shielded and films improved so that, on the whole, there were seldom any apparent effects of exposure: with the prevailing idea that there was a threshold, there was no incentive to reduce doses.

One of the incentives might have been inter-comparison but this depended on having some standard methods for measuring and reporting doses—a not insignificant issue given the variety and number of radiographs that were being taken.

The first international review by UNSCEAR in 1958 was concerned principally with genetic impacts, so many of the conclusions they drew seem less relevant today. However, they did consider the mean bone marrow dose (called the "dose" in what follows) that resulted from various examination procedures so, since this is reasonably close to the quantity that would concern them later (effective dose equivalent), following it gives some insight into how things have changed. There were at least two uncertainties: there was little knowledge of where the active bone marrow actually was but this was dwarfed by the great variations in doses found between different countries and hospitals.

If we exclude for the moment fluoroscopy and mass surveys, the procedures that contributed most to collective dose in the 1950s were examinations of the gastrointestinal tract. Examinations of the upper and lower tract both required several views and resulted in mean marrow doses of 5-7 mGy (in mrem in the original) for the set. Excluding dental x-rays, chest examinations were in fact the most common(almost 1/10 of the population of advanced western countries were thought to have one each year) but resulted in only around 0.4 mGy per examination. Fluoroscopy examinations were conducted on between 4 and 31% of the population in the three countries for which data were reported. Although some fluoroscopy doses may have been in the range below 10 mGy, doses to patients could, in extreme cases, be "very high". This probably means that they were significant fractions of a Gray. Fluoroscopy doses could be reduced significantly if an image intensifier (invented in 1948 by Coltman) were used. The first of these to be used in routine radiography became available in the early 1950s and a number of systems, mainly designed for use during surgery, began to appear around the middle of the decade. They had the potential to reduce doses to patient and operator by a factor of around 1000.

The 1950s were perhaps the peak of mass chest x-ray screening for tuberculosis and it was the time when effective drugs became widely available. Between 10 and 20% of the population of many western countries experienced an examination. In many countries this was performed by photo-fluoroscopy where the fluorescent screen was photographed and these images were later scanned (a mass miniature system had been available since the 1930s). The doses were about 1 mGy per examination. In some countries the approach was rather different: the examinations here were conducted by straightforward fluoroscopy (a method abandoned elsewhere) and the doses were considerably larger. In 1962 UNSCEAR estimated the doses at 10-20 mGy per examination in Austria, France and Spain. In France over half the population were being examined by this means each year with doses estimated at 12 mGy each time.

There may well have been some decrease in dose per examination for specific examinations over the next two decades in normal radiography but in 1982 UNSCEAR [12] (who now began to use effective dose for medical exposures) commented that there had been little or no reduction in general medical exposures in spite of faster films and the more sensitive rare-earth intensifying screens that had become available. There had been though a significant move away from mass fluoroscopy to radiography: in France the number of fluoroscopy installations had fallen from 13,000 in 1976 to 5000 in 1982 with an expectation that they would be eliminated by 1985. By then, anyway, the incidence of TB was declining and the other possible reason for chest screening—early detection of lung cancer— did not seem good enough to justify continuing the exercise. So while 10-30% (or more) of the population in many countries were still screened, in the UK and the US the proportion was down to 5%.

Patient doses required for mammography examinations fell by factor of about ten during the 1970s as techniques and equipment improved. On the other hand, the procedure became more popular: examinations doubled in Sweden in just two years in the late 1970s. But the most significant development was probably the introduction of computed tomography.

One of the dreams of x-ray pioneers was to obtain a three-dimensional view of the inside of the body. As early as 1896 Elihu Thomson published a paper in the *Electrical Engineer* entitled *Stereoscopic Roentgen Pictures*: two images were produced with two tubes in different positions (later special double tubes were used) and viewed through a stereoscopic viewer. It was possible to locate foreign objects quite well and this proved useful during the First World War. Stereography, where film is used to record the images for later viewing, remained in use throughout the century but, from the 1930's, an alternative was available: tomography.

The principle of tomography is simple enough. If an x-ray tube is moved backwards and forwards above a patient while the film is moved back and forth in an exactly opposite way, then the image of just one layer of the body will fall in precisely the same place on the film all the time. In principle there will then be a sharp image of that layer when the film is developed. Other layers will be more or less blurred. By adjusting the relative movements of the tube and film, different layers can be brought into sharpness and made to stand out against a blurred background.

The principle seems to have been appreciated and used just before and during the First World War by one or two radiographers but it was the Frenchman André Bocage who really developed it and published a patent for a practical device in the early 1920s (two other Frenchmen patented something very similar but apparently not quite as good at about the same time). Bocage was unable to actually construct his device so the credit for the first practical device usually goes to a Dutch engineer who had turned to medicine: B G Ziedes des Plantes. Unaware of Bocage's work, he produced a prototype for planigraphy (as he called this form of tomography) in 1931 and the first commercial version appeared in 1936. There were several other people who produced, about the same time as Ziedes des Plantes, apparatus based on the same principle and shortly after this several variants appeared; in one, for example, the tube and film moved around the patient on arcs rather than linearly. The technique was popularised in the UK by E Twining and in the USA by Kieffer, who patented a variant called the laminagraph.

TAMING THE RAYS

The plain tomograph, as it is sometimes called, more or less in the basic forms established in the 1930s reigned (although devices with complex movements were developed to give better pictures) until the 1970s and the appearance of computerised tomography. It resulted in larger doses to patients than straight radiography. In chest examinations for example—for which it was widely used—doses could be around ten times higher. But, of course, the technique provided much more information.

Godfrey Hounsfield invented computerised assisted (later "axial") tomography working at the British company EMI in 1972 (it is said that some of the development money came from the record sales of the Beatles). The method depended on sending a large number of pencils of x-rays through a slice of the body at various angles. The attenuation of each ray depended on what was in its path. With a large number of pencils criss-crossing the slice it is possible to untangle the results, using a computer and a suitable software algorithm, to find the attenuation caused by each element of the body, giving a picture of the structure of the slice. Hounsfield received a Nobel Prize in 1979 (which he shared with Allen M Cormack who had independently developed many of the theoretical techniques and experimented without producing a machine) and the CAT scan became an essential part of medicine, revealing the internal structure of the body in images (later in 3D) of great complexity, beauty and diagnostic value. By 1973 head scanners were being used clinically around the world; whole body scanning became available at the end of the decade. The typical dose from a whole body scan was less than 1 mSv.

Over the 1980s there was some progress in reducing patient doses from chest x-rays [13] so by 1990 doses had fallen to 0.1 mSv per examination although because of the large numbers of people examined, they remained the major contributor to medical collective dose. Mammography doses continued to fall. When in 1992 they reviewed patient dose in the UK in the preceding 30 years, Hart and Le Heron [14] found that there had been a significant reduction in dose per examination but that the number of examinations had increased (by a factor of four between 1945 and 1983) to give a collective dose that had decreased rather little.

CAT scanning was being used more widely by the late 1980s and the average dose per examination and was estimated by UNSCEAR as 5 mSv and increasing. In the UK the collective dose from

4 Natural, Medical and Other Sources

conventional radiography had remained roughly steady at 16,000 manSv in the decade to 1993 [15] but the contribution from CT (where approaching 1 million scans were being made annually) had risen to 4500 manSv. Doses from body scans in the UK were closer to 10 mSv than UNSCEAR's 5 mSv.

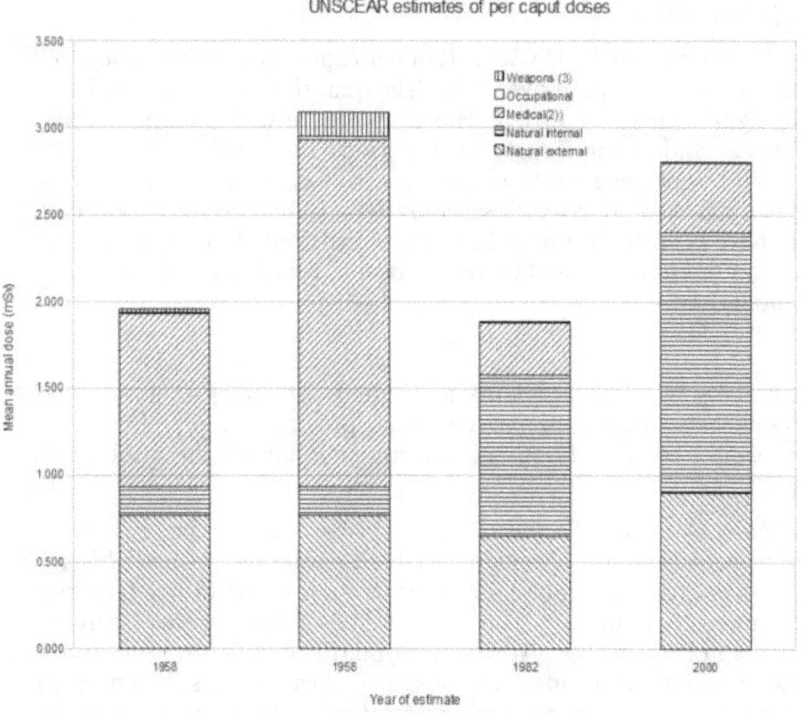

Figure 6 Per caput dose from different sources (UNSCEAR)

A significant event in the UK was the publication in 1990 of a joint report from the Royal College of Radiologists (RCR) and the NRPB on patient dose reduction[16] The report reviewed clinical practice to see how many routine x-ray examinations were "clinically unhelpful, repeated and unoptimised" and concluded that it might be possible to cut collective dose by nearly 50% without any impact on clinical care. This would require both improved radiographic equipment (rare-earth intensifying screens should be more widely used for example) and procedures (a reduction in the number of radiographs and introduction of better quality assurance). Tighter criteria for referral to radiology departments were drawn up by the RCR [17] and the NRPB began to publish data on doses achieved by better performers so that radiology departments could benchmark their own performance.

The estimates made by UNSCEAR of per caput doses from all sources are not easy to compare over time. The quantities have changed and their significance has varied. However Figure 6 summarises three estimates made over a forty year period. The 1958 estimates are lower and upper estimates of mean marrow dose; the later ones are best estimates of effective dose equivalent. Whatever the variations there have been (radon would not have contributed much to marrow dose even if known about) there has been a remarkable constancy in the total.

The more significant accidents associated with nuclear power and weapons have been summarised in the previous chapter but there have been a number of others, affecting employees, the public and patients

Accidents at irradiation facilities, where either large gamma sources or accelerators provide very high gamma or x-ray fields for research or sterilisation of equipment, have resulted in five fatalities and numerous injuries [18]. The five fatalities all occurred at gamma facilities: in Italy in 1975, Norway in 1982, El Salvador in 1989, Israel in 1990 and Belarus in 1991. The dose in one case at least was over 20 Gy. The high exposures at accelerators tended to be of parts of the body (notably hands, feet and arms) and resulted in amputations being required.

A company sterilising medical products had been operating near San Salvador in El Salvador in spite of the civil war going on[19]. However, by 1989—after ten years of war—the gamma facility they used was seriously degraded. The activity of the Co-60 source had decayed to 660 TBq, there had been no maintenance worth speaking of and the installed gamma monitor that prevented entry to the shielded enclosure if the source was exposed had been removed. The three people trained when the facility was set up had left soon after and subsequent training had been "informal and oral". The

4 Natural, Medical and Other Sources

conditions in the country meant that there was no regulatory regime in place and barely any radiological protection expertise available.

The source was made up of a series of cobalt pencils arranged in a rack in the form of a grille. The products to be irradiated were placed in boxes and these were shuffled around the source by a series of pistons to give a fairly uniform irradiation. Between irradiations the source was lowered into a pool of water by a cable system.

In the early hours of the morning of 5 February 1989 the operator's coffee break was interrupted by an alarm which showed that the source was in transit between the irradiation and storage positions: it persisted indicating that the source was stuck. The source position indicator light was flashing. This was not unusual because boxes had become jammed in the past and interfered with the source operation. The operator did what had been done before and went up to the roof and pushed and pulled the source hoist cables to free the source. In doing so he tripped the micro-switch so that the position indicator light now indicated that the source was down.

He then opened the door into the enclosure (without checking radiation levels with the portable monitor provided) by a process which had become standard since the installed gamma monitor had been removed: he quickly cycled the buttons on the control panel, tricking the system into allowing the door to open. He then waited a few minutes for the ozone f by the source to clear and went in. He found that some of the boxes were indeed jammed and, when he removed them, he saw the source jammed by one of the hoist cables he had pushed and pulled. He then tried to release the source and when he found this impossible he went off and asked two other workers—not connected with the irradiation facility—to lend a hand, explaining to them that it was quite safe. Between them they struggled with the cable and did manage to lower the source.

Shortly after, the operator began to vomit violently and was taken off to hospital. The other two workers also became ill. All three were formally diagnosed with food poisoning and given sick notes.

The facility continued operating for several days but, because the source rack had become distorted during the accident, a source pencil fell out into the radiation room. This was subsequently dealt with—although there were further significant exposures—but it drew attention to what had happened earlier. It was then realised that the men were suffering from radiation sickness after a massive dose of radiation. The operator's condition deteriorated—partly as result of inappropriate medical treatment—and he died on 20 August, 197 days after the accident. His mean body dose was estimated at 8.1 Gy. One of the other workers had been heavily irradiated around the legs (> 30 Gy) and both of these were amputated but he and the other helper survived after mean doses of 3.7 and 2.9 Gy.

In this example lack of training (to the extent of total ignorance of the danger the source presented) and removal or routine bypassing of the safety systems are the likely basic causes of the accident. These seem to be features of many of the accidents that have happened. With over-familiarity, overconfidence and time-pressure they make a devastating combination.

In their review[20] Ortiz et al referenced eighteen incidents involving radioactive sources—usually large radiography or radiotherapy sources—which had either killed or seriously injured members of the public. Their common feature was that the sources fell into the hands of people who did not understand the dangers they presented. The seven incidents that resulted in deaths were (fatalities in brackets) in: Mexico in 1962 (4), China in 1963(2), Morocco in 1984(8), Brazil in 1987(4), China in 1992(3), Estonia in 1994(1) and Georgia in 1997(1). The accident in Goiania, Brazil can perhaps stand for all of them[21].

A private radiotherapy partnership in Goiania, Brazil moved to new premises and a radiotherapy unit, containing 51 TBq of Cs-137, was left behind as a result of a dispute over who owned it. The old clinic was subsequently partly demolished and used by vagrants as a shelter and the word seems to have spread that some of the contents looked valuable. Between 10 and 13 September 1987 some local men found the unit and dismantled it, taking the source home—exposing themselves to about 4.6 Gy/h at 1 m. In dismantling they broke open the source capsule and the caesium chloride salt was released. Over the next two weeks the remains of the unit were sold to scrap-yards and, since it glowed so beautifully, friends and family were invited to admire the source and were given pieces of it. Although several people fell ill it was not until 28 September that the authorities were made aware.

Twenty people were found to be seriously ill and four of them, with doses in the range 4.5 to over 6 Gy, died. Two more people with

4 Natural, Medical and Other Sources

doses in that range survived. Surveys showed that 249 other people were contaminated internally or externally and contamination was also widespread in the environment. Twelve locations were found covering in all around a square kilometre with dose rates as high as 2 Gy/h at 1 m, forty two houses were demolished and some 3500 m³ of radioactive waste was generated before the cleanup was over. It has been estimated that, by then, about 44 TBq of the initial 51 TBq had been recovered.

Radiotherapy depends on very large doses (perhaps 10s of Grays) being delivered to localised parts of the body to kill cancer cells. It uses penetrating radiation in a beam from a radioactive source (or sometimes an accelerator) or places a source in contact with the region to be treated to generate localised strong fields (brachytherapy). Treatment plans are often complex and the equipment must be carefully calibrated because the margins of safety are not large. Ortiz et al found seven major examples of overdoses being caused to groups of patients varying in number from just one person to almost 1000. They found 36 fatalities altogether. Most of the accidents arose from errors made during calibration of Co-60 sources but a Spanish accident in 1990 overdosed 27 patients and killed 18 of them through errors during and after the maintenance of a clinical accelerator: whatever electron energy was set 36 MeV electrons were generated. ICRP Publication 86[22] added accidental exposures in the USA and Canada between 1985 and 1987 when three people died as a result of treatment from an accelerator because of a software problem. One death, in the USA in 1992, resulted from a brachytherapy source being left inside a patient for several days through a fault in the equipment and in spite of high radiation readings near the patient. There have also been cases of underdosing which can, in some circumstances, be as harmful as over-dosing: ICRP estimated that of the 1000 patients under-dosed between 1982 and 1991 in the UK almost one half died as a result of local recurrence of cancer because of the under-dosing.

All the accidents reported in the sources occurred in western Europe and North America. It seems unlikely that the situation elsewhere is any better.

Notes Chapter 4

1. (Kathren R L, 1984)
2. (Hess V F, 1965)
3. (Hultqvist B, 1956)
4. (UNSCEAR, 1958)
5. (UNSCEAR, 1962)
6. (Shapiro J, 1956; Schraub A, Aurand K and Jacobi W, 1957)
7. (UNSCEAR, 1993)
8. (Haque A K M M, Collinson A J L and Blyth-Brook C O S, 1965)
9. (Cohen B, 1995)
10. (Darby S, Hill D et al, 2005)
11. (UNSCEAR, 2000)
12. (UNSCEAR, 1982)
13. (UNSCEAR, 1993)
14. (Hart D and Le Heron J C, 1992)
15. (Kevles B H, 1997)
16. (NRPB/RCR, 1990)
17. (Royal College of Radiology, 2007)
18. (Ortiz P, Oresegun M and Wheatley J, 2000)
19. (IAEA, 1990)
20. (Ortiz P, Oresegun M and Wheatley J, 2000)
21. (IAEA, 1988)
22. (ICRP, 2001)

---

5 Interactions and Cell Death

The first detailed study of the nature of particle tracks seems to have come from W H Bragg. As President of the Australasian Association for the Advancement of Science he was required to give a talk on recent developments in some branch of science to their meeting in Dunedin, New Zealand in 1904. He chose as his topic "On some recent advances in the theory of the ionization of gases". Until this point Bragg's career was remarkable only for his youthful appointment as Professor of Mathematics and Physics at the University of Adelaide in 1896: he had concentrated on being a teacher and shown only passing interests in x-rays and wireless. By 1904 he had been in the post for 18 years, was 42 years old and had made no significant contributions to research on anything. However, his preparation for his talk convinced him that this was a fruitful area of study and, specifically, that measuring the ranges of alpha particles could be a way of measuring their energies. He returned to Adelaide and started work with a young student R D Kleeman and, within a year, produced two key papers on the tracks and energy loss of alpha-particles[1]. They provided both the profile of the ionization intensity along the track (the Bragg curves) and the relationship between ranges in different materials (the Bragg-Kleeman rule).

This thread of research was to lead to the understanding of the ionisation caused by particles but there was to be striking confirmation from another quite different approach: the cloud chamber. C T R Wilson had been developing this device since 1896 and using it to study ions generated in it. Early on he had seen that x-rays and "Uranium-rays" caused ionisation in the chamber but it was not until 1910-1911 that he saw individual tracks of alpha-particles and electrons. They were striking confirmations of the conclusions that Bragg had already reached on the nature of the tracks [2] and that understanding was to be a significant factor in untangling the effects that radiation had on living tissue.

In the early enthusiasm for radiation, any number of biological systems were irradiated ranging from bacteria to man. The results were variously unreliable, inconsistent and, for man, unintentional and tragic. Simply measuring the dose delivered to an experimental subject was hard enough—with all the problems of reproducibility that followed. But the essential problem came from the unfathomable complexity of biological systems and little real progress could be made until simplified systems for study could be developed and some standardisation in the effects to be studied agreed. So, while the immediate interactions of radiation with living tissues might be understood it was when scientists turned to viruses, bacteria and other simpler organisms cultivated *in vitro* that a body of understanding began to build up. And it was when the effect studied became cell killing that reproducible results began to appear. Variously called cell killing, inactivation, sterilization, disinfection (in the early days) and loss of clonogenic capacity (to reflect the measurement technique) this referred to the loss of the cell's ability to reproduce.

Three key features of cell killing were taken as the important ones to be explained by any model of radiation action: dose dependence, the effects of ionisation density and that of dose rate or fractionation.

The usual way to present dose dependence was (and is) as a plot of the logarithm of the fraction surviving against the dose delivered—the so-called survival curves. On such plots many of the simple systems showed a response that plotted as a straight line—or close to it—as in Figure 7 from Lea [3] for bacteriophages.

5 Interactions and Cell Death

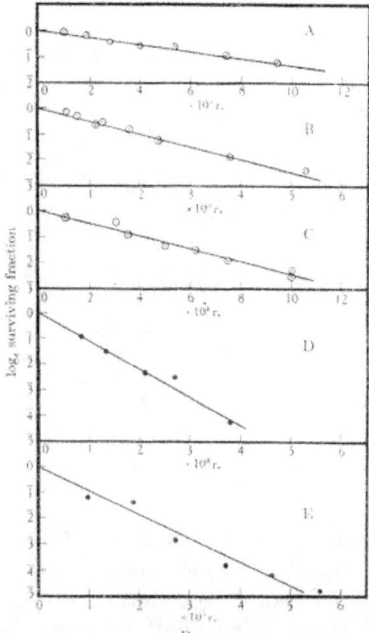

Figure 7: Exponential survival curves of bacteriophages

A graph, in the 1961 review article of Powers[4] of the survival of *Serratia marcescens*, a bacterium able to produce a deep red pigment and suspected as the cause of communion bread turning red in mediaeval monastic safes in apparent transubstantiation miracles, is itself remarkable. The surviving fraction is shown to be exponentially related to dose over an impressive 9-decade range, its behaviour marred just by another feature of interest: the straight line portion at higher doses projects back, on a logarithmic plot to intercept the y-axis not at 1.0, as a true exponential would, but at a higher value. It is, in the terminology a "shouldered" curve. The explanation of the exponential form and the appearance of the shoulder—seen better in the survival of mammalian cells in Figure 8[5] —was a major aim of radiobiology for much of the century.

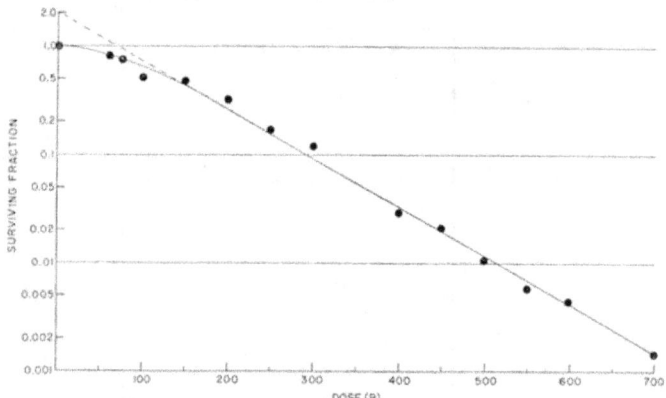

Figure 8: Survival of mammalian (HeLa) cells after x-ray exposure

The effects of ionisation density on cell killing were probably first studied by Zirkle in 1935 [6] and 1940 [7]. He used filtered alpha particles to expose nuclei of fern spores to different ionisation densities using the Bragg curve and found different effectiveness for several biological effects including cell division. All the effects studied showed an increased efficiency with increasing ionisation density and Zirkle put a formula to it: biological effectiveness was proportional to $I^{2.5}$ where I is the ionisation density.

He also found that mould spores and yeast showed increased effectiveness with increased LET while *E coli* (he used the older name of *B coli*) were just the reverse. Other workers found a similar decrease in effectiveness with increase in LET with other bacteria and viruses.

So, from around the 1930s the main differences to be seen between radiations of different LETs:

- For simple organisms like viruses and bacteria, high LET radiation is less efficient at cell killing than low LET radiation

- For more complex organisms (including mammalian cells when the data became available) the reverse is true

When Lea reviewed the rather scanty data available on dose rate effect in 1942 [8] he was unable to see any effect on the killing of simple organisms.

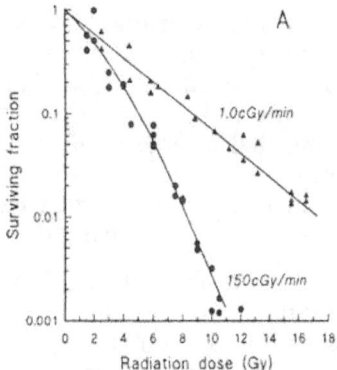

Figure 9: Effect of dose rate on survival of human melanoma cells

Later experiments on mammalian cells have shown that there is a generally decreased effectiveness as the dose rate decreases or—perhaps better put—as the time over which a fixed dose is delivered is increased. Figure 9 [9] shows a modern demonstration of the dose rate effect.

A basic question that had to be answered was "How could so little energy have such a great effects?": the energy absorbed from damaging x-ray exposure corresponded to a temperature rise of only a fraction of a degree. Dessauer[10] suggested in 1922 and 1923 that the harm arose because the thermal energy was being concentrated in very localised regions of the protoplasm, killing minute parts that decomposed and released substances that poisoned the cell. This was his "point-heat" theory.

Blau and Altenburger [11] in 1922 proposed that each cell or organism died when it had absorbed a particular number of x-ray photons. Although the theory was flawed, they appear to have been among the first to suggest that all cells were equally sensitive to radiation so the survival curve arose from hit statistics. They were first to state some of the key mathematics that was to become central to target theory in general.

Crowther used target theory in 1924[12] in his interpretation of the results of Strangeways and Oakley[13]. His 1926 paper [14] reported some careful measurements on the effects of x-radiation on the protozoan *Colpidium colpoda* where he found sigmoid survival curves, interpreted as showing that a large number of x-ray quanta (more than 40) were required to kill one organism. He may have been wrong about the mechanics but the curves he saw were simply a result of statistics—the essential feature of the target theory.

When Charles Packard reviewed the field in 193 [15] he could point to a range of experiments on *Drosophila* eggs, *Colpidium* (by Crowther), *Ascaris* eggs and with tumour tissue that had been minced, irradiated and injected into rats to culture. The survival curves were all distinctly different but showed the general sigmoid (or, as he put it, ogee) shape—interestingly the *Drosophila* results lay on a lognormal curves. Crowther's multi-hit theory had been developed more by Condon and Terrill[16] to accommodate quanta of different energies and this thread—the effectiveness of x-rays of different wavelengths—continued to be controversial. Packard thought that this might be resolved once better measurement of dose delivered became standard—as was by then possible.

Mayneord[17] in 1934 progressed a version of target theory that depended on a particle track passing through cells to cause biological change. He made detailed calculations of ranges and energies of photo-electrons and Compton electron produced by X and gamma-rays and deduced a number of survival curves on the basis of single and multiple hits being required.

Douglas Lea and his co-workers[18](this was Lea's first paper in radiobiology), suggested in 1936 that, although much had been published, "The mechanism of disinfection ... remains obscure". However, in Lea's elegant prose, they disposed of Dessauer's heat idea and of the notion that bacteria killing could be the result of some poison produced by radiation. The exclusively exponential survival curves they saw were best explained by target theory with each bacterium having just one sensitive region and with one hit (with a

5 Interactions and Cell Death

hit being the passage of an energetic particle through the region). The idea that the survival curve could be the result of varying sensitivity among the organisms was discussed and rejected. Ionisation and its effects were the key to the damage done by radiation. The authors' extension of the measurements to gamma-rays[19], where exponential survival curves were found, was sidetracked into an investigation of the long threads produced by *E coli* when irradiated. They concluded that these were bacilli that had continued to grow after irradiation but failed to divide. It illustrates how, even at this relatively late date, experiments on simple organisms were still producing surprising results.

Crowther, when he delivered the Silvanus Thompson Memorial Lecture in 1937[20], still preferred his random ionisations explanation to the other target theories that had emerged in the 1930s: Glocker[21], Mayneord[22] and Lea[23]. These three were all based on treating radiation in terms of particle tracks so that when a track passed through a sensitive area a hit is registered. He also disposed of a variation, down to Holweck and Lacassagne, that assumed a hit was somewhere where a photon was absorbed inside a sensitive volume [24]. Like Lea he had little time for the poison hypothesis: radiation produces some unspecified poisonous substance that diffuses into the sensitive areas of the cell and damages them. He also quoted J J Thomson, his mentor, in relation to the target hypothesis: "A hypothesis is a policy, not a creed".

The next word perhaps belongs to Douglas Lea who made target theory central to his 1946 book *Actions of Radiation*[25]. However, he did restrict its application. He was convinced that it applied to inactivation of small viruses and the production of chromosome aberrations and thought it probable that it applied to the killing of larger viruses and bacteria and the production of gene mutations. As he saw it, three factors should be taken into account in deciding whether it could be applied: the shape of the dose response, the effect of dose rate and the dependence on radiation type and wavelength. With these guidelines he focussed on two aspects: cases where the biological effect was supposed caused by a single ionization (which covered most of his cases) and the production of chromosome aberrations by the passage of a single densely ionising particle. He specifically excluded detailed discussion of multi-hit theory, where several ionising events were required from different particles, and implied that the experiments that pointed to it through a dependence of effect on dose rate or fractionation might be better explained by the existence of a repair process.

Target and hit theory can be distinguished but this is not done here because they are really subsets of a more general theory. Behind both of them is a proposition that cell killing (or some other end point) can be modelled by assuming that within each cell there are a number of targets and that when a number of these targets are "hit" by radiation the result is death (or some other effect). This general situation would be a multi-target multi-hit situation. In Lea's original work he found that the results of cell killing experiments on bacteria and viruses could be explained if it were assumed that there was one target per cell and that it could be deactivated by just one hit. A multi-target model might require (and this is the most common) that all the targets in a cell are hit before the biological consequence is triggered but, in principle at least, it might be enough for just some of them (even one) to be hit. Conversely, it is conceivable that there is just one target that needs to be hit several times before an effect is seen.

The major achievement of the single-hit target theory, for Lea, was the explanation of the differing efficiencies of different radiations. For single-hit events—i.e. those leading to exponential survival curves—the efficiency of more densely ionising particle would be expected to be lower that that of sparsely ionising ones because, the ions being close together, it is more likely that several will be deposited in a sensitive volume when one would be enough. The wastage that resulted would make alpha-particles and neutrons less biologically effective per unit dose. Precisely the reverse would be true where several ionisations close together are required for an effect: so for chromosome breakages densely ionising particles would be expected to be more effective.

However, the real attraction of the single hit theory was the possibility it gave for estimating the size of the sensitive volume. Lea tried four methods for doing this using data on the inactivation of a bacteriophage and found reasonable consistency, with his favoured method giving a target diameter of about 16 nm for gammas, x-rays and alpha-particles alike. Other systems gave diameters of 4-40 nm.

There might, he thought, be some circumstances where a multi-target theory was needed—where there were multiple target each of which, if hit, could cause an effect. "Lethal mutations" was one such example where genes (the targets) were strung out along a chromosome. Larger viruses and bacteria might also require this approach.

The book was pivotal since it clarified much that had gone before by insisting that the single-hit model, with the multi-target caveat, could account for most of the data seen to date on prokaryotic and simple organisms. Sadly, Lea died in 1947 but the book was revised by L H Gray and a second edition was published in 1955.

A second book that developed the target theory, by Timoféeff-Ressovsky and Zimmer, appeared in 1947 [26] and, with Lea's, provided the basis for much of the thinking for the next decade or so. However, by the time the second edition of *Actions of Radiations* appeared there were several important developments that were to have at least significant and at most profound effects.

The first, in 1951, was the description of the cell cycle by Howard and Pelc that gave a foundation for the known variation of the radio-sensitivity of cells as they passed through the cycle. The second was the identification by Watson and Crick in 1953 of DNA as the carrier of genetic information and their unravelling of its double helix structure. In 1956 there was a significant experimental advance when Puck and Marcus [27] found a way to prepare, irradiate and assay mammalian cells in vitro for the first time. Suddenly experiments that had been possible only with bacteria and yeast could be performed on mammalian cells. The cells, known as HeLa cells, were derived from ones taken (seemingly without her permission) from a patient Henrietta Lack (hence "HeLa") with cervical cancer and their descendants are still used today. Indeed, although the lady died in 1951, it has been estimated that more than her bodyweight of the cells still exists.

Puck and Marcus found that the killing of mammalian cells gave quite different survival curves from those seen before: the curves were shouldered on a logarithmic plot rather than a straight line. This was interpreted as evidence for a two-hit process using the equation for multi-hit theory derived by Luria[28] since the extrapolation back to the y-axis gave a value very close to 2. The two hit process, the small inactivation dose of 96 r and the ability of the cells to reproduce a limited number of times all pointed, they thought, to a lethal effect in "the genetic apparatus" rather than the operational parts of the cell. It could not be a simple single gene inactivation but the "locus of the action could be chromosomal".

Barendsen[29] in 1962, in an extensive study of cultured human cells with a range of radiations found that the results could not be modelled with a hit or target approach. Instead he proposed that a high LET particle might cause enough damage in passing through a nucleus to inhibit reproduction. Low LET radiation might not deposit enough energy locally to cause lethal damage but separate particles might produce two events close enough together to be effective. As he put it: "a survival curve of the type derived from our experiments can be explained on the basis of pure physical considerations, without postulating anything about the biological phenomenon".

Bender and Gooch[30] in 1962 addressed some of the uncertainties in experimental procedures and concluded from their results that the survival curves for mammalian cells were rather better fitted by a model—they called it a multiple process model—where there were a number of processes that could lead to cell killing. Although they thought that the simple two-hit model with one target could be eliminated they commented that: "Probably many of the cell's structures are, in fact targets. The work of Barendsen et al (1960) with alpha-particle irradiation suggests that important targets are located in the nucleus. Beyond this, however, we can only speculate."

Kellerer and Rossi's 1972 work[31] was driven by developments in microdosimetry and was prompted by a desire to explain the differences seen in response to high and low-LET radiations. They developed their theory of dual radiation action. It was assumed that radiation produced a number of sub-lesions and that it was through the combination of these that lesions with biological consequences were formed. In the original formulation (sometimes called the "site" model) it was assumed that sub-lesions within a certain distance (typically around 0.4 microns) of one another would always combine and that those beyond this critical distance would not.

It led to the result that more lesions would be formed per unit dose from high-LET radiation than low-LET radiation, as seen for many biological consequences.

5 Interactions and Cell Death

The model led to the so-called linear-quadratic formula for the dependence of lesion (whatever they might be) numbers on dose:

$$E(D) = k(\zeta D + D^2)$$

with the linear component down to lesions arising from sub-lesions caused by a single particle track and the quadratic to those fro sub-lesions from different tracks.

It was not long before some challenges to the model emerged (summarised by Goodhead[32]. Some were based on pre-existing inconsistencies: the value of k was found to depend on radiation quality and the quadratic dependence was not always seen where it was expected. However, new sources of data emerged and demanded a rethink. One was the results from irradiation of mammalian cells with ultra-soft x-rays (a fraction of a keV). Such x-rays produce secondary electrons with ranges of a few nanometres creating around a dozen ionisations each—with a low probability of more than one sub-lesion arising from a track. So they should, if the "site model" is correct, be much less efficient than higher energy x-rays or gamma-rays in producing biological effects. In fact the precise opposite was found experimentally.

Kellerer and Rossi generalised the theory in 1978 [33] as the "distance model" to include, among other refinements, an interaction probability function that expressed the variation of likelihood of lesion formation with sub-lesion separation better than the crude "site model" did. Critics complained that this still did not address the long-standing question about k depending on radiation quality and also that the long-range interaction still assumed to take place (it was needed to justify the quadratic term) was, in fact, not required by the experimental evidence.

Chadwick and Leenhouts in their 1978 paper [34] and subsequently[35] were more specific than some workers before and after, identifying the critical (sometimes lethal) molecular lesion as a double-strand break (DSB) in DNA. This could arise either directly from the passage of a highly ionising particle or indirectly from the combination of two single-strand breaks. The single-strand breaks could be caused by two different particles—giving a squared relationship to dose. The consequence is a linear quadratic dependence of DSB numbers— and hence biological consequences—on dose.

Another strand of models was based on the repair mechanisms and began in 1960 when Elkind and Sutton[36] threw some doubt on the solely chromosomal nature of the damage and introduced a third important factor: the cells they used were able to repair rather rapidly the damage caused by radiation.

Powers [37] in his review of target theory in 1961 interpreted the results of split dose experiments as possibly being due to the depletion of the "pool of substances that protect against part of the radiation damage" and replenishment—or some other kind of recovery— between irradiations. The idea of the "pool" was to be developed by Orr and his co-workers in the early 1970s [38].

Calkins[39] in 1971 derived survival curves on the assumption that there is a time (the fixation time) beyond which lesions caused by radiation cannot be repaired. The rate of repair of lesions depends on how many lesions are present and so the number repaired before fixation depend on the initial number—and thus on the dose. Drawing on theories of enzyme action, Calkins was able to produce very respectable shouldered survival curves.

Kappos and Pohlit [40] in 1972 proposed what they called a cybernetic model for cell killing based on two possible consequences of the interaction of radiation with "essential molecules"– which they provisionally identified as DNA in the cell nucleus. One type of interaction damaged the molecule irreparably and lethally but the damage caused by the other might be repaired or the molecule might go on to the lethally-damaged state. If the molecules were DNA then these alternatives would be the creation of DSBs and SSBs. Some seemingly-straightforward simultaneous linear differential equations were set up but it was found that the rate constants could not, in fact, be constants. The rate constant for producing repairable damage (mainly due to indirect action) was supposed dose dependent – possibly because of competition for radicals by the essential molecules and others. A time dependent repair constant was found necessary to fit repair experiments. With these adjustable parameters they found a good fit with survival curves.

5 Interactions and Cell Death

In 1980 Tobias et al [41] suggested a further model that introduced non-linearity into the system but this time by considering the dynamics of repair of "uncommitted" (or U) lesions in macromolecules. They were seen as metastable defects in the molecules that the cell attempted to repair. The repair process develops in two ways: either the individual lesions are repaired (more or less successfully) in so called linear repair or, while one lesion is being repaired, another one comes along and causes a misrepair (quadratic misrepair). When allowance is made for the statistical fluctuation of the number of U lesions, the Repair–Misrepair (RMR) model appears to give a reasonable description of survival curves. It can be extended to cover chromosome aberrations and mutations if it is assumed that some of the quadratic misrepairs are not fatal for the cell.

Stanley B Curtis accommodated many of the features of earlier models in his LPL Unified Repair Model of cell death in 1986 [42]. He disregarded interactions between tracks and considered just the biological lesions caused in a cell by a single track. These were presumed to be of two types: lethal ones that would inevitably lead to cell death and potentially lethal ones that might. Several things might happen to a potentially lethal lesion: it might be repaired (in a time of 90-120 minutes), it might interact with another potential lesion to cause a fatal one (he called this binary misrepair) or it might suffer "fixation" by cellular processes and cause it to become a fatal one. With these a assumption he set-up a non-linear differential equation and showed that the system it described had many of the characteristics seen in shouldered survival curves and led to the linear-quadratic approximation. Curtis tentatively identified the potentially lethal lesion with a DNA double strand break.

To accommodate the effect of LET it was necessary to look back a step to the biochemical "prelesions" that cause the lesions. When these prelesions are close together, they lead to a lethal lesion; if isolated they may lead to a potential lesion. Thus a high-LET particle would lead to more lethal lesions per unit dose than a low-LET one and this would give the observed dependence of survival on ionisation density. Prelesions were identified as prelesions in DNA.

Goodhead, in a sequence of papers from the end of the 1970s[43] proposed what he called "saturable repair". This did not depend on either a dose-dependent depletion of enzymes as considered by Calkins or on invoking a process where sub-lesion interaction caused interference with repair. Indeed, he did away with the whole "dual action" idea. Instead, he thought, that there is a limited pool of repair enzymes which is depleted each time a repair is made. If repairs are not made before the damage is "fixed" then the damage is permanent.

Some simple assumptions led to a mathematical model that predicted rather well many of the radiobiological observations. The model was not unlike that of Orr and his colleagues except that Orr assumed that the fixation mechanism continued to act in competition with repair rather than coming into play only (and conclusively) after a fixed time.

There have been a variety of mechanisms put forward to explain chromosome damage. Some are termed exchange aberrations—rejoining models:where there is contact between chromosomes, a radiation hit and then some kind of exchange of chromosome material. This, the "contact first" model was an early idea from Serebrovsky and Muller but, by the late 1930s there was evidence that the breakage came before contact and the breakage-and reunion model, where the chromosomes joined after breakage, was promoted by Sax and by Lea. The idea was that the chromosome material was broken by a radiation event and that the two frees ends could either rejoin, join up with another free end from another chromosome or simply remain broken. Such a model could explain most of the observed chromosomal damage seen and its relationship to dose and LET. It remained the basic model for the rest of the century, perhaps because, as various authors pointed out, it is the easiest one to explain and visualise. The model was challenged in the late 1950s by the so-called exchange theory of Revell. The exchange theory proposed that radiation caused "lesions" in the chromosome material rather than breaks. Each lesion had a limited life but, if it came close to another lesion during this time, there might be an exchange of material. Revell was able to explain all the aberrations seen: a deletion for example would result from the interaction of two lesions forming a loop which would become detached, removing genetic material.

Sax, Lea and Revell were all working at a time before the structure of the chromosomal material and the central role of DNA were understood. At the time it was thought that it had a defined backbone (called the chromonema), made mainly of protein, with a diameter of about 100 nm. DNA was just a liquid in the nucleus with no role in the structure of the chromosome; it was merely adsorbed onto the chromosome as it prepared for mitosis.

Now we know that the essential material, carrying the genetic code, is DNA and that this is wound around a series of histone beads to form a microscopic (but rather long) necklace). The whole structure is coiled and folded by different degrees at different parts of the cell cycle. The chromosomes are formed by the 30 nm fibre looping itself around a protein scaffold immediately before mitosis.

By the time of the molecular theory of Chadwick and Leenhouts the nature of chromosomal material was much better understood. They proposed that all aberrations originated in single DSBs (with their linear-quadratic dose response). To achieve this the authors had to assume that a broken end could join up with a telomere, an unbroken chromosome end, and this led to "some bizarre aberration shapes, and expectations that are contrary to established fact", as Savage put it much later [44]. An alternative explanation based on repair mechanisms was offered by Resnick in 1976. The relative merits of the theories remained in some doubt and, by the end of the century, new techniques for studying chromosomes (such as FISH-painting) were showing that aberrations that were thought to be simple were actually rather complex and that many arise from multiple breaks in more than one chromosome.

Also the understanding of the complex supporting structure for DNA means that a double strand break does not lead to loose ends waving around wildly waiting to join to other similarly fractured structures (like chopped up spaghetti in a plastic bag in a well-used simile). Instead, DSBs are usually quite tightly constrained. There is also structure at a higher level in the nucleus that restricts contact between different chromosomes: each chromosome is, more or less, restricted to its own domain in the nucleus during interphase and even the two arms of the same chromosome are not necessarily side-by-side. There is thus no intimate mingling of chromatin: broken ends of different DNA strands do not usually meet. There is some limited contact between chromosomes since there seems to be some DNA ("cables") that stretches through the regions between domains and loops that are associated with nuclear pores[45]. It raised the possibility that DNA comes into contact in a limited but perhaps very specific way and that, when it does, it is sensitive to the effects of an ionising particle. It takes us, in a way, back to the contact-first idea of early in the century.

There is an enormous literature of chromosome aberrations (and they are known to be associated with radiation-induced and other tumours) but they have had no great direct significance in radiation protection terms other than their use in the retrospective assessment of doses following accidents. However, like the study of cell killing (which is after all significant only at extremely high doses) their importance lies in what they can tell us about cases where cells are not killed and chromosomes may not be grossly damaged. The two consequences of central importance are hereditable changes caused by radiation and changes that transform cells and lead to their uncontrolled proliferation. The first of these, the genetic effects, is reviewed in Chapter 7 on genetic effects and the second, carcinogenesis, in the following one.

The chemical action of radium on water was seen as early as 1901 by Curie and Debierne: there was, they found, a steady generation of oxygen and hydrogen. Lea in 1946 summarised what was known of indirect action. It was agreed that normal water decomposed into hydrogen and oxygen and some hydrogen peroxide and that at moderate doses there was rather less oxygen and more peroxide. But, as Lea said, there were "still some obscurities concerning the production of chemical change by ionizing radiation". In solutions he was confident that there was an indirect action (in the sense that the molecules of the solute had not been excited or ionised directly but by their reaction with excited or ionised solvent molecules) because the yield of excited solute molecules was independent of its concentration. This was true of solutions of enzymes as well as less complicated molecules. The yield of solute ions (the number reacting per ion pair produced by radiation) was between 0.1 and 2.0 suggesting that activated water was formed at the rate of about one molecule per ion pair and then a fraction was lost/deactivated before it encountered a solute molecule.

Hydrogen peroxide formation could not account for all the results seen but Weiss [46] had suggested, in 1944, that the key mechanism actually revolved around the formation of free H and OH radicals. These would, Lea agreed, produce many of the results seen and he embraced the explanation. He presented a theoretical treatment of diffusion of the radicals as they recombined and managed to explain the decline in reaction yield at low solute concentrations. (His diffusion distances were of the order 10 microns). However, he then appears to have lost interest in indirect action: the remainder of the book is devoted, more or less, to direct action and the target theory.

By 1955, when Bacq and Alexander wrote their monograph [47], the idea that free radicals were an essential element of the interactions was in all theories (there were two or three of these) but there was still some differences in views on how they were formed, in particular the roles of excited molecules and ions. The details would take us too far into chemistry but the different mechanism of formation did result in different spatial distribution of the radicals—with significant implications for the differences between particles of different LET.

One of the uncertainties was the fate of any free electrons generated. While there were suggestions in the 1950s that these could have an extended life of their own, it was 1962 before the hydrated electron (e^-_{aq}) was identified by Hart and Boag. This gave three radicals of real consequence H·, OH· and e^-_{aq} that create an extremely complex pattern of reaction among themselves, with their own reaction products and with water molecules. It is their reaction with significant organic molecules that causes damage by "indirect action".

Many cellular effects of radiation other than cell killing have been studied and a history of radiobiology would properly review these. However, this is a history of protection and we have seen enough radiobiology to know how complex the interactions with biological systems are and how there remain differences in interpretation of even relatively crude destructive effects. It is appropriate now to look at higher organisms and see how they are affected by radiation through more subtle changes. Before we can do that, we need to look at how radiation dose has been characterised.

Notes Chapter 5

1. (Bragg W H and Kleeman R D, 1904, 1905)
2. (Wilson C T R, 1927)
3. (Lea D E, 1955)
4. (Powers E L, 1962)
5. (Puck T T and Marcus P I, 1956)
6. (Zirkle R E, 1935)
7. (Zirkle R E, 1935; Perry R T, 2005)
8. (Lea D E, 1940)
9. (Steel G G, 1996)
10. (Dessauer F, 1922, 1923)
11. (Blau M and Altenburger K, 1922)
12. (Crowther J A, 1924)
13. (Strangeways T S P and Oakley H E H, 1923)
14. (Crowther J A, 1926)
15. (Packard C, 1931)
16. (Condon E U and Terrill H M, 1927)
17. (Mayneord W V, 1934)
18. (Lea D E, Haines R B and Coulson C A, 1936)
19. (Lea D E, Haines R B and Coulson C A, 1937)
20. (Crowther J A, 1938)
21. (Glocker R, 1932)
22. (Mayneord W V, 1934)
23. (Lea D E, Haines R B and Coulson C A, 1936)
24. (Holweck F and Lacassagne A, 1930)
25. (Lea D E, 1955)
26. (Timofeeff-Ressovsky N W and Zimmer K G, 1947)
27. (Puck T T and Marcus P I, 1956)
28. (Luria S E, 1947)
29. (Barendsen G W, 1962)
30. (Bender M A and Gooch P C, 1962)
31. (Kellerer A M and Rossi H H, 1972)
32. (Goodhead D T, 1982)
33. (Kellerer A M and Rossi H H, 1978)
34. (Chadwick K H and Leenhouts H P, 1978)
35. (Chadwick K H and Leenhouts H P, 1981)
36. (Elkind M M and Sutton H, 1960)
37. (Powers E L, 1962)
38. (Laurie J, Orr J S and Foster C J, 1972; Orr J S and Laurie J, 1975)

39 (Calkins J, 1971)
40 (Kappos A and Pohlit W, 1972)
41 (Tobias C A, Blakely E A, Ngo F Q H and Yang T C H, 1980; Tobias C A, 1985)
42 (Curtis S B, 1986)
43 (Goodhead D T, 1985)
44 (Savage J R, 1998)
45 (Savage J R K, 2000)
46 (Weiss J, 1944)
47 (Bacq Z M and Alexander P, 1955)

6 Dose Concepts and Models

Much of the theory of measurement and many of the instrumental techniques of dosimetry originate in radiology with the aim of either protecting the radiography patient from excessive exposure or—perhaps more significantly—delivering radiation dose accurately in radiotherapy. Radiation protection of workers and members of the public—our subject—has benefited from much of this work but it is concerned with radiation levels that are generally chronic and often orders of magnitude lower than those relevant to patients. In some accounts of the earlier measurement systems this distinction is occasionally lost and this may be the case from time to time in what follows.

The challenge in defining units and quantities has been to find something that is measurable and at the same time relevant. We have already mentioned in Chapter 1 the pastilles and film strips that were used to measure diagnostic and therapeutic x-rays. They measured (very high) doses without there being a clear understanding of how. Their relevance was therefore solely judged, initially, through the significance of the effect used to calibrate them: the epilation and later erythema dose. Progress really came with the adoption of ionisation in gases as the measurable quantity because it was fairly basic and reproducible. Such a basic process as ionisation was recognised early on as central to the effects of radiation on people, even if the mechanisms of damage were not understood. So, although it was to be a very long time before full relevance could be established, measuring ionisation was a key event in measuring

6 Dose Concepts and Models

radiation. Later the energy absorbed by materials from radiation was seen as a more important quantity for radiation protection but even this was frequently measured, in practice, through the effects of the ionisation produced.

To retain some structure the quantities have been grouped under the headings: exposure, absorbed dose, kerma and dose equivalent. These have been, or may be, transient names but they have been widely enough used and understood to make them useful.

The first suggestion that a unit might be defined in terms of the ionising power of x-rays came from Belot in 1906 and in 1907 both C E S Phillips and C L Lennard proposed a unit based on the same principle. However it was Villard in 1908 who made the first clear quantitative proposal suggesting the unit as the liberation of 1 esu of charge per cc under normal conditions of temperature and pressure. Other definitions of a unit as the radiation setting free 10^{12} ions came from Christen in 1913 and Szilard in 1914 but Villard's suggestion was taken up by Kronig and Friedrich in 1918, after a comprehensive study, as the e-unit. The e-unit then formed the basis for the definition by Behnken in 1924 of the R-unit or German Röntgen. There was also a French R-unit (proposed by Solomon) based on the ionisation produced at a fixed distance from a standard radium source but it was the German R-unit that was adjusted somewhat and adopted by the Second International Congress of Radiology (ICR) at Stockholm in 1928 as the internationally agreed unit, the Röntgen (r):

> The quantity of X-radiation which, when the secondary electrons are fully utilised and the wall effect of the chamber is avoided, produces in 1 cc of atmospheric air at 0°C and 76 cm of mercury pressure such a degree of conductivity that 1 electrostatic unit of charge is measured at saturation current [1].

The only practical difference between the R-unit and the rontgen was that the temperature was specified as 18C in the German unit and it was therefore about 7% bigger than the r. The standard was realised with several different designs of free-air ionisation chamber and international inter-comparisons made by US, German and UK workers meant that the Third ICR in Paris in 1931 could be assured that the rontgen was an accurate basis for measurement of x-rays up to about 200 kV.

The rontgen was defined specifically for x-rays and there were several analogous units in use for gammas. The most widespread at the time were the D-unit proposed by Mallet and Proust in 1927 and Sievert's Imc of 1923. Both were related to the ionisation produced by radium sources but the use of the rontgen for gamma-ray measurements was clearly desirable. Following the Paris ICR, it was recognised that cavity chambers with air equivalent walls, relying on the Bragg-Gray theory (see below), offered potential as secondary standards. It was soon established that cavity chambers with a suitable air-equivalent wall material gave results for x-rays in good agreement with free-air chambers. However, the results from comparisons between free-air and cavity chambers using radium rays were less encouraging than for x-rays and it was clear that the extension of the rontgen to gamma-rays would not be simple. The problem lay in the much higher photon energies and the much greater ranges of the secondary electrons produced. In the small ionisation chambers used, these electrons were not "fully utilised" and electronic equilibrium was not being achieved. The 4th ICR was unable to extend the definition to gamma-rays.

The ranges of the electrons generated by gamma-rays proved to be about 4 m in atmospheric air and two different approaches were adopted in the UK and USA. At the UK National Physical Laboratory (NPL) a room-sized free-air ionisation chamber was constructed; at the US National Bureau of Standards(NBS) a pressurised free-air chamber was used. The results from these, with the good agreement now achieved with thick-walled cavity chambers, showed that the rontgen could be extended to gamma rays. At the 5th ICR in 1937 the definition was modified to:

> The rontgen is that quantity of X- or gamma radiation such that the associated corpuscular emission per 0.001293 g of air produces, in air, ions carrying one electrostatic unit of quantity of electricity of either sign. It is to be noted that 0.001293g is the mass of 1 cm^3 of dry atmospheric air at 0°C and 760 mm mercury pressure.

6 Dose Concepts and Models

The theory of the cavity chamber—Bragg-Gray theory—was developed about this time and is important to the development of dosimetry so we briefly trace its history here. W H Bragg[2] in 1912 considered using the measurement of ionisation in a small air-filled volume to deduce the electron fluence in the surrounding material. He concluded that the electron fluence under photon irradiation was equal to the product of the consequent number of electron emitted per unit volume and the electron range. This, he argued, would be proportional to the ionisation in the small air volume. This was independently stated quantitatively by L H Gray in 1929[3] in his first published paper, developed and explored experimentally by him[4] and became known as Bragg-Gray theory. The principle states that the ionisation per unit volume in a small air-filled cavity in a medium such as tissue is proportional to the ionisation per unit volume in the medium. The constant of proportionality is the ratio of the linear stopping power (dT/dx) of air for secondary electrons to that of the medium and this can be calculated and applied as a correction. Gray assumed that the stopping power ratio was independent of electron energy but this restriction was removed with the work of Laurence in 1937 although it was still necessary to assume that slowing down is a continuous process. Subsequently the theory was extended further analytically in 1955 by Spencer and Attix and by Burch to include the effects of delta-rays. Further theoretical work to calculate the correction factor has generally used Monte Carlo techniques[5].

The value of air-equivalent cavity chambers as secondary standards, mentioned above, was one important implication of Gray's work. The other was that it gave a sound basis for measurement of the ionisation caused—and the energy deposited—in materials other than air.

The rontgen was a measure of the interaction of photons with air. It became apparent that there was a need to measure interaction with other materials, notably tissue, and to do this for radiation other than photons. While there were several possibilities for the quantities that could be measured—which could be energy or ionisation related—attention focussed on energy absorbed in a material. In his 1937 paper, for example, Gray stated that:

> All material changes brought about by radiation may be correlated with the radiational energy actually absorbed, so that

from the physical standpoint the most natural unit of dosage is the absolute increase of energy per unit volume (ergs/c.c.) of the absorbing medium immediately consequent upon radiation.

A number of units appeared in the literature: tissue rontgen, nominal rontgen, equivalent rontgen, rhegma, rontgen equivalent, gram rontgen and Gray's energy unit. Each embodied the idea of absorbed energy per unit volume or per unit mass. While the importance of absorbed energy appears initially to have been more widely recognised by radiologists in the UK than in the USA, the unit that was to become the basis for the next step forward in practical radiation protection was first used in the Manhattan Project in 1943: Herb Parker's rep [6].

Parker, in 1948, described the rep as a practical energy absorption unit and recommended that "one rep represents an energy absorption dose in irradiated tissue of (exactly) 93 ergs per gram". Since slightly different definitions of the unit had emerged as its use spread through the MED and beyond (so that it was variously 83 ergs per gram, 83 ergs per cubic centimetre), Parker thought authors should make it explicit just which version was being used. One of his concerns in defining the unit was that there should be continuity and he was well aware that many radiologists would be happy to continue to work with the rontgen. The choice of 93 ergs per gram reflected two experimental observations:

- One rontgen of x-rays and gammas resulted in about 83.8 ergs per gram absorption in air and this corresponded to about 83.8 ergs per gram of wet tissue
- One rontgen of hard gamma resulted in about 93 ergs per gram energy absorption in water

The second of these was the basis for Gray's energy unit and gave Parker the 93 figure and a degree of compatibility with the rontgen. The differences between 84 and 93 were not significant. Parker noted that "Very little radiobiological work to date is accurate enough to make results sensitive to the difference between 83 and 93".

He was keen that the unit name should begin with the letter "r" and rep came from "rontgen equivalent physical", a name clearer from an alternative definition: the quantity of radiation which, absorbed in the body, would liberate the same amount of energy as 1 roentgen of x-rays or gamma rays would.

The 6th ICR in London in 1950 then took the step of recognising the importance of energy absorbed and defined the quantity absorbed dose as the quantity of energy absorbed per unit mass of irradiated material at the point of interest. The 7th ICR in Copenhagen in 1953 modified the definition of absorbed dose slightly to make it the amount of energy imparted per unit mass and defined the unit as the rad, 100 ergs per gm. The quantity absorbed dose has, for most practical measurement purposes in radiological protection, not changed significantly since that definition.

The concept of absorbed dose has remained fixed and is a well-defined quantity for most practical purposes; given a particular material it is possible to measure the absorbed dose within it. However, a practical problem remains in, for example, survey measurements. That is, which absorbed dose in what material should be measured for protection purposes? This was addressed in ICRU19 and its Supplement in 1971 through the introduction of Index Quantities and one of these was the Absorbed Dose Index (D_I):

> The absorbed dose index at a point is the maximum absorbed dose within a 30 cm-diameter sphere centred at this point and consisting of material equivalent to soft tissue with a density of 1 g cm^{-3}.

The sphere, later called the ICRU sphere, was defined to have a mass composition of 76.2% oxygen, 11.1% carbon, 10.1% hydrogen and 2.6% nitrogen.

A quantity related to absorbed dose that has specialised but important use is kerma, the kinetic energy released per unit mass.. This quantity was introduced in ICRU Report 10a in 1962 after a 1958 proposal by Roesch as:

> The quotient of ΔE_K by Δm, where ΔE_K is the sum of the initial kinetic energies of all charged particles liberated by indirectly ionizing particles in a volume element of the specified material, and Δm is the mass of the matter in that volume element.

The definition is similar to absorbed dose (and is measured in Grays) with the difference that the energy of the liberated charged particles is not lost through ionisation in the volume element. It is carried away to be lost elsewhere. Kerma is significant because it can be calculated readily even when the liberated particles have long ranges and is thus useful with, for example, gamma or x-rays of energies greater than a few MeV. Of course, if the material dimensions are large enough and the radiation field uniform enough kerma and absorbed dose are equal: the liberated particles escaping from the volume element are exactly replaced by those coming in.

As well as the ideas behind the rep described above, Parker also suggested in 1948 that the different biological effectiveness of different radiation could be accounted for, in chronic exposures, by defining the rem:

> One rem is that dose of any ionising radiation which produces a relevant biological effect equal to that produced by one roentgen of high-voltage x-radiation, other exposures being equal.

He then presented a scale of relation:

X-rays, gamma rays	1 rep = 1rem
Beta rays	1 rep = 1 rem
Protons	1 rep = 10 rem
Fast neutrons	1 rep = 10 rem
Slow neutrons	1 rep = 5 rem
Alpha rays	1 rep = 20 rem

This, Parker realised, ignored the differences between low voltage x-rays and gamma-rays but he was content that these uncertainties could be controlled, in radiation protection, by the application of suitable safety margins.

He considered application of the rem concept to both acute exposures and therapeutic applications. The acute case he recognised as different requiring, at least, the separate consideration of individual organs. As for therapy "no profitable application is known to the writer"; more precise dosimetry in reps was required before "extension to the biological equivalence unit should be attempted". In the discussion following the presentation of the paper Parker addressed the question of whether one unit could be defined to cover all biological effects. He thought it impossible: it would mean, he said, that "you could write the whole subject of experimental radiobiology in one sentence".

The original name for the unit was the reb (rontgen equivalent biological) but it was changed to rem to avoid confusion in speech (as Parker himself pointed out). Ron Kathren[7] gives a little more detail:

> The unit (the rem) was originally called the reb (roentgen equivalent biological), but during one of his early presentations of the new unit, Parker was suffering from a cold, which led to difficulty in differentiating it from the rep. Accordingly, the name of the unit was changed to rem.

Rem was an acronym for rontgen equivalent man or, possibly, rontgen equivalent mammal.

The most significant development after the Copenhagen ICR was probably the development of the idea behind Herb Parker's rem unit: the definition of a quantity which was an indicator of the damage caused by all ionising radiation. In its 1957 report ICRU used the idea of Radiobiological Effectiveness and recommended a quantity "RBE dose" calculated as the absorbed dose x RBE and gave a RBE v LET relationship for some radiations. This was subsequently criticised since RBE had a specific meaning in radiobiology and took on a range a values in different conditions and for different endpoints. When the subject was reviewed a few years later by a joint ICRU/ICRP working party[8] "RBE" was replaced by "quality factor", a new term to acknowledge that the factor was now a compromise, conservative one specifically chosen for use in protection..The quantity "absorbed dose x quality factor" was now called the dose equivalent. The unit was the rem.

The dose equivalent concept allowed some unification of the treatment of different radiations: in principle a measuring device could measure the dose equivalent rather than just dose. Just as for absorbed dose, the ICRU in 1971 defined a dose equivalent index (H_I):

> The dose equivalent index at a point is the maximum dose equivalent within a 30 cm-diameter sphere centred at this point and consisting of material equivalent to soft body tissue with a density of 1 g cm^{-3}.

In the 1980 ICRU Report 33 the Sievert was used as the special SI unit for dose equivalent and two further indexes were defined: the shallow and deep dose equivalent indexes.

> The dose equivalent at the respective depths of 0.007 cm (shallow) and 1.0 cm (deep) in a sphere of soft tissue of a density of 1 g/cm^3 and a diameter of 30 cm.

ICRU 39, in 1985 introduced three Operational Quantities:

> Ambient dose equivalent: $H^*(d)$
> The dose equivalent that would be produced by the corresponding aligned and expanded field, in the ICRU sphere at a depth, d, on the radius opposing the direction of the aligned field

> Individual dose equivalent, penetrating, $H_p(d)$
> The individual dose equivalent, penetrating, $H_p(d)$, is the dose equivalent in soft tissue (defined as in the ICRU sphere) below a specified point on the body at a depth, d, that is appropriate for strongly penetrating radiation.

6 Dose Concepts and Models

> Individual dose equivalent, superficial, $H_s(d)$
> The individual dose equivalent, superficial, $H_s(d)$, is the dose equivalent in soft tissue (defined as in the ICRU sphere) below a specified point on the body at a depth, d, that is appropriate for weakly penetrating radiation

In ICRU 51 in 1993 two new operational quantities, directional dose equivalent and personal dose equivalent were defined

> The directional dose equivalent $H'(d,\Omega)$, at a point, is the dose equivalent that would be produced by the corresponding expanded field in the ICRU sphere at a depth d on a radius in a specified direction Ω.

Directional dose equivalent was considered of particular use in the assessment of dose to the skin, when d=0.07 mm, or eye lens, when d=3 mm. It was to be used for weakly penetrating radiation and the ambient dose equivalent, as defined in 1985 with d=10 mm, i.e. $H^*(10)$, for strongly penetrating radiation in the assessment of area monitoring results.

The other new quantity, personal dose equivalent $H_p(d)$, replaced the two individual dose equivalents as the basis for personal monitoring:

> The personal dose equivalent $H_p(d)$, is the dose equivalent in soft tissue, at an appropriate depth, d, below a specified point on the body. $H_p(d)$ can be measured with a detector which is worn at the surface of the body and covered with an appropriate thickness of tissue-equivalent material.

For penetrating radiation, a depth of 10 mm is employed; thus the notation is $H_p(10)$. For weakly penetrating radiation, a depth of 0.07 mm for the skin and 3 mm for the eye are employed; thus the notations are $H_p(0.07)$ and $H_p(3)$, respectively.

As ICRU continued to interpret the dose equivalent concept from a measurement viewpoint, the ICRP used it in the 1977 recommendations of Publication 26 and the 1990 recommendations of Publication 60 although for 1990 there were some changes in nomenclature. The "dose equivalent" was re-named the "equivalent dose" and the "quality factor" became the "radiation weighting factor" without any significant change in meaning.

By 1977 it was clear that cancers could be caused by radiation in a number of organs and tissues and that the fatal risk associated with them varied considerably. Some way of accounting for this was essential if non-uniform irradiations—as resulted from intakes of many radionuclides—were to be properly dealt with. ICRP's solution was to assign to each important tissue a "weighting factor" proportional to the fatal cancer risk per unit dose equivalent to the organ. The weighting factors were then normalised so that when the products "weighting factor x organ dose equivalent" were added up for all the organs the result would be the dose equivalent which, if uniform over the body, would give the same fatal cancer risk. This dose was called the "effective dose equivalent". The concept survived bouts of criticism (for example, that it was a constructed rather than a measured quantity which would change as weighting factors changed) and, with a change in name to "effective dose" in Publication 60[9], stayed at the centre of radiation protection. The weighting factor was renamed at the same time: it became the "tissue weighting factor" to avoid confusion with the renamed quality factor.

The earliest internal dose standard was that for radium and was based on setting a level for the body burden well below that at which any clinical effects had been seen from human exposure. This was used as the basis for the first plutonium standards and the evolution of these is discussed in Chapter 12 on standards. Here we will trace the development of approaches which are of more general applicability in setting standards for the wide range of nuclide that have been created since the 1940s. They have replaced other techniques for all nuclides.

By the late 1940s the need to define standards for other nuclides had become clear. A number were being produced by reactor and accelerators which could not be dealt with by (relatively) simple extrapolation from radium data. So, when USA, Canada and the UK came together for the Tripartite Conference in 1949[10], the permissible levels of nuclides where these could be derived from radium data

6 Dose Concepts and Models

were considered: U-233, Pu-239, Po-210, Sr-89,Sr90 and Th-234. However, permissible amounts of other nuclides were also an important topic and body burdens for seven more nuclides were agreed: H-3, C-14, Na-24, P-32, S-35, I-131 and Co-60. The values were all derived from some knowledge of what was the critical organ and the rate of clearance from it. So, while several of the isotopes could be treated as distributed uniformly around the body, some were known to concentrate in particular organs: P-32 in the red bone marrow, Co-60 and S-35 in the liver and I-131 in the thyroid. Then, with knowledge of the energy characteristics of the decays, the concentration in the critical organ that would lead to exposure at the dose limit (at the time 0.3 r per week), could be estimated. With the mass of the organ, the total allowable body burden for workers (it was called the "best estimate of safe dose" at the time) could be calculated. With some simple assumptions the maximum permissible air and water concentrations for continuous exposure were calculated. The air concentrations for Ar-41, Xe-133 and Xe-135 were also calculated, based on external dose as the limiting factor. The approach was adopted by ICRP for their 1950 recommendations[11] where the the permissible levels quoted were essentially those from the Tripartite meeting. There was a tentative quality about the "maximum permissible amounts" quoted. The Commission was not "in a position to make firm recommendations regarding the maximum permissible amounts of radioactive isotopes that may be taken into, or retained in the body" but "reasonable calculations" could be made as had been done by the Tripartite partners.

By 1960 the ICRP was ready to publish, as Publication 2[12], a comprehensive set of internal standards based on metabolic data and derived from the current dose limits. The calculations were based on the critical organ idea and so were made for several organs to decide which one was limiting for a particular nuclide. The methodology allowed for continuous entry of a radionuclide into the organ with an exponential removal (the rate of removal was proportional to the concentration in the organ) by both decay and biological elimination. Two entry routes were considered: the respiratory and gastrointestinal tracts. The lung route took into account the solubility of the compound by assuming it was either "soluble", when it would immediately enter the bloodstream, or "insoluble" when it would be removed from the lung with a biological half-life of 120 days to pass out through the gut (the gut doses resulting were not allowed for in this model). This classification reflected the work done ten years earlier in the USA and briefly reported at the Tripartite conference. The GI tract was represented simply with a mean residence time and a parameter f_1 describing the fraction of material transferred from the gut to the bloodstream. The results were presented as maximum permissible concentrations (MPCs) in air and water and maximum permissible body burdens—actually the burden there would be after 50 years of continuous exposure to the MPC rather than a control quantity. There were adjustments and extensions in ICRP Publications 10 and 10A[13].

The calculations were based on Standard Man, a concept developed in the late 1940s by Herman Lisco at the Argonne National Laboratory. Standard Man was a compilation of parameters designed to represent a man (and indeed woman) for use in dosimetry calculations. It included height, weight, organ masses and radii and other relevant factors like breathing rate, water ingestion and output rates. Standard Man survived until 1975 when he was replaced by Reference Man, to form the basis for the landmark ICRP Publication 26[14] published in 1977. With Publication 26 the tolerance dose ideas and critical organs were swept away to be replaced by a risk-based approach although threshold effects remained through the specification of non-stochastic limits these had no great effect on internal dose calculations other than for the actinides where the dose to bone surface was the limiting factor.

Internal dose calculations using the new approach were presented in the multi-volume ICRP Publication 30[15] which appeared over the next few years. Since, in effect, the dose had to be calculated for every major organ in the body, the scheme was more complex than that that required for the "critical organ" approach. The respiratory tract model now contained 10 compartments and included transfer between compartments, clearance to the gut and absorption into the "transfer compartment" (the body fluids). Four compartments described the GI tract, with clearance by excretion or to the "transfer compartment".

Each organ had its own transfer parameters for each element and, transfers were assumed to be exponential in nature (although an exception was made for the alkaline earths for which clearance was taken as a power-law relation). So the model for almost all nuclides came down to a large set of simultaneous linear first-order differential equations and required a computer to obtain solutions.

6 Dose Concepts and Models

There remained the question of precisely which organ doses to calculate and this was largely decided by the practicabilities of control: it was intake which could be limited, not dose. The quantity that was calculated was the dose equivalent that would inevitably follow an intake, conventionally over the following 50 years: the committed dose equivalent. For assessment against any potential stochastic effects it was then possible to use the organ weighting factors to calculate the committed effective dose equivalent from an intake. The Annual Limits on Intake (ALIs) were then evaluated by calculating the intakes that would lead to committed doses equal to annual dose limits. These ALIs, for inhalation and ingestion, and Derived Air Concentrations calculated from them were presented in Publication 30, published in four Parts with Supplements and an Addendum between 1979 and 1989[16].

The scheme remained much the same following ICRP Publication 60 in 1991. The names of some of the quantities were simplified and the models (which included a new respiratory tract model in Publication 66[17]) were extended to cover children and the foetus and they became much more refined and more complex still. They allowed for so-called "cross-fire" doses from more penetrating radiations, where radionuclides deposited in one organ could irradiate others. Extensive lists of dose coefficients (dose per unit intake) for individuals of a range of ages were published[18] and ALIs[19] but there remained concerns that the microdosimetry associated with very short-range particles was not taken into account. To some it seemed that a complex structure was being built around a core of substantial uncertainty in model structures with, sometimes, a set of parameters with limited and uncertain provenance[20].

Notes Chapter 6

1. (Mould R F, 1993, 1995)
2. (Bragg W H, 1912)
3. (Gray L H, 1929)
4. (Gray L H, 1936, 1937)
5. (Carlsson G A, 2008)
6. (Parker H, 1950)
7. (Kathren R, 1986)
8. (ICRP-ICRU, 1963)
9. (ICRP, 1991a)
10. (Warren S et al, 1949)
11. (ICRP, 1951)
12. (ICRP, 1960)
13. (ICRP, 1968, 1971)
14. (ICRP, 1977)
15. (ICRP, 1979)
16. ibid
17. (ICRP, 1994)
18. (ICRP, 1989b, 1993, 1995, 1996b, 1996a)
19. (ICRP, 1991b)
20. (Goodhead D (Chairman), 2004)

---

7 Genetic Effects

The discoveries of radiation and radioactivity were about as close to Charles Darwin's *Origin of Species* (published in 1859) as we are to America putting a man on the moon. By the time of Röntgen and Becquerel the theory of natural selection was widely accepted but there was still little understanding of the process of inheritance and of the source of the variation needed for natural selection to work on. Darwin's own pangenesis theory of inheritance held some sway. It assumed that each organ produces microscopic "gemmules" that passed around the body and entered the sex cells, each of them carrying the properties of the organ and cells it originated in. As the embryo formed, the gemmules from mother and father were supposed to interact with its new cells to create cells and organs of the appropriate kind and pass on some particular properties from the parents. The theory accounted quite reasonably for many of Darwin's observations, particularly on cross-breeding, but even he thought of it only as a working hypothesis.

It was 1900 before the rules of inheritance began to be widely understood. In one of the more dramatic rediscoveries of science, three scientists independently doing plant breeding experiments found that there were relatively simple rules for the transmission of characteristics between generations. These rules, observed by Hugo de Vries, Carl Correns, and Erik von Tschermak, had actually been discovered (and published) more than 30 years before by the Augustinian monk Gregor Mendel after a meticulous series of experiments on edible peas at his abbey in Brünn (now Brno in the Czech Republic).

Mendel found that the transmission of many characteristics of the peas could be predicted if it were assumed that each pea plant had just two genes (to use the modern term) that determined whether a particular characteristic appeared or not. If there was a gene R that caused round seeds and another r that caused wrinkled seeds then Mendel found the following rules applied:

- Each parent had two genes associated with a particular characteristic and passed one of these, chosen at random, to each offspring
- If the offspring received an R gene from each parent or one R and one r (i.e. RR or Rr) they would have round seeds; when offspring inherited two r genes (so rr) they would develop wrinkled seeds. The R gene was thus dominant (only one was needed) and the r gene recessive (two were needed if a change was to be seen).

With these simple rules Mendel found that he could predict the statistical outcome of crosses of various types (and with seven different characteristics) with remarkable accuracy.

His results were reported in a lecture to the Natural History Society of Brünn in 1865 and published the following year but by the time the principles were rediscovered in 1900 he had been dead 16 years. His ideas were not immediately accepted even when rediscovered and it was largely through the conviction and energy of the British geneticist William Bateson that, by 1910, Mendelism was accepted as the conceptual basis for inheritance. Although there are, as we shall see, many instances where it is more complex than Mendel thought (even he appears to have developed doubts about its universality), his ideas remain key insights of genetics. Mendel had no idea how his rules were realised; a reasonable understanding of the mechanics behind them was more than a hundred years away.

Even while Mendel remained undiscovered, another strand of genetics was developing. Cells had been seen through the earliest microscopes but, enabled by the developments in microscopes and staining techniques for visualising cell structures, the division of cells in mitosis and the chromosomes (the term was coined by the German Wilhelm Waldeyer to reflect the way they stained) were first seen independently by three biologists in 1873.

7 Genetic Effects

Mitosis, the normal way in which cells proliferate, was described in salamander larvae with just about the detail we see in textbooks today by Walther Flemming in 1882 and the process of sexual reproduction began to be understood soon after. The fusing of sperm and ovum (the sex cells) in fertilization of the sea urchin had been seen in 1876 by Oskar Hertwiga and shortly after this several workers, studying Ascaris, a parasitic worm with just four chromosomes, realised that the sex cells had half the number of chromosomes (were haploid) of normal (diploid) cells. By 1890 the way in which sex cells were produced in meiosis was broadly understood—although there were important surprises to come.

As early as 1866 Ernst Haeckel had speculated that the nucleus might be the seat of inheritance. But it was not until nearly 20 years later, in 1884, that four German scientists working independently all suggested that that chromosomes might be the carriers of information. So, by the end of the century it was a reasonable hypothesis that the chromosomes had something to do with inheritance. It was advanced greatly by the work of Theodor Boveri— who showed that a full set of chromosomes was necessary for normal development of sea urchin embryos—but it was Walter S Sutton's work on grasshoppers, published in papers in 1902 and 1903, that brought general acceptance. Sutton showed that the chromosome structures persisted across fertilization and that each chromosome was probably linked to some specific characteristics—they followed the logic of Mendel's genes. Or as Sutton himself put it[1]:

> I may finally call attention to the probability that the association of paternal and maternal chromosomes in pairs and their subsequent separation during the reducing division as indicated above may constitute the physical basis of the Mendelian law of heredity. To this subject I hope soon to return in another place.

So by the early years of the twentieth century and within a few years of Röntgen and Becquerel's discoveries, the notional genes of Mendel and the chromosomes seen through a microscope were drawn together. However, it was the second decade of the 20th century before Mendelism and chromosomes were truly reconciled.

TAMING THE RAYS

In 1910 the American geneticist Thomas Hunt Morgan, working at Columbia University, chose a small fly with a taste for ripe fruit, *Drosophila melanogaster,* as the subject for genetic experiments. This small fly, about 3 mm long, proved a fortunate choice. When a pair was sealed in a small bottle with some simple fruity food they produced several hundred young within just two weeks; experimenting was cheap and the flies were large enough to study with a hand lens.

Linked genes—ones that were located on the same chromosome and were therefore expected to follow one another across generations—were soon encountered in the *Drosophila* work. Sutton had already predicted this but the analysis of the results of crosses showed that the linkage was not invariable—the genes on the same chromosome were not always inherited together—suggesting that there was a transfer of genetic material between chromosomes. The formation of cross-like structures by chromosomes during meiosis had been seen by the French biologist F A Janssens in 1909 (he called them chiasma) and he had speculated that this corresponded to breaking and rejoining (crossing over) of chromosomes with exchange of genetic information. With the *Drosphila* experiments Morgan was soon able to show that this fitted in with the behaviour seen in linked genes. This was an important step in understanding how nature injected genetic variation through sexual reproduction but it also gave a way to begin mapping the genes because the closer together two linked genes were on a chromosome, the more likely they were to stay linked in offspring. This phenomenon allowed one of Morgan's students, Arthur Sturtevant, to begin to map the genes linearly onto the chromosomes.

All this strengthened the hypothesis that chromosomes were the hereditary material but it was Calvin Bridges, another of Morgan's protégés, who published results in 1914 and 1916 showing that, for characteristics known to be sex-linked, the genes were located on the sex chromosomes. This was generally regarded as convincing proof that the chromosomes were the physical embodiment of the genes but the clinching cytogenetic evidence for crossing over (i.e. actually seeing it take place) did not come until 1931 with the work of Barbara McClintockMcClintock and Harriet Creighton on maize and that of Curt Stern on Drosophila.

The term "mutation" was proposed by the Dutch botanist Hugo de Vries between 1901 and 1903 to denote the abrupt changes he saw in the Evening Primrose (*Oenothera biennis*) from generation to generation. While his studies encouraged the idea that mutations could be studied systematically, it was found subsequently that these mutations were not the result of changes in the genes. The Evening Primrose genetics proved difficult to work out but it was eventually clear that the changes were rearrangements of existing genes rather than the creation of new ones. However, de Vries's term stuck and came to mean a hereditable change in a gene. Morgan became interested in whether there were mutations in his fruit flies and he soon found one: a male with white eyes, rather than the normal red ones, suddenly appeared. Experiments showed that the white-eye gene followed the X chromosome: more evidence that chromosomes carried the genetic information.

It was the further development of techniques for investigating the occurrence of such spontaneous mutations in *Drosophila* that led to the first demonstration, in 1926, that mutations could arise in offspring because of irradiation of a parent—a true genetic effect.

As we saw earlier, scientists were not slow to study the effects of radiation on living things. While many experiments were unstructured and looked at the effects on the tissues, there were scientists looking at effects at the cellular level and in the cell nucleus. For example, in 1903 Bohn was able to conclude that the main effect of radium treatment on sea urchins was damage to the chromatin. In the following year, after studying the effect of radiation on developing eggs of *Ascaris*, Perthes suggested that the chromosomes of developing eggs were fragmented. This was confirmed in 1905 by Koernicke who treated *Lilium* (lilies) with radium[2].

So, the fact that radiation could damage chromosomes and therefore disrupt subsequent cell division was known by the end of the first decade of the century. It was clear that rapidly dividing cells were more susceptible (Bergonie and Tribondeau[3]) and it seemed likely that the chromosomal damage was the cause of the somatic radiation effects that were by then known in people. It was Bardeen, in 1906, who was the first to show that irradiated toad sperm led to fertilized eggs that, after a seemingly normal start, failed to develop properly[4]. Since the spermatozoa are entirely nuclear material this showed that it was the nuclear material—the chromosomes— damaged by the radiation that led to the deleterious effects. The results were quickly confirmed with other species such as frogs and rabbits[5].

However, it was the observations of Mavor in the early 1920s that showed convincing evidence of radiation damage to chromosomes. In 1921 and 1922 he showed that doses of x-rays increased the frequency with which chromosomes failed to separate in mitosis and in 1923[6] he showed that there was also an effect on the frequency of crossing over. This was confirmed by Anderson who, in 1925, also found a female fly with her two X-chromosomes attached to one another after irradiation[7].

Now while this meant that the damage to the chromosomes did
affect the individual irradiated (and probably produce the obvious somatic effects) and might persist through a number of normal cell divisions, it did not show a true genetic effect. While the effects of radiation could be seen microscopically on chromosomes as breakages and other gross (in chromosomal terms) damage, there might be other more subtle changes that affected individual genes and could be passed on to offspring. The distinction between chromosomal damage and mutations of individual genes was to be an important, although occasionally fuzzy, one for the rest of the century and the understanding of the ability of radiation to cause mutations made its first major leap forward in 1926 with the work of Hermann J Muller[8].

Muller was a product of the Columbia University laboratory of T H Morgan and his interest was predominantly in mutations. After his doctoral work at Columbia he took a position at the Rice Institute with the British biologist Julian Huxley. He returned to Columbia in 1918 but his two year appointment was not extended and he left in 1920 for the University of Texas and it was here that he developed an ingenious technique for studying mutations based on a particular strain of *Drosophila* called ClB[9]. Using females from this strain Muller was able, through relatively simply chosen crosses, to detect lethal mutations on the X chromosomes of sperm from normal males: when these were present, after two crosses, there were no male progeny. Simply by counting the number of males it was possible to estimate the mutation rate. The method was sensitive enough to determine the spontaneous mutation rate (about 1 in 1000 for lethal mutations on the X chromosome) and detect a temperature dependence. The temperature dependence arose, he thought, because[10]:

...the mutations ordinarily result from submicroscopic accidents, that is, from caprices of thermal agitation, that occur on a molecular and submolecular scale

and this led him to consider that X-radiation might also cause mutations. He used the ClB technique to show quite clearly the effects of irradiating the fly sperm with x-rays: after some 5000 r the mutation rate increased more than 100 times[11]. Writing much later[12], Muller suggested as noteworthy:

...(a) that the induced mutations, like the natural ones, gave evidence of arising in punctiform fashion, in just one or two identical genes in a given cell and (b) that, having arisen, the mutant genes proved to be highly stable and were inherited in the regular Mendelian, chromosomal manner.

In our context the importance of the results was that they showed that radiation could cause mutations—something that was to become a dominant concern of radiation protection years later. At the time, Muller thought it was a least as significant that he had found a means to alter and measure the mutation rate (he saw more mutations in a few months than had been previously seen in all the studies of *Drosophila*) and therefore might have provided geneticists with an important tool to study the genes and inheritance. He was wrong in this; the major advances in genetics were to come through the identification of DNA as the genetic material and the understanding of its structure and chemistry.

Muller just scraped in as the discoverer of radiation-induced mutations. Lewis Stadler, at the University of Missouri, started work on mutations at about the same time as Muller but he choose barley as his subject and, barley being an annual plant, his results were not available until after Muller's. He is remembered as merely confirming Muller's findings[13]. Muller was awarded the Nobel Prize for Physiology and Medicine in 1946 for his discovery and subsequent exploitation of the production of mutations by x-rays. His life after the discovery was ever turbulent: his left-wing views led him to leave Texas for Berlin in the early 1930s and then, as Germany fell into the grip of the Nazis, to Russia. Here a productive few years were brought to an end by the rise of T D Lysenko who brought Lamarkian views to dominance in Soviet genetics. He eventually managed to return to the USA during the war where he finally obtained a post at Indiana University. He remained there until he died in 1967.

A key question posed by Muller's work was the relation between the dose and the number of mutations that resulted. Hanson and Heys, by the end of the 1920s, showed that there was a proportionality between dose and mutations in *Drosophila* and that this was true for betas, gammas and x-rays[14]. Oliver, in 1930, demonstrated this over a 16-fold range so that, in his Nobel lecture in 1946[15] Muller was able to summarise the work on radiation-induced mutations by saying that the number produced was proportional to dose down to as low as 400 r and rates as low as 0.01 r per minute for X and gamma rays.

He believed that there was no threshold dose and that the individual mutations resulted from individual "hits" that affected genes in their vicinity. The nature of these hits was undecided: they might be individual ionisations or clusters of ionisations either at the end of electron tracks or at their side branches. Whatever the hits were, the mutations resulted from "disturbances on a microscopically localized scale".

The experimental work of Muller and others on *Drosophila* was the foundation of genetics in general and radiation genetics in particular and the fruit fly continues even today to illuminate the basic physics and biology of inheritance and radiation effects. However, by the time Muller received his Nobel Prize, there was intense interest in the degree of genetic risk posed to humans by radiation and the investigation of this required something closer to us biologically than a fly. It was thus the mouse that was to provide most of the data for risk estimation in the second half of the century and the key players were the Russells at Oak Ridge, with what became known as their mega-mouse experiment, and T C Carter in England.

William Lawson Russell (known as Bill) was an English geneticist who went to the USA in the early 1930s to work on *Drosophila* and guinea pig genetics. In the late 1940s he moved to the Biology

7 Genetic Effects

Division at Oak Ridge with his second wife and co-worker Liane Braunch and a plan to research mouse genetics. With the growing concern about fallout from weapons testing, he was encouraged to investigate the effects of radiation by the division head Hollaender—supported by H J Muller—and to think big. An entire floor of an old factory was turned over to a mouse house and the Russells started a programme on an unprecedented scale. The first results came in 1951[16] when just over 50 mutations were seen in the progeny of 48,000 irradiated males against two in a slightly smaller control group. The mouse house was promptly tripled in size and, with the Russells in a leading role, continued to provide the majority of the data for the assessment of genetic risk from radiation into the 1990s. It has been calculated that around seven million mice were used in the programme. This was, as Alvin Weinberg the Director of Oak Ridge said, Big Biology with "big institutes, big experiments, big money and, one hopes, big ideas." Bill Russelll died in 2003.

The key to Russell's success was not just the scale of the experiments; he also developed a simple method for detecting recessive mutations in first generation offspring of irradiated mice known as specific locus assay. This involved mating the irradiated mice with a special strain of mice homozygous for seven autosomal recessive visible mutations. Mutations of the marker loci caused by radiation will result in visible effects in the progeny.

The seven loci chosen to be followed in the specific locus test were defined by recessive mutations with visible homozygous phenotypes that were easily distinguished in isolation from each other, and had no effect on viability or fertility. The seven loci are agouti (a is the recessive non-agouti allele), brown (b), albino (c), dilute (d), short-ear (se), pink-eyed dilution (p), and piebald (s). A special 'marker strain' was constructed that was homozygous for all seven loci[17].

A similar large-scale programme was set up after the war at the MRC Radiobiological Research Unit Edinburgh Scotland under the direction of T (Toby) C Carter before moving to Harwell in England.

In parallel with the increasing understanding of the effects of radiation, there was improved knowledge of naturally-occurring genetic disease inherited in a Mendelian way. It started with the sex-linked diseases (those from genes associated with the X chromosome) where a report on the inheritance of colour blindness was made to the Royal Society as early as 1779. In 1820 the sex-linked nature of haemophilia inheritance was recorded and somewhat later that of Duchenne muscular dystrophy.

In 1902 Archibald Garrod[18], working at the Hospital for Sick Children, Great Ormond Street, London identified the condition alkaptonuria that led to a metabolic variation that caused the urine of sufferers to turn black. Other clinical symptoms developed in later life—among them something akin to arthritis. The condition was more common among the offspring of first cousins. With advice from William Bateson he suggested that it was an autosomal recessive trait due to single recessive gene but it was not until six years later, after further cases were studied, that this was confirmed. Garrod believed that albinism, cystinuria and porphyrinuria had similar origins and subsequently it was found that cystic fibrosis and sickle cell anaemia were further examples. In the meantime the first autosomal dominant condition in man (brachydactylyl—in which sufferers had short stubby fingers) was identified by a Yale PhD student William C Farabee in 1903[19]. As the century progressed many more diseases were added to the list of conditions inherited in a Mendelian fashion.

The concern about the effects of bomb fallout in the early 1950s created an urgent need to assess the actual genetic risks associated with radiation. Since there were no data from humans, it was necessary to use the information from animal experiments and of spontaneous occurrence of genetic diseases in the human population. The single most important parameter for the next 50 years, one which linked these two sources, was the doubling dose (DD), the radiation dose that caused the mutation rate to double from the natural background rate.

The evolution of the ideas and the data in the second half of the 20th century is best traced through the reports of the various national and international bodies who examined the question of the genetic risk of radiation. This does not always acknowledge the origins of some of the contributions (although these can be traced in the individual reports) but illustrates how the informed and authoritative mainstream of thinking developed. Of course, it means also that we may not give enough weight to more controversial ideas that have come and, usually, gone.

7 Genetic Effects

The Medical Research Council (MRC) Committee on the Hazards to Man of Nuclear and Allied Radiations reported in June 1956[20] and based its consideration of genetic effects on the DD method, starting its analysis by considering the effects that a doubling of mutation rates in one generation would have on the burden of genetic disease. Three example diseases were presented in detail based on the work of L S Penrose (described in appendices) and the results for these are summarised in Table 2.

Disease type	Sample disease	Effect of doubling of mutation rate
Dominant	Achondroplasia	80% increase in generation 1 then falling back over next few generations
Sex-linked	Haemophilia	29% increase in generations 1 and 2 then falling back over next few generations
Recesssive	Phenylketonuria	1% increase in generation 1 then slow increase and decline over many generations

Table 2: Estimates of effects of doubling mutation rate

The effects on a number of other mental conditions were also looked at. In its discursive style the report then gave an impression of the social impact of such rises.

It was only after this that the Committee turned to radiation and an estimate of the DD. Here they initially noted that the DD had to lie above 3 r, the cumulative dose to the gonads from natural radiation before reproduction generally ended. The human data were, of course virtually non-existent. However, it was known that radiation at the background level caused about 0.01% of the spontaneous mutations in *Drosophila*. Allowing for the longer timescales with humans and the suspected differences in spontaneous rates between man and the fly they estimated that about 2% of spontaneous mutations in man were due to radiation. There was though another correction to apply: mouse genes were known to be ten times more sensitive than fly genes. This pushed the fraction up to 20% giving a lower limit for the DD of 15 r. The upper limit, 150 r, was based on the 2% fraction.

The doubling dose information that existed for other organisms was reviewed and it was concluded that much of this pointed to values between 25 and 60 r. Combining both these sets of information the Committee put forward a best estimate for the DD of between 30 and 80 r.

At about the same time the Genetics Committee of BEAR[21] set up by Bronk reported. Its members included Hollaender, Crow, C C Little, Muller, Neel, Russell and Sturtevant and the report was prepared as the first information on the effects of radiation on mammals—notably mice in Russell's experiments— began to appear.

The main conclusions were:

- Mutant genes are usually recessive
- A small but not negligible part of the harm appears in the first generation
- There is a proportionality between dose and increased mutation rate and the effect is cumulative

Three aspects of genetic hazard concerned them:

- Risk to offspring and descendants of people receiving larger doses
- Effect of average doses on population as a whole
- Because of the increased death rate that might arise from prolonged exposure "the population, considered as a whole, would decline and eventually perish"(page 17)

Two methods for assessment were adopted: the doubling dose one and a more direct alternative. The "responsible" suggestions to the Committee for the size of the doubling dose ranged between 5 and

150 r and they settled for a statement that it almost certainly lay between these values but that it "may very well be" between 30 and 80 r. So the conclusions on DD were, reassuringly or disconcertingly, similar to the British MRC values (in fact there seems to have been collaboration). The spontaneous rate of genetic disease was estimated from the observed 4-5% of birth defects seen in the USA and it was taken that about half of these had a simple genetic origin and appeared prior to sexual maturity. There then seemed to be an assumption that almost all of these were due to background radiation because it was deduced that a DD to all the Americans then alive would lead to an increase of between two million and four million in the eventual number of defective children in subsequent generations. If the US population were subject to 10r each then there would be 50,000 defective live births in the first generation and half a million in all generations.

While this may seem comprehensible enough to us, it apparently did not satisfy all the geneticists on the Committee as meaningful. In an alternative approach six members of the Committee set out to calculate the total number of "mutants" that would result from the 10 r to the gonads. They agreed remarkably on a central value of 5 million and this was consistent with the DD calculations but they were honest enough to suggest that they could all be a factor of ten in error either way: the number could be between half a million and fifty million. This was, as the Committee put it, "disappointingly vague". Nonetheless, the numbers were big enough to suggest that genetic effects should be the major concern—a theme that was to dominate radiological protection thinking for decades to come.

By the 1960 MRC Report[22] there was more information from Japan and much more from animals. While there were a number of discoveries—for example of the effect of dose rate on mutation rate—overall it was clear that the previous conclusions were not far out. The rather surprising discovery in 1956[23], after the first report, that man had 46 rather than 48 chromosomes had no effect.

The 1962 UNSCEAR report[24] reviewed spontaneous rates and concluded that one rad/generation would cause an increase of between 1/100 and 1/10th the spontaneous frequency, so keeping broadly the same DD estimate. The theme continued and in its 1966 report[25] UNSCEAR concluded that between 2 and 3% of live-born children are affected by severe disabilities arising from dominant mutations. Other conditions may also be genetic but their frequency could not estimated.

The rate of mutations induced by radiation was estimated from experiments with mouse germ cells because the human data were "meagre" and it was estimated that there would be two mutations per 1000 male gametes per rad when the radiation was delivered as a single acute dose. While the resulting disabilities would be harmful, like the spontaneous ones, it was not possible to give an estimate of their frequency. The authors limited themselves to a statement that a dose of one rad per generation would add something like 1/70th of the total number of mutations arising spontaneously in a generation —within the previously estimated range. While the 1/70th might also apply to hereditary diseases in man it had to be remembered that there were many complexly inherited conditions also.

UNSCEAR reported again in 1972[26] and in the same year another authoritative source of genetic risk assessment appeared: the first BEIR report from the US National Academy of Sciences[27]. In both reports the Doubling Dose method of assessing mutation risks was used: in UNSCEAR it supplemented the direct method but in the BEIR report it was adopted as the principal assessment method for gene mutations. UNSCEAR took the DD to be 100 rad for male germ cells and made no estimates for females while BEIR took a range of 20-200 rad.

UNSCEAR gave two estimates of recessive point mutation induction based on radiation-induced mutations rate of mice spermatogonia : the difference in the values obtained (a factor of 40) could be explained away on technical grounds. The increase in disease with a mutational cause that might be brought about by radiation was estimated with a DD methodology using the spontaneous incidence data from the earlier reports. BEIR used a similar indirect approach to mutations using their range of DDs.

UNSCEAR gave estimates for males but, because of the lack of data, were unable to give risk estimates for females for recessive lethals, dominant visibles and skeletal mutations. BEIR gave results for both using, where necessary, the conservative assumption that female and male risks were the same. UNSCEAR restricted themselves to the first generation after exposure while BEIR, with some further assumptions, made estimates for both the first generation and equilibrium.

7 Genetic Effects

By 1976 UNSCEAR was ready to review the mouse data and its methods again[28]. The direct method was used to estimate the number of recessive and dominant mutations that might be produced by irradiation. Recessive mutation rates were estimated as 60×10^{-6} per gamete per rad for irradiation of sperm and much lower for female germ cell irradiation. The reduction factor of three, used in their previous report to allow for the low dose rate applicable to human exposure compared with that in the mouse experiments, was confirmed and used again.

The risk of dominant mutations that might lead to serious handicap was estimated from the rate of induction of skeletal mutations in the offspring of irradiated mice. The risk of these skeletal mutations (about 4×10^{-6} per gamete per rad) was adjusted to allow for about 10% of dominant mutations in man being associated with skeletal abnormalities and about 50% of these dominant mutations causing serious handicap. Overall this led to a dominant mutation risk associated with serious consequences of about 20×10^{-6} per gamete per rad for male irradiation. No estimates were made for female irradiation but the risk was considered low.

The doubling dose method was also used for estimating the increased incidence of Mendelian diseases taking spontaneous rates as 0.1% for recessive diseases and 1% for dominant and X-linked diseases. With the doubling dose of 100 rad derived from mouse experiments (and now indicated as a minimum value by data on the mortality of children born to bomb survivors) they estimated the impact of 1 rad per generation of low-LET radiation. There would be, at equilibrium, about 100 additional cases per million births of dominant and X-linked disease. In the first generation, per million live-born, there would be about 20 additional cases: not much different from the direct estimates. However, they did comment (para 637) that one of the main assumptions of the doubling dose method, that of the proportionality between spontaneous and induced rate, "still remains to be proved".

In the report published in 1982[29] UNSCEAR essentially confirmed its conclusions in 1977. The knowledge the occurrence of spontaneous mutation diseases was more firmly established and the animal data had expanded considerably both in terms of species and irradiation conditions. The lower sensitivity of female germ cells was confirmed and there was support for the basic proportionality assumption of the doubling dose method from fruit flies and bacteria. The estimates from the doubling dose method of the increased equilibrium incidence of dominant and X-linked disease remained unchanged from 1977 but the number expected in the first generation reduced slightly. In summary—and in the new Gray units—the incidence of mutation-related diseases in a population of one million exposed to 1 Gray per generation would increase to 1500 in the first generation and 10,000 at equilibrium. Such diseases would then account, as we will see, for some two-thirds of the total genetic impact.

The 1986 UNSCEAR report[30] saw some further adjustments to the assessed mutation risks. Most significant perhaps was a somewhat lower dominant mutation risk for males as indicated by further mouse data. It was also possible to make an estimate of the risk from induced autosomal recessive mutations (which in previous reports had been assumed negligible). For our one million population and a single dose of 1 mGy, there would be no extra cases in the first generation and possibly one extra one in the following ten generations.

The 1993 UNSCEAR report[31] pointed to three conditions that confused inheritance. Mosaicism, when both normal and mutated germ (and somatic) cells can be present in the same individual, genomic imprinting, where the expression of genes sometimes depended on whether they had been inherited from the mother or father and uniparental disomy, when all the chromosomes came from one parent.

The 2001 UNSCEAR report[32] was the first to take advantage of the advances of the late 20th century in molecular biology which, in many cases, complicated the treatment of Mendelian diseases. The one-to-one relationship between genetic make-up (the genotype) and the physical make-up (the phenotype) had broken down: there were mutations that could lead to several diseases; some similar diseases came from quite different gene mutations and genes that were dominant in some people were recessive in others. Allelic expansion (where a gene expands as it is transmitted between generations) is also an example of effects that greatly increased complexity. It was also recognised that the division of genetic damage into gene mutation and chromosomal damage was a fairly arbitrary one: point mutations were at the molecular end of a continuum that extended up to the damage visible through a microscope. Although the

7 Genetic Effects

distinction was formally maintained, in fact chromosomal disease was subsumed under Mendelian diseases and congenital disease.

The key risk estimates were made using the doubling dose method based on the spontaneous disease rates in man (rather than the mouse) but taking the doubling dose itself from mouse experimentation. The concept of the Mutation Component (MC), initially developed by Crow and Denniston[33] in the 1980s, was developed as a mathematical model in the 1990s with the MC defined as the fractional increase in disease per fractional increase in mutation rate.

With this additional factor the risk equation became

Risk per unit dose = P x(1/DD) x MC

Where P is the background incidence of the disease.

For the dominant Mendelian diseases MC is related to the selection coefficient, s, the probability that an individual with a genetic disease will go on to reproduce. For these diseases the Mutation Component value can be estimated for a variety of situations using standard population genetics models. For example, for a one-time increase in mutation rate (a so-called burst) the Mutation Component for the subsequent gth generation is:

$MC(g)=s(1-s)^{g-1}$

The value chosen for s by UNSCEAR[34] was of the order of 0.3 and this defined the MC for first generation risks. This value was used in making the risk estimates for autosomal dominant and X-linked diseases in their 2001 Report and the following BEIR. The value for autosomal recessive diseases is expected to be near zero.

The 1990s saw the beginnings of the resolution of one of the persistent problems of genetic risk: the squaring of mouse data with the fact that, as Sankaranarayanan put it in a 2002 paper[35], "no radiation-induced germ-cell mutation, let alone an induced genetic disease has been found in humans!". The results from the children of bomb survivors[36], at the beginning of the decade, began to suggest that the doubling dose of about 1 Gy derived from the extensive mouse experiments, might be several times too low. A better understanding of the nature of mutations made it clear that the source of the discrepancy might lie in essential differences between spontaneous mutations and those cause by radiation. They were different in nature: those caused by radiation were largely deletions of single or multiple genes while spontaneous ones were a mixture of types. They were differently caused: radiation damage arose at random while the spontaneous mutations were often closely related to DNA sequence organisation. And they had different effects: the radiation-induced changes generally caused a loss of gene function while the spontaneous ones sometimes resulted in a gain of function. These new complexities made the simple concept of the doubling dose a good deal shakier than before. Specifically, they suggested that the mutation caused by radiation, that might lead to genetic disease, might also frequently lead to the non-viability of the genetic material. If this were so, the genetic changes would never be expressed in live births and they would be lost. In the jargon, they would not be "recoverable".

This interpretation seemed to offer an explanation of the differences between the mouse experiments and the experience with humans. The doubling dose had been deduced from the mouse experiments and most recently from work on recessive mutations in just seven genes. But these were subject to a bias since these genes were chosen for study because they were not essential to survival of the animal and were in a non-essential region of the genome: they were chosen just because they were "recoverable".

It was clear that it was necessary to estimate just what the impact of this was and this was done, in the late 1990s, by Sankaranarayanan and Chakraborty[37]. They estimated, in a detailed study based on experience with mice, the expected impact of a radiation-induced deletion on 63 autosomal and X-linked human genes that between 15 and 30% might be recoverable and be capable of causing genetic disease. They termed this the Potential Recoverability Correction Factor, PRCF and thus extended the basic risk equation to:

7 Genetic Effects

Risk per unit dose = P x(1/DD) x MCx PRCF

The approach was adopted by both UNSCEAR[38] and BEIR.

Because of the better understanding of the nature of genetic disease, its spontaneous occurrence was recognised as being higher than previously thought: overall, estimates nearly doubled to 240 cases per 10000 live births from the figures used previously. However, the more detailed mathematical treatment led to estimates for autosomal dominant and X-linked diseases that were not very different from those of 1993. With the 1 Gy doubling dose, the estimates for parental exposure of 1 Gy to low-dose low-LET radiation are 750- 1500/million in the first generation and 1300-2500/million progeny in the second. The risk of recessive diseases was put at zero. The Committee's estimates were limited to the first two generations to recognise the uncertainties in the demographics and the various parameters used in the risk assessment over longer period of time.

The evolution in the last half-century of the estimates of the risks of Mendelian diseases due to radiation exposure from the different sources are summarised in Table 3. While a few assumptions have been made here in deducing single figures to represent the outcome of complicated arguments and there have been changes in classification over the last fifty years, the consistency is striking. Perhaps it should not be surprising though; the doubling dose has remained about 100 r (in old units) and the estimates of background rates have not really not changed significantly.

	Risk per million liveborn/Gy per generation		
	Autosomal dominant and X-linked		Recessive
	1st generation	Equilibrium	1st generation
BEAR 1956			
UNSCEAR 1962		1000-10000(c)	
UNSCEAR 1966		1300(b)	
UNSCEAR 1977	2000 2000(a)		
UNSCEAR 1983	1500 1000(a)	10 000	
UNSCEAR 1986	500-1000		0-
BEIR-V 1990	500-2000 severe 100-1500 mild	2500 severe 7500 mild	<100
UNSCEAR 1993	1500		~5
UNSCEAR 2001	750-1500(d)	No estimate made	0

a) based on direct method

b) 0.1%(para 4) x 1/70 per rad (para 25)

c) based on 0.1% incidence and 1/100-1/10 natural incidence per rad

d) includes some diseases previously classified as chromosomal

Table 3: Mendelian disease risks from radiation

The Mendelian effects discussed to this point have arisen from mutations and these are associated with changes in just a single gene. It is the fact that some such changes lead to viable offspring who can pass on the mutation that allows it to persist. Such point mutations are, however, just one end of the spectrum of damage that radiation can cause in chromosomes.

Changes in entire sets of chromosomes in plants had long been known and these had confused early workers (as noted earlier on the origins of the word "mutation"). However, more relevant from our point of view were the results that emerged from *Drosophila* in the 1920s where, to explain the observed gene linkages, workers were forced into assumptions about the behaviour of chromosomes. Three broad types of gross damage were postulated to occur spontaneously: deficiency (a length of chromosome was lost), inversion (a piece was removed, turned around and put back) and translocation (a piece was transferred to another chromosome).

To some these invisible changes were just speculation and an easy way to explain anomalous results. However, in 1933, T S Painter discovered the gigantic chromosomes in the salivary glands of *Drosophila* larvae that were large enough to study under the microscope and make detailed maps of the banding. Soon it was possible to tie the locations of genes predicted by purely genetic means to the bands on the chromosomes and, in a striking confirmation of the insight of the earlier workers, the various presumed types of changes to chromosomes could now actually be seen. They were real. It was later found that the giant chromosomes had been known to cytologists in 1881; it had taken 50 years for geneticists to find them and exploit them in these critical observations.[39]

The mechanism by which radiation created these changes was not really understood until the 1930s. Early work assumed that, after a chromosome was broken, the "centric part" (that with the centromere) could be passed down through cell divisions as a fragment. These and other fragments would join up in different ways to cause the postulated damage. It was H J Muller in 1932 who developed the picture that fitted the majority of cases: the *Drosophila* chromosome was actually broken in two places and these ends were then rejoined in the different ways. The original free ends of chromosomes were always free; they were specific structures, telomeres, defined by the genes there. Muller's idea was based on a wide range of experimental data and by 1938 he was convinced that this was the general model of structural change[40] and he was also sure that the breaks that were the source of the damage were independent—based on the dependence of the frequency of their occurrence on the 1.5th power of the radiation dose that caused them. Similar results came from Sax with plants[41] and Carlson[42] with grasshoppers.

On the surface a squared relation with the dose was to be expected but Muller explained his results by the increased cell killing that occurred at higher doses and the various susceptibilities among the exposed cells. He also noted that where the radiation was densely-ionising fast neutrons the relation was proportional, suggesting that two breaks were produced close together by the same track and joined.[43]

By 1956 the MRC Report[44] was able to summarise the effects of chromosomal damage:

> 1. Structural changes are caused most readily by large acute doses of radiation such as x-rays and from atomic bombs; extended exposure to low intensity x-rays only rarely causes them although neutrons and alpha particles are more effective
>
> 2. Chromosome breakage usually results in cell or embryo death
>
> 3. But, if fragments reunite in new patterns, these may be capable of passing through cell division
>
> 4. The resulting structurally changed chromosomes may be transmitted to apparently normal offspring and one type of change may lead to repeated abortions or malformations in descendants
>
> 5. Major chromosomal changes may bring sterilisation or abortion but do not, as a rule, cause abnormalities

It has become apparent that spontaneous chromosome damage is not uncommon. A high proportion of abortions and about 0.5% of live births show some kind of chromosome anomaly with the most common being trisomies followed by unbalanced translocations. Trisomies are extra chromosomes which result from the failure of chromosomes to separate properly (non-disjunction) in meiosis and occur in several chromosomes. Down's syndrome results from the presence of an extra chromosome 21 (Trisomy 21), but Trisomy 18, Trisomy 13, Turner syndrome (a missing or incomplete X-chromosome in women), and Klinefelter syndrome (an extra X chromosome in men) are also seen in live-born infants.

However, the conclusion of the 1956 MRC committee has remained broadly valid: "chromosomal structural changes are likely to be of comparatively little importance among the radiation hazards to man" (para 125)

In spite of the time that has been spent in this chapter on Mendelian diseases, most traits or diseases with a genetic component are not inherited in such a simple way; they are associated with more than one gene (they are polygenic). Their expression is also not determined solely by genetics but by environmental factors and they tend to run in families. Normal traits that behave like this include stature, intelligence and skin colour. Abnormal conditions such as common congenital disorders, diabetes mellitus, rheumatoid arthritis as well as some psychiatric conditions like schizophrenia are inherited in this way. While in multifactorial inheritance (as it became known) the individual genes are inherited in a Mendelian fashion, their combination and the folding in of environmental factors is far more complex and requires a different approach

In normal traits the variation caused by multifactorial inheritance shows up as a normal (or Gaussian) distribution of an attribute such as stature but for abnormal conditions it is the likelihood of occurrence of the condition which is important. To predict this Falconer[45] in 1965, using earlier concepts of Wright[46], defined a liability: an unobservable continuous parameter determined by an additive combination of genetic and environmental factors. This liability is then expected (on the basis of the central limit theorem) to be normally distributed and, in the so-called threshold model, should its value be greater than a certain threshold value, then the individual will develop the abnormality or disease.

The application of these ideas to risk estimation has been fraught ever since: it has been difficult to establish which diseases have a genetic component and just how big that component is and, beyond this, there have been varying estimates of spontaneous rates. Since the spontaneous rates have generally formed the baseline for the assessment of any enhancement of risks from radiation exposure, these assessments have usually been restricted to congenital abnormalities where there is some agreement on the background occurrence. If there is an overall pattern it is of early confidence followed by pessimism and then a cautious and measured optimism.

UNSCEAR 1966 and the BEIR 1972 report[47] both made estimates of spontaneous occurrence of multifactorial diseases and by 1977 UNSCEAR[48] was confident enough to make some estimates of radiation risk based on a doubling dose of 1 Gy and a mutation component of 5%. These were reiterated a few years later in 1982 but by the time of its 1993 report[49] there was a rethink:

> ...the uncertainties did not justify continuing this procedure, given the higher estimated incidence of these conditions of varying seriousness that can arise throughout a lifetime (para296)

It seemed to say that while the methodology was satisfactory while the results were rather insignificant, now the prevalence of MF disorders was understood to be high it could not be trusted. Estimates of radiation risks for these disorders were not made by UNSCEAR between 1982 and 2001 but evidence showing that multifactorial diseases were rather common continued to grow

The UNSCEAR 1986[50] report divided multifactorial disorders into congenital abnormalities and others. The background rate for congenital abnormalities used in the 1982 report[51] (and based on British Columbia data) was 430/10,000 live births. By 1986 it was possible to re-evaluate this using data from Hungary and the United States that had not been fully and systematically taken into account in the earlier report. The use of data from a variety of sources proved, as might be expected, difficult and the conclusions were rather complex. However, broadly, the prevalence was between about 430 and 850/10,000 live births depending upon exactly which categories of abnormalities were included. The Committee decided upon 600/10,000 , derived from the Hungarian data but representing a realistic average of the available information.

From the British Columbia data, a rate of 470/10,000 had been derived by the workers Trimble and Doughty but this was really only for disorders that showed up before the age of 21 and was therefore clearly an underestimate. Hungarian work, just becoming available, suggested a much higher value: 6000/10,000. The proportion of people affected was actually less than 60% since many suffered from more than one defect. The Committee was unable to make any comment on the effects of radiation.

The 1988 report[52] could add little to the previous one and UNSCEAR 1993[53] did not adopt an optimistic tone in discussing multifactorial disorders. Rather it detailed the many uncertainties and complexities that stood in the way of an adequate treatment. The spontaneous rates were essentially unchanged, the availability of mathematical models for multifactorial disorders was noted but the conclusion on radiation was not encouraging:

> ...in view of the complex genetic basis of most malformations and of multifactorial diseases, it is impossible to estimate increases caused by radiation...(para 296)

The 1990 BEIR V report[54] arrived at a lower value for spontaneous congenital abnormalities than the figure adopted by UNSCEAR in 1986: 20-30 per 1000 live births rather than 60. However, there was an enormous increase in the number of other multifactorial diseases(or as BEIR preferred "disorders of complex etiology") from 600 per thousand to 1200 per thousand. These were broken down into heart disease (600), cancer(300) and "selected others" (300). This means that there are more disorders than births.

The conclusions at various time on spontaneous rates are shown in Table 4.

In contrast to UNSCEAR, BEIR did make some estimates of the impact of radiation—but only on congenital disorders. They concluded that there would be one additional case seen in the first generation per thousandlive births per Sv per generation and between 1 and 10 per thousand in each generation at equilibrium. A doubling dose of 1 Sv and a mutation component of 5-35% was used.

Estimated spontaneous burden of multifactorial disorders (per 1000 live births)		
Source	Congenital Abnormalities	Other multifactorial
Stevenson 1959	10	10
UNSCEAR 1966	25	15
BEIR 1972	15	25
Trimble & Doughty	43	47
UNSCEAR 1977	90	
Carter 1977	24	-
BEIR 1980	90	
UNSCEAR 1982	43	-
Czeizel & Sankaranarayanan 1984	60	-
UNSCEAR 1986	60	600
UNSCEAR 1988*	60	600
BEIR V 1990	20-30	1200
UNSCEAR 1993*	60	650
UNSCEAR 2001*	60	650
BEIR VI (draft)*	60	650
From BEIR V Table 2-5 except * from original documents. All rounded to nearest whole number. UNSCEAR 1977 and BEIR1980 are combined values		

Table 4: The background of multifactorial disorders

The work through the 1990s on Mutation Component and, latterly, PRCF that contributed to the understanding of the risks of pure genetic diseases finally made it possible to construct some estimates of the enhanced risks from multifactorial diseases caused by radiation. The ICRP Task Group set up to consider multifactorial diseases was able to report in 1999[55] with a structured approach to making the estimates.

7 Genetic Effects

To estimate MC they used a hybrid model that became called the Finite Locus Threshold Model (FLTM) which combined the threshold model for liability with a mechanistic population genetic model. The threshold model described above assumed that the number of genes involved was very large so that the liability could be expected to have a normal distribution. The Task Group took a rather different approach: they assumed that there were just a few genes (actually five) responsible for the genetic component. The environmental contribution was assumed to be normally distributed.

The model was then subjected to what was essentially a sensitivity analysis to see how, for a plausible range of parameters (such as s and the threshold level), the Mutation Component varied with heritability. The results were complex but some simplifying principles emerged. The most important was the discovery that, in the early generations, MC was very much less than 0.02 for heritability between about 30% and 80%. Multifactorial diseases were thus predicted to be "far less responsive to induced mutations than Mendelian diseases". The factor 0.02 was adopted for risk estimation for chronic multifactorial diseases by UNSCEAR[56] and BEIRV[57]. Conclusions are summarised in Table 5.

Source	Effect of 1 Sv (low-LET) per generation/10^6 live births/generation	
	First generation	Beyond
UNSCEAR 1977	used DD 1 Gy+MC 5%	
UNSCEAR 1982	used DD 1 Gy+MC 5%	
BEIR V 1990	1000(n/e)	1000-10000(n/e) at equilibrium
UNSCEAR 1993	n/e(n/e)	n/e(n/e)
UNSCEAR 2001	~2000(~250-1200)	2400-3000(~250-1200) in second generation

n/e: not estimated. Main figure for congenital disorders, "other multifactorial" in brackets

Table 5: Radiation risk estimates for multifactorial diseases

Chromosomes were known from rather early in the century to be made of proteins and DNA, a long stringy molecule that was thought to be some kind of constructional material. Until 1944 almost all scientist thought that the genes were somehow located on the proteins but, in that year and after decades of incremental research work, a team at the Rockefeller Institute for Medical Research in New York led by Oswald Avery showed that they were in fact carried by the DNA. It was not an instant success as a theory, partly because the work was actually on the transformation of bacteria and partly because DNA seemed too simple a molecule for this awesome task. For another ten years the matter was not settled; the crucial illumination of the molecular basis of the genes had to wait for the development of the science of x-ray crystallography and the imaginative genius of James Watson and Francis Crick. Their double helix model of DNA, published in 1953, answered two key questions: how is the genetic code carried and how is it passed on? One of the great insights of 20th century science, it has explained in molecular terms all the observations of genetics in the previous century and has been a fertile agent of new ideas ever since.

It has allowed us to track down the molecular changes that cause (or predispose us towards) many diseases. But it has also shown us that nature is even more complex than we thought and much more messy. The DNA molecule is largely composed of apparently meaningless repetitive sequences interrupted by seemingly randomly-located stretches that carry instructions. It was, of course, created and modified by natural selection and, as Professor Steve Jones puts it in his book *The Language of the Genes*[58]: "Natural selection has superb tactics, but no strategy". It is a process that has acted "like a handyman rather than a craftsman" leaving "products that are badly—not to say extravagantly –planned and roughly made".

Nonetheless, we now know a great deal about the molecular nature of our genes, how they are passed between generations and how they are used to construct the proteins that make up our bodies. We have witnessed one of the great revolutions of science but, by the end of the century, the molecular nature of the genes had had surprisingly little impact on the understanding of the action of radiation in genetic terms. Our assessments would have been little different even if our knowledge had been much less.

Notes Chapter 7

1. (Sutton W S, 1902)
2. (Sturtevant A H, 1965)
3. (Bergonie J and Tribondeau L, 1906)
4. (Bardeen C R, 1907)
5. (McGregor J H, 1908; Regaud C and Dubreuil G, 1908)
6. (Mavor J W, 1924)
7. (Sturtevant A H, 1965)
8. (Meggitt G C, 2016)
9. (Moore J A, 1972; Crow J C and Abrahamson S, 1997)
10. (Muller H J, 1964)
11. (Muller H J, 1927)
12. (Muller H J, 1958)
13. (Stadler L J, 1928)
14. (Hanson F B and Heys F, 1929)
15. (Muller H J, 1964)
16. (Russell W L, 1951)
17. (Silver L M, 1995)
18. (Garrod A E, 1902)
19. (Farabee W C, 1903)
20. (MRC, 1956)
21. (BEAR Committee, 1956)
22. (MRC, 1960)
23. (Tijo J H and Levan A, 1956)
24. (UNSCEAR, 1962)
25. (UNSCEAR, 1966)
26. (UNSCEAR, 1972)
27. (BEIR Committee, 1972)
28. (UNSCEAR, 1977)
29. (UNSCEAR, 1982)
30. (UNSCEAR, 1986)
31. (UNSCEAR, 1993)
32. (UNSCEAR, 2001)
33. (Crow J F and Denniston C, 1981)
34. (UNSCEAR, 2001)
35. (Sankaranarayanan K, 2002)
36. (Neel J V et al, 1990)
37. (Sankaranarayanan K and Chakraborty R, 2000)
38. (UNSCEAR, 2001)
39. (Moore J A, 1972)
40. (Muller H J, 1958)

TAMING THE RAYS

---

41 (Sax K, 1938)
42 (Carlson J G, 1938)
43 (Giles N H Jr, 1940; Muller H J, 1954)
44 (MRC, 1956) p28
45 (Falconer D S, 1965)
46 (Wright S and Eaton O N, 1923; Wright S, 1934)
47 (UNSCEAR, 1966; BEIR Committee, 1972)
48 (UNSCEAR, 1977)
49 (UNSCEAR, 1982, 1993)
50 (UNSCEAR, 1986)
51 (UNSCEAR, 1986)
52 (UNSCEAR, 1988)
53 (UNSCEAR, 1993)
54 (BEIR Committee, 1990)
55 (ICRP, 1999)
56 (UNSCEAR, 2001)
57 (BEIR Committee, 1990)
58 (Jones S, 1993)

---

8 Somatic Effects

The radiation injuries among pioneers were terrible: dermatitis, epilation, degeneration of blood vessels, changes in the bone marrow, necrosis and anaemia were all reported. Cancer associated with an x-ray ulcer was reported as early as 1902 and leukaemia was described in five workers in 1910.

The effects were summarised in the first international recommendations of the International X-ray and Radium Protection Committee(IXRPC) in 1928[1]:

> The effects to be guarded against are (a) injuries to superficial tissues, (b) derangements of internal organs and changes in the blood.

They were then, and for decades afterwards, seen as the results of chronic overexposure: keep the exposures down and the consequences would virtually disappear. So Mutscheller could write in 1925[2]:

> ...it seems that under present conditions and standards accepted at present, it is entirely safe if an operator does not receive every thirty days a dose exceeding 0.01 of an erythema dose...

The search for the precise value of such a safe level—what was soon called the "tolerance dose"—continued for decades and is described in detail in Chapter 12. The belief that somatic effects of significance would appear only after a threshold level had been exceeded dominated radiation protection for many years. If we leap forward twenty of them from Mutscheller to Cantril and Parker[3] writing in 1945 we will find statements that:

> The majority of radiation effects are thought to be of the threshold type

> Fatal anaemias which do not respond to any form of treatment have appeared in a considerable number of radiation workers after long continued overexposure. This is a manifestation of bone marrow exhaustion in which repair has not been able to keep up with the continued insult produced by overexposure.

By the 1949 Tripartite talks between the USA, the UK and Canada[4] the terminology "tolerance dose" had been dropped in favour of "permissible dose" but thinking was still based on there being a threshold level.

It was agreed to regard the blood-forming organs as the critical tissue when the body is exposed to hard x-rays or gamma-rays, while the skin is to be considered the critical tissue in the case of soft x-rays or beta rays.....Damage to the blood-forming organs caused by over-exposure over a period of many years is manifested by the development of leukaemia, while epitheliomas are the end result of skin over-exposure.

In the notes from the US Pre-meeting[5]:

> ..it may be expected with confidence—justified by clinical experience—that if the exposure to radiation is insufficient to cause perceptible skin changes, no cancer attributable to radiation will develop at any time.

> It may be confidently expected, therefore, that when the exposure level is low enough to prevent perceptible changes in the blood count, no leukaemia will develop.

Mutscheller had realised in 1925 that the factual basis for assuming a threshold was not very strong. No doubt applying the tolerance levels did limit the effects that were being looked for but some doubts crept in that suggested a more precautionary approach. One of the more significant was that there might be different susceptibilities in a population to these effects. This biological variability and the limitations of current knowledge led Neary to write[6], in 1952:

8 Somatic Effects

... agreed exposure levels ...involve a risk which is small compared to the other hazards of life. The figures arrived at are therefore usually referred to as maximum permissible levels rather than tolerance levels, to emphasise the desirability of avoiding any unnecessary exposure.

By the late 1950s information was beginning to appear from the bomb victims (and from the work of Court Brown and Doll[7] that there might be no threshold for the induction of leukaemia but rather that the risk increased in some continuous way with dose. It began to seem that there were two quite different kinds of late radiation effects. The first, which generally became called deterministic, where the severity increased with dose and where there often seemed to be some kind of threshold—however blurred it might be. The second where radiation appeared to act as a kind of trigger for leukaemia (and later other cancers), increasing its likelihood. With what was known from the human and animal data, backed up by theories of cancer and radiobiology, there was no good reason to think that the risk, albeit decreasing, did not extend down to the lowest levels of dose. This was already accepted for genetic effects.

In their first report in 1958[8] UNSCEAR summarised what was known of the effects of radiation and reflected the new understanding of tumour induction:

> ...it has been believed that there is a minimum (threshold) dose of radiation causing the induction of tumours. Such thresholds vary from organ to organ and with the age of the organism. Owing to limitations in experimental methods ....and the physical background radiation, the possibility remains that there may not be a true threshold. The situation would then be analogous to that obtaining in the case of genetic changes.

They then went on to discuss the implications of the somatic mutation theory:

> the somatic mutation theory would postulate that each increment in radiation above the natural background would

carry with it a proportional probability of tumour development (linear response).

By the next UNSCEAR report in 1962[9] the incidence of bomb leukaemias had levelled off but there were now indications of other forms of cancer. They considered two hypotheses to explain this. First the somatic mutation theory (as they had already noted) pointed to a proportionality of incidence with dose—although it might suggest, given the possibility of cell killing, that lower doses would be relatively more effective at inducing tumours than high ones. On the other hand, if tumours somehow arose from somatic mutations generated during the body's attempts to repair damage caused by radiation, there might be a threshold effect. The committee could thus see reasons for thinking that the dose response might be linear (proportional), supra-linear or sub-linear.

By 1964 ICRP would write in Publication 6, their update[10] of the 1959 Publication 1 [11]:

> It appears possible on some theoretical and experimental grounds that when either the total dose or dose-rate is very low any effects will be directly proportional to the total dose.

Other effects have been associated with radiation—including non-specific life shortening and the effects of large exposures apparent from the earliest days—but cancers of various kinds emerged as the most significant potential consequences of exposure. The growing awareness of this and the struggle to express and manage the associated risks became the dominant theme of radiation protection.

The earliest descriptions of cancer date back to Egyptian times, perhaps 1600BC: the treatment for breast cancer was cauterising with the terrifying "fire drill". The recorded speculation on causes starts with Hippocrates (who used the term "carcinoma") around 400BC. He believed that the body contained four fluids or "humours": blood, phlegm, yellow bile and black bile. In a healthy body these four were balanced and disease came from any disturbance of this balance; cancer was an excess of black bile that collected at sites in the body. The ideas were endorsed and expanded by Galen in the 2nd century and persisted until the 18th century when they were undermined by Harvey's discovery of the circulation of the blood, the microscope revealing the cellular structure of living things and other advances, such as the discovery of the lymphatic system. Abnormalities in this system became the next candidates for causes of cancer: Stahl and Hofman suggested that cancer arose from degenerating and fermenting lymph. The theory was widely accepted and the great John Hunter (1723-1792), often called the father of scientific surgery, subscribed to it. Hunter saw tumours as growths from lymph and was perhaps the first to suggest rationally that surgery might be worthwhile for some.

The systematic and microscopic study of body tissues that followed revealed that cancers were made of cells, not lymph. Johannes Muller, who discovered this in 1838, thought the cancer cells arose by a budding process from between normal cells—the so-called "blastema" theory—but it was quite quickly apparent from work by Muller's student Rudolph Virchow that all cells arise from other cells. Virchow proposed chronic irritation as a source of cancer, which arose from the connective tissue, and thought it spread like liquid. Karl Thiersch showed that metastasis was due to cell movement.

Towards the end of the 19th century the theory that cancer was caused by trauma or irritation took hold and persisted to at least the 1920s. Indeed, there were many cases of compensation being sought for cancer arising from injuries at work until well into the 20th century—in spite of no real experimental evidence to back up the idea[12]. It was one of the reasons that the possibility of low doses of radiation might cause cancer was not taken seriously until rather late in the century. A related theory that cancer was due to parasites seemed to receive respectability with the award of the 1926 Nobel Prize to Johannes Fibiger partly for showing that a parasite nematode caused a cancer in the stomach of rats. The experimental outcome could never be repeated and the parasite theory—and indeed poor Fibiger—fell by the wayside.

The modern era of cancer research began in 1910 when Peyton Rous found a viral agent that caused cancer in chickens—Rous sarcoma virus. Rous had to wait until 1966 to receive his Nobel Prize and it has often been suggested that this unprecedented delay might have been due to the Fibiger experience.

TAMING THE RAYS

But is was also clear from early epidemiological studies that environmental agents and lifestyle played some part in cancer. In the early 18th century it was reported that Italian nuns experienced hardly any cervical cancer but had a relatively high rate of breast cancer. In London in 1775 cancer of the scrotum in chimney sweeps was associated with soot and, even earlier in 1761, there were indications that tobacco in the form of snuff caused cancer. However, it was not until 1915 that a definite experimental causal link was established when, after many had failed, Yamigawa and Ichigawa at Tokyo University induced skin cancer on rabbits' ears by rubbing them with coal tar. This was seen as giving support to an irritation theory and it was not until 1934 that the specific carcinogenic ingredient in the coal tar was identified by Kennaway and co-workers.

Of course, cause was one thing but it was clear that cancer development was another and it was really the insight of Rous, with his colleague Beard, in 1939 that separated the initiation of the cancer from its promotion: these were potentially quite distinct elements needed to produce a clinical cancer. It was found in animals that some agents had no effect unless preceded by an initiating event. Initiators were found usually to be mutagens but promoting agents less so: wounding an area that had been subject to an initiating agent could, in some laboratory experiments, produce tumours (suggesting that trauma could have some role). The notion of progression was extended from the laboratory and became established in the more general sense of a multistage cancer process.

The notion that viruses were causes was supported by Ludwik Gross—who found in 1951 that mouse leukaemia was carried by a virus—to the extent that other causes were largely dismissed by him in his theory of "vertical transmission". While several other cancers associated with viruses were found before and after Gross's discovery, it was to become clear that even viruses had an impact through genetics.

And it was genetics in one way or another that was to provide the dominant theory for all but the very early years of the century. Theodor Boveri was one of the first to show (in the first decade) that chromosomes were important to development and started to form the link between them and heredity. In mitosis of normal cells the chromosomes divide equally between the daughters but Boveri saw that, in tumour cells, this division was unbalanced (what we now call aneuploidy) and, while many of the cells would be non-viable, some might go on to uncontrolled proliferation. In his 1914 book[13] he clearly saw (and he seems to have been the first person to do this) cancer as a cellular problem originating from a single cell with defective chromosomes. The book was written in German and Boveri died, in 1915, soon after it appeared. It perhaps had more influence after his wife published an English translation in 1929 but it seemed more important in retrospect when it was re-discovered in the 1970s. In retrospect too his ideas have been widely interpreted as supporting the idea that aneuploidy, which continued as the cell line evolved (and now seen as a genomic instability) was a result of mutation in individual genes. Boveri was not alone in seeing "mutation" as the source of cancer. Ernest E Tyzzer in his 1916 paper on tumour immunology[14] introduced the idea that mutations might be their source: "From the evidence in the biological character of tumors of a permanent modification of somatic tissue, it appears logical to regard a tumor as a manifestation of somatic mutation".

It is not clear that Tyzzer had the modern ideas of mutation in genetic material in mind; the term was applied more loosely at the time. Certainly similar ideas were current in the early 1920s[15] but it is the book published (in German) by Karl Heinrich Bauer in 1928[16] that is usually regarded as setting out a comprehensive theory of the genetic origins of cancer. This was, of course, just a year after H J Muller established that radiation could cause mutations[17].

Just two years later in 1930 McCombs and McCombs were to publish a short paper in Science[18] that both stated clearly the somatic mutation hypothesis, referencing just Boveri's work, and making a tentative link with radiation-induced cancer:

> It is a well-known fact, genetically, that mutations experimentally can be speeded up tremendously by exposure to stimulating amounts of radiation and the x-ray effect, where greater doses are destructive. Perhaps the frequently observed "skin-cancers" in Roentgenologists are due to such mutations occurring from stimulating exposure to the x-ray effect (the release of the cathode ray).

Although in the following twenty years the theory continued to be promoted, it seemed to some to stumble in explaining the rapid variation of cancer incidence with age and the long latent periods. The resolution of this problem began in the early 1950s with the suggestion by H J Muller and Carl Nordling that several mutations, perhaps as many as seven, in sequence might be necessary to cause an observable cancer. Nordling summarised these thoughts in 1953[19] and this was picked up by Peter Armitage and Richard Doll, working at the Medical Research Council Statistical Research Unit in London in what was to become a landmark paper[20]. They extended Nordling's data: where he had considered total cancer deaths in men, they look at cancers at various sites for both men and women—17 in total. They found for many of them a power law order around 5 to 6 fitted the increase in incidence with age and that generally those that did not fit this were associated with endocrine secretions—a plausible confounding factor. Their analysis then suggested that a six or seven stage process would account for this and might be an explanation for latent periods. The analysis was presented mathematically and this was probably the first time this was done.

The paper was widely cited 50 years later when a short commentary[21] by (then Sir) Richard Doll addressed one of the striking presentational points of the paper: they steered away from assuming that the cellular changes leading to cancer were mutations. Why did they do this? :

> It is difficult after so long a time to know what we had believed at the time, but I am pretty sure that we had mutations in mind. We did not want to describe the changes as such, however, as we did not want to put off the many cancer specialists, who were not happy with the mutational theory...

Elsewhere in the commentary he wrote, with a touch of indignation:

> Even as late as 1960 it was possible for Austin Brues, a distinguished American scientist, to write a Critique of mutational theories of carcinogenesis.

In fact Armitage and Doll recast their model quite dramatically in 1957[22] when they found that no evidence had emerged to support the idea that cancer involved more than two stages. While the multistage theory would be compelling if incidence increased as the simple power law they had assumed, in fact the increase was not in that form across the whole age range so other models could fit it too. In this paper they followed up a suggestion made by Platt in 1955[23] and developed a two-stage model. This had a first event as a change in a cell that confers on it a faster rate of multiplying than that of the unchanged cells and an exponential increase in their number with time. A second event would then cause a clinical cancer. Armitage and Doll showed that such a process could be consistent with the observed cancer incidence rate and latent periods. By then they could suggest that one or both of the changes "could be of the nature of somatic mutations".

The clonal nature of cancers—which was the observation behind Platt's suggestion—had been suspected for some time (Boveri had suggested it) but it was in the 1950s that hard evidence began slowly to accumulate. By 1976 Nowell could write[24] that:

> This thesis is still being elaborated but, in general, it has been supported for many neoplasms by evidence obtained over the last two decades.....

In his paper Nowell was able to incorporate the monoclonal nature of tumours into a theory of their development. He did not define the initiating event that created the first neoplastic cell (that "remains a basic problem in cancer research") but appeared to favour genetic damage of some kind—whether visible chromosome damage or not—caused by agents such as radiation, chemical carcinogens or viruses. It was the progression of a tumour once the initial event had occurred that was the key topic of his theory and this depended on both the clonal nature and an associated genomic instability. The instability— most plainly apparent in the aneuploidy of tumour cells —had been seen by Boveri much earlier. Nowell thought that the first altered cell must have some growth advantage and would hence set it on the route to cancer. As the cell divided, whatever change had taken place might result in further genomic damage and see cell variants arise with even greater growth ability. These variants would develop into sub-populations that would come to dominate the evolving tumour before themselves varying. There would thus be an evolving group of cells, with changing genetic structure, that became progressively more aggressive as they grew into a clinical tumour with the considerable individuality, genetically and biologically, of advanced tumours. The ideas were developed in the early 1980s, notably by Knudson and Moolgavkar[25] in what became widely known as the MVK model.

The first direct evidence for a multistage process had come somewhat earlier from Alfred Knudson[26]. He found that the occurrence of the disease retinoblastoma could be accounted for if two mutations were required to cause it and one of them was hereditary. The patients who inherited a mutation developed the disease earlier in life and were more likely to have more than one tumour.

Support for the idea that mutation is at the centre of cancer came also from the work of Bruce Ames. In the early 1970s Ames developed a test for carcinogenic potential based on the observation that carcinogens are often also mutagens. The test was based on the effect of suspect materials increasing the mutation rate of prepared strains of *Salmonella* bacteria and was widely used, because of the expense and time involved in cancer tests on rodents. The justification to use it rests on the good general correlation with the rodent tests that are done.

Howard Temin had been promoting the idea that some viruses—the viruses with an RNA genome later called "retroviruses"—could alter cellular DNA since 1964 but this was not accepted until 1970 when he [27] (and independently David Baltimore[28]) found the enzyme "reverse transcriptase" in retroviruses and this was quickly confirmed by many other workers. It could turn the viral RNA into a strand of DNA—thus reversing the "central dogma" of molecular genetics which claimed that DNA could produce RNA but not the reverse. This could the be incorporated into cellular DNA and be passed on through cell divisions. Temin and Baltimore received the Nobel Prize in 1975.

The discovery opened up the field of genetic engineering and provided a mechanism to support the virogene-oncogene theory of cancer then being developed by Robert Huebner and George Todaro, at the National Cancer Institute in the USA. This proposed[29] that, in the course of evolution, portions of viral RNA had become incorporated into the genome and were transmitted "vertically" from animal to offspring and cell to daughter. These contain "oncogenes" that are normally suppressed but can be activated by a carcinogen to lead to the development of a cancers.

Proto-oncogenes were discovered by Harold Varmus and Michael Bishop in the mid-1970s while studying the Rous sarcoma virus, puzzled by where the SRC oncogene responsible for the cancer came from[30]. They found that the cellular counterpart of SRC was present in chickens, in all the bird species they looked at and then in the mammals they tested. So it seemed that not only had the retroviruses picked up the gene through the process of transduction, but that it lurked unseen in all of us. Varmus and Bishop speculated in 1976 that the proto-oncogene might be "involved in the normal regulation of cell growth and development" and this proved to be correct. More examples of proto-oncogenes were found so that by the time Varmus and Bishop received their Nobel Prize in 1989 more than 100 were known.

It subsequently became clear that oncogenes were generally changes responsible for hyperactivity in the cell and three classes were distinguished. A point mutation or deletion in a proto-oncogene (and there was some preference for calling these "cancer-critical genes") might cause it to produce normal amounts of a protein but in some much more active form. Otherwise, for some reason several copies of a cancer-critical gene might somehow be created on the genome. This "gene amplification" would lead to excessive amounts of the normal protein—and again over-activity. Serious chromosome re-arrangement, on the other hand, could have significant consequences in either or both ways.

It now seems that there are several ways that proto-oncogenes can be transformed into oncogenes. One mechanism involves retroviruses (although this is not common in humans) and can happen in two ways. The first is that when the virus interacts with the cellular DNA, as well as inserting a length of DNA from its own genome, it receives a section of DNA from the cell (transduction). The second way that cancer may result is from the disturbance caused by the transfer of viral DNA into the cell's genome. It is possible that such viral actions could transfer a damaged proto- oncogene between hosts in an infection process. Otherwise transformation may arise through mutation—caused by radiation or some chemical means.

However, we know that genome damage is not uncommon and that just a tiny fraction of what does occur is ever expressed as cancer. This is a result of the body's ability to react to genetic damage to prevent it developing into cancer and this is based on set of genes known as tumour suppressors. If these genes are transformed then the suppression function may be lost and the pathway to cancer opens up. The BRCA1 and BRCA2 genes produce proteins responsible for DNA repair; the protein from mutated genes may be unable to do this. Mutations of the BRCA1 and BRCA2 genes are known to predispose people to breast and other cancers. The Rb gene responsible for retinoblastoma is a tumour suppressor: its inheritance pattern as a recessive shows that one functioning gene is enough to maintain the suppression function. The protein p53 is responsible for regulating the cell cycle and also acts as a tumour suppressor. The gene responsible for encoding it is often found mutated in human tumours, presumably compromising the role of the protein.

The picture of cancer that was generally accepted towards the end of the century[31] was of a slow multistage process originating from a single cell. A somatic mutation causes some change that encourages the cell to divide in an uncontrolled way. If the normal systems that suppress this are disabled by the same or another mutation, a clone of cells will begin to grow—perhaps promoted by other agents. The clone that grows from this cell is genetically unstable—particularly susceptible to mutation—because of earlier damage. Further mutations make some progeny better able to overcome other barriers (such as lack of oxygen at the centre of the clone as it gets larger) to expansion of the clone than others and eventually, through an evolutionary process, these progress as a tumour and undergo metastasis.

There were a few dissenting voices. Just how important were epigenetic changes—changes in the chromatin other than in the DNA sequence—and how might they modify gene expression? Was the genomic instability a cause or an effect? Was the theory based on somatic mutations too reductionist a view, not taking full account of the complexity of cellular and inter-cellular interactions? However, in general scientists were continuing to investigate the extraordinary

8 Somatic Effects

details of the cancer process (just sketched here) within the somatic mutation paradigm.

The injuries suffered by many of the early workers with radiation have been mentioned in the early chapters. Of these, radiation dermatitis was the most common condition and this frequently was followed by highly malignant squamous carcinomas. Of the "martyrs" whose names appear on the memorial erected by the German Röntgen Society in 1936 no less than three-quarters of the deaths (96/130) recorded are due to skin cancer. Something over 10% of the deaths were attributed to anaemia and two out of the 130 were recorded as having died of leukaemia.

While the association of skin cancer with radiation was clear within a decade of Röntgen's discovery, there was a perception that it was closely associated with the more obvious damage. If the obvious damage could be avoided then skin cancer would not follow.

For much of the first half of the century the possibility that radiation caused leukaemia was either disregarded or was controversial[32]. While a few cases were reported before the 1930s, it was only in 1931 that Aubertin reported five cases of myeloid leukaemia in radiologists when he had seen only one in other medical practitioners. Even after this the link was not regarded as conclusively established and when Colwell and Russ reviewed radiation injuries in 1934[33] (the year Marie Curie died) they could not decide whether leukaemia was one. It was March's study of US radiologists in 1944, described later, that gave the first solid evidence.

Two other indications of radiation's hazards were the fate of many miners and the radium dial painters. The unfolding of the risks from radon to the miners in the Harz mountains of Germany began in the early years of the century and by mid-century the link was largely accepted; the terrible story of the radium dial painters was first told in the 1930s. We will return to both these issues later as we cover internal exposure because it was the risks from x-rays and other external sources that were to be understood first and this began with H C March's radiologists.

There had since the early days been concern about the effects of x-rays on radiologists and there was an understanding that forms of cancer could arise but it was thought that these would follow obvious and serious injury. The first indication that there should be concern, even when there was no apparent radiation damage, came from the study by H C March in 1944[34]. March found that the death rate from leukaemia of US radiologists—using death notices in the *Journal of the American Medical Association*—was about 10 times that of other physicians. A similar results came from Ulrich in 1946[35] who reviewed the obituaries of radiologists published in the *New England Journal of Medicine* and found that leukaemia mortality was eight times that of other doctors. March continued his work into the 1960s[36] (and others e.g. Dublin and Spiegelman[37] made similar studies) but the most significant subsequent contribution was perhaps by E B Lewis[38] in the late 1950s. After eliminating the possibility that diagnosis was better among radiologists than others (a recognised possible confounding factor), he concluded that there was indeed a significantly higher death rate from leukaemia, aplastic anaemia and multiple myeloma.

In the preparation of the Medical Research Council report[39] in 1956, it was appreciated that there were no extant data on humans that would allow an estimate to be made of the risks associated with radiation. Some information was emerging from Japan[40] but, as we will see, this was incomplete and could not yet form the basis for a risk estimate. It was against this background that W M Court Brown was asked by the Committee to review the information from the radiation treatment of patients with ankylosing spondylitis where there were already three reports of leukaemias.

Ankylosing spondylitis is a disease which affects the joints, particularly those of the spine. It causes severe pain and the joints become steadily more immobile until, sometimes, they become rigid. One treatment which is effective is irradiation of the spine; for some patients at least it can reduce pain and increase mobility.

Court Brown and his associate Richard Doll obtained the records of some 13,000 patients who had been treated between 1935 and 1954[41]. Of these, 49 were found to have developed leukaemia, aplastic anaemia or myelofibrosis (thought to be a variant of leukaemia). Twenty-eight had already been certified as having died of leukaemia, 13 from aplastic anaemia and 1 from myelofibrosis. Four more deaths were added which had been certified as from other causes. With some corrections the leukaemia deaths were finalised as 25—and this was to be compared with an expected number of just 2.4.

Using information on treatment regimes obtained from a smaller sample, the incidence of leukaemia (there were 37 cases) was calculated for the men as a function of the estimated doses to the whole body (in Mgm.r) and the maximum dose to the marrow of the spine in roentgen. In both cases it showed a steady increase. No threshold was seen with either dose measure (although the doses were high by protection standards) but it was not possible to deduce anything more about the shape of the curve: the spine marrow approach suggested a linear relationship, the other an ever- increasing curvilinear one.

This was the first demonstration of a dose effect relationship—even if the form of the relationship was unclear—and it was to be followed up and reanalysed several times in the future. However, it was to be the analysis of the atomic bomb survivors that was to provide the main body of radiation risk data.

Early in the morning of 6 August 1945 the *Enola Gay* dropped the gun-type uranium weapon codenamed Little Boy high over Hiroshima and three days later the implosion-type plutonium bomb, Fat Boy, was dropped on Nagasaki from another B-29, *Bockscar*. Large areas of the two cities were flattened by the blasts and the devastating fires that followed destroyed yet more. Over 100,000— perhaps over 200,000— people died immediately or within a few months and there were a similar number of serious casualties. The destruction of major hospitals and the death of many medical staff meant that minimal help was available making infection a major problem; at Hiroshima the problems were compounded by a typhoon on 17 September that caused further damage to the infrastructure and slowed recovery.

Within a few days of the Japanese surrender on 2 September a small team—William Penney, Robert Serber and G T Reynolds— arrived to estimate the yield and survey the physical damage. By 12 October a Joint Commission for the Investigation of the Effects of the Atomic Bomb in Japan had been set up with about 60 US staff and 90 Japanese. One of their tasks was to estimate the number of casualties that could be attributed to radiation rather than burns and blast. Based on a survey of 13,503 people known to have been alive 20 days before the bombings and data collected by the Japanese workers, it was estimated that about 30% of the deaths were from radiation. When it reported in 1946 the Joint Commission urged further study[42].

As a result the National Research Council was asked to advise on this and, through James Forrestal, the Secretary to the Navy, President Truman was prompted to issue in November 1946 a Presidential Directive requiring further long-term study. It was to review several specific concerns[43]:

> Cancer, leukaemia, shortened life span, reduced vigour, altered development, sterility, modified genetic patterns, changes in vision, "shifted epidemiology", abnormal pigmentation and epilation

The Atomic Bomb Casualty Commission (ABCC) was formed the following year with funding through the Atomic Energy Commission but direction coming from the Committee on Atomic Casualties set up by the NRC and it then initiated several studies. This, of course, was against a background[44] of rather little information about the effects of radiation. Some studies had been made before the war and the Manhattan Project had sponsored some experiments on animals —but the results of the latter were not widely known. It was not known if the effects were in some way proportional to dose and the idea that there was some kind of threshold dose was current. What was known was that the illnesses caused by radiation were, as far as was known, present in some degree anyway and that separating radiation as a cause might be difficult.

A haematological study in 1947 and 1948 of 1000 heavily exposed survivors established that their blood-forming systems had recovered but it was an early priority, given that the genetic effects of radiation had been seen in experiments on plants and animals, to assess the genetic impact. In March 1947 a study was begun, under the direction of William J Schull and and James V Neel, of the impact of the bombs on children born to exposed parents to try to establish whether there were any genetic effects and whether they were likely to have long-term implications for the population. With the help of midwives and local Japanese doctors, children were examined for malformations at birth and then, in some cases, 10 months later. Between 1948 and 1954, when the study was ended, autopsies were

8 Somatic Effects

carried out on 750 infants. There was no clear effect of radiation other than a trend in the sex ratio.

However, a study of women who had suffered miscarriages or stillbirths showed that this was more likely after radiation exposure during the pregnancy. By 1952 [45] it was clear that radiation could also cause severe damage to the developing embryo (an effect already known from radiotherapy) with the main effect being microcephaly (reduced head size) and associated mental retardation. The early published results from the *in utero* group, available at the time of the MRC report[46], showed that there were eleven mentally retarded children born to mothers who had been between 700 and 1200 metres from the hypocentres. Ten of the women had acute effects from the explosions. All of the children had suffered from microcephaly. The most sensitive time for the effect, judging from the Hiroshima data, was between 12 and 18 weeks of gestation.

Early studies aimed at finding any chromosomal damage suffered by survivors were unsuccessful. The procedure adopted—a painful needle biopsy from the testes—gave chromosomes but these were difficult to study. There was even some doubt about the number of human chromosomes. Better techniques were developed and by the mid-1950s damage was being seen.

The development of children who had been exposed to radiation was also studied from the start by the ABCC. Greulich found some evidence for radiation having affected growth when he reported in 1950 but it was difficult to disentangle the effects of radiation from those due to general conditions and nutrition. A follow-up study by Earl Reynolds also found some growth retardation effects but the reception the study got was tainted by his plans to photograph naked teenage girls for record purposes. The subsequent study of Sutow and Emory West used x-rays of the children but showed no discernible effects[47].

The first cancers were found, not by the ABCC's researchers but by a local doctor Takuso Yamawaki who noticed a higher than expected incidence of leukaemia among survivors. All the nine cases he saw in 1949 came from within 1.5 km of the hypocentre. In a study by Folley, Borges and Yamawaki over the next two years[48] 19 more cases were found among nearly 100,000 survivors at Hiroshima and 10 among a similar number in Nagasaki. By 1953[49] Moloney and Kastenbaum found 50 cases in Hiroshima with a clear relation to distance from the hypocentre. The increased leukaemia risk was found to be about one in 13,000 at 2.5 km and beyond; within 1 km it was near one in 80.

In its 1956 report the MRC Committee reviewed what was made available to them on leukaemia incidence from Japan. By then, in the 98,000 survivors studied, there were 61 confirmed and four suspected leukaemias in Hiroshima. The ratio of observed cases to expected went from 1.2:1 for people at 3 km or more from the hypocentre to 100:1 for those with 1 km. Raised incidence was also seen at Nagasaki but little detail was available. Alarmingly the rate of leukaemia occurrence was not decreasing. There was no real possibility of reliable risk estimates at this stage because of the lack of information on doses, the best available data released by the US in 1950 (*The Effects of Atomic Weapons*[50]) was rather generic and took no account of shielding. While there was clearly an increase in incidence with dose the Committee found it could not "infer with certainty whether the relationship between dose and the incidence of leukaemia is a curvilinear or linear one"(para 60).

The first solid tumours were found, in the Life Span Study (see below), by Gensaku Ohu in 1956 and this was to be the focus of a major effort for the rest of the century, as we will see.

The first hydrogen bomb was exploded in June 1954 with a yield a thousand times greater that the Japanese bombs. It distributed radioactive fallout around the world and prompted great concern about the long-term effects the relatively small individual doses might have. At the National Academy of Sciences it led to a realisation that the current ABCC studies might not give the information needed for the understanding of the effects of these small individual fallout doses. This led to the first major programme review by the the Francis Committee in 1955. The Committee, in its report, recognised that much important and relevant work had been done but in a somewhat haphazard fashion; if credible risk data were to be obtained (as was needed) a more structured approach was essential. The Committee therefore recommended to NRC that three well-defined cohorts should be set up and studied. The Life Span Study (LSS) should follow up 100,000 or so people in a mortality study, the Adult Health Study (AHS) should continue the clinical examinations of around 20,000 survivors (in four groups with a range of exposures) and the Pathology Study should collect and analyse post-mortem results from a cohort that was initially about 70,000 people. This unified study programme was initiated and formed the basis for the next 20 years although there were changes.

The changes to the LSS were in the late 1970s when first a further 10,000 proximally exposed people were added and then 11,000 more from Nagasaki. This was at least in part to ensure that as many highly exposed people as possible were being studied. As a result there were about 120,000 people in the LSS of whom about 94,000 were in the cities at the time of the bomb. The AHS was increased by just over 1000 people in 1977 who were judged to have doses >1 Gy matched with a similar number who were not exposed. In 1975 about 1000 of the *in utero* group were also added so that they could be offered the biannual physical examination.

Not all the studies formally used these cohorts. The essentially *ad hoc* work that had led to the discovery of the effects of radiation on the embryo mentioned above was formalised somewhat when the clinical cohort PE86 was set up with some 1600 members in 1956.

The clinical studies of offspring (such studies became called rather bleakly "F1 studies") ended in 1954 but data collection on sex ratio and neonatal death lasted until 1968. The genetic work then continued with cytogenetic studies but these showed, once more, no clear chromosomal effects due to radiation.

By 1973 the US was unwilling to continue funding the ABCC and was seeking a cost-sharing arrangement with Japan and this led to the setting up, in 1975, of the Radiation Effects Research Foundation (RERF), a jointly managed and funded research organisation. However, before the ABCC was to hand over to the RERF, the Crow Committee reviewed the programme and, in its 1975 report, recommended some changes. The LSS and the AHS should continue but the AHS should move away from clinical studies to laboratory ones. The *in utero* studies should be formally brought into the AHS. As we have seen, no radiation effects had been seen in the F1 clinical studies designed to detect genetic effects in first generation children but the Committee recommended that the cohort studied should be formally defined (the F1 cohort) and that the work should focus on laboratory and mortality studies for the cohort. The first generation studies cohort was initially about 50,000 strong but this grew to near 90,000 with almost all available for mortality work; subsets of perhaps 20,000 were used for individual laboratory studies. The Committee recommended that the post-mortem programme be run down (this had been modified significantly since set up in the 1950s). So overall there was a move from clinical studies to mortality and laboratory ones.

Although the clinical and mortality studies of offspring had shown no significant effect of parental exposure up to the end of the programme in the 1970s, the knowledge that radiation produces mutations in all other species tested led to a continuation of the programme in the form of extension of the cytogenetic (chromosomal) studies begun in 1967. By the end of the century these too had shown no effects of radiation either (other perhaps than a lower level of chromosome damage in the offspring of survivors than in others). However, from the 1980s, techniques for the study of DNA developed rapidly and these promised much greater sensitivity by detecting damage at the molecular level and by the end of the century pilot studies were in progress using the leading DNA technologies. The parallel cytogenetic studies of survivors continued to show effects—as they had done since the mid-1950s— and it was established that chromosomal damage increased with estimated dose. The techniques developed were used in radiation accident dosimetry.

The occurrence of leukaemia, established by the earlier work, peaked around the early 1950s and then steadily decreased (although it stayed higher than expected in Hiroshima in the early 1980s). By 1974 more solid cancers than leukaemias were being seen and by 1978 the incidence of leukaemia among those exposed at Nagasaki was the same as in the control group[51].

The finding of solid cancers in 1956 led to tumour registries being set up by local medical societies in Hiroshima in 1957 and in Nagasaki the following year. These were funded by the ABCC and later the RERF and were the route for continued discoveries of solid cancers at a number of sites. The results from the registries and from the mortality study of the LSS documented the increase in solid cancers. First thyroid cancer increased in 1960-1961[52], then breast cancer in 1966-1968 and then lung cancer (about 1968). Stomach cancers—the most common cancer among Japanese—was first documented in 1977 along with others associated with the digestive and urinary organs.

There were searches for the entire follow-up period for causes of death other than cancer but the findings were almost completely negative until around the turn of the century. A possible increase in risk of cardiovascular disease in highly exposed survivors was first seen around 1990.

The interpretation of many of these results depended upon an understanding of the doses that survivors had been exposed to so, while the studies of the health effects were undertaken, there were parallel exercises to estimate doses. This was necessary if the results were to be used for the assessment of risks from radiation.

The first dose estimates were made at Oak Ridge in the AEC-sponsored Ichiban project. Determinations were made of the kerma in tissue, free-in-air as a function of distance. Based on the so-called York curves[53] and with allowance for shielding, these were presented as a function of distance from the hypocentre as the tentative 1957 dose estimates (T57D). These estimates were not widely used (and were not accepted by the ABCC) because of concerns about their reliability. The first widely accepted estimates were those of 1965, the T65D dose estimates, which were based on data from weapons test explosions in Nevada, the BREN experiments (with an unshielded reactor or large cobalt source as the radiation source) and other measurements with fixed sources. The agreement with independent estimates from thermoluminescence dosimetry of roof tiles and bricks to give estimates of gamma doses and activation of steel reinforcing bars for neutron doses was reasonable. The T65D system was used through the 1960s and 1970s but a presentation by Harald H Rossi to the NCRP in 1976 prompted a reassessment. Rossi had adjusted the T65D data to calculate critical organ doses and, taking account of new evidence of neutron RBE, concluded that neutrons made a much larger contribution to the health effects than previously thought. This prompted the NCRP to set up a task group under Harold Wyckoff to review the T65D data— but they were unable to do this because much relevant material was classified. The DOE were then prompted to fund George Kerr at ORNL to update the calculations. The update, using more advanced weapons and radiation transport codes showed that there were indeed fewer neutrons at Hiroshima than previously thought; the doses were several time lower. The overall agreement with TLD and activation measurements (which had both significantly improved) was better.

	Comparison doses T65D and DS86	
Hiroshima	gamma	Up x 2 to 3.5
	neutron	Down x 0.1
Nagasaki	gamma	Down x 0.7 to 0.9
	neutron	Down x 0.3 to 0.5
From DS86 report Vol 1		

Table 6: Comparison of T65D and DS86

The reassessment was reviewed in several workshops over the following years and was finally adopted in 1986 as DS86. Table 6 shows the changes from T65D.

In the decade after the DS86 system was adopted, concern began to grow that there was a discrepancy between it and the various activation measurements: higher neutron doses were being indicated. It was also apparent that there was a discrepancy between the measurements and the information about the bomb. In the 1990s a further revision was started that would take advantage of the improved techniques for the measurement of very low levels of residual activity and the advances in weapons and radiation transport calculations. The results, revealed as DS02, in 2004 were, in fact, close to the DS86 systems for Hiroshima and the differences for Nagasaki were overall less than 20%.

While the bomb data is by far the most significant source of information about radiation risks, a number of other studies have been very significant in specific ways. They range from reviews of cancer mortality to people subjected to intakes of radioactive material for accidental and medical reasons.

The 1958 Court Brown and Doll[54] study was of increased death risk for 1377 British radiologists who had joined a radiological society before 1955. It was found that those who had registered before 1921,

8 Somatic Effects

when the first British XRPC recommendations had been published, did indeed have a significantly increased risk of dying of cancer. The excess was confined to the skin and pancreatic tumours (6/0.6 and 6/1.6 observed over expected in social class I respectively) and possibly leukaemia (2/0.5). The work was continued in 1981[55] when the follow-up period was extended to 1977. By this time all those who had registered before 1921 had died and they were found to have had a death rate from cancer 75% higher than that of other medical practitioners—the comparison group. Those who had registered later had a total mortality from cancer that was not significantly different from the comparison group—although there was an indication that risk increased with the number of years since registration. These results were confirmed by a further study[56] when the follow-up was taken to 1997. There was no evidence of increased cancer mortality for those who first registered after 1954. There was considerable uncertainty about the doses received but the later results suggested that the risks were significantly (several times) lower than those coming from the bomb Life Span Study. It was also found, again in conflict with the LSS, that there was no evidence of radiation causing other diseases even for the very early practitioners.

In 1956, in the Oxford childhood survey[57], Alice Stewart found a doubling of risk of cancer for children who had been irradiated *in utero* as part of diagnostic procedures and received about 0.02 Gy in total. The results were supported by the Tri-State study in the 1970s. MacMahon (in 1962 and 1984) and others also found an increased risk from leukaemia but not always for other cancers. In a study of childhood cancer in twins born in Connecticut between 1930 and 1969, Harvey found relative risks of 1.6 for leukaemia and 3.2 for other cancers for doses of about 10 mGy. The excess risks were not statistically significant but the study added further weight to the notion that there was an effect[58]. One of the reservations throughout was that such effects had not emerged from the Japanese bomb data; when they were seen in the late 1980s they were at a much lower level than expected and they were solid cancers rather than leukaemia. Although the results were not fully consistent and some of the work could be criticised for flaws in methodology, by the end of the century it was widely accepted that there was an effect. Doll in 1995[59] suggested that the risk of childhood cancer from foetal irradiation was 6% per Gy and of childhood leukaemia was 2.5% per Gy.

The use of luminous paint—zinc sulphide with added radium or mesothorium (Ra-228)—seems to have begun in Germany in the 1900s and by the beginning of the First World War it was taken up in America. The US Radium Corporation became a leading supplier of luminous watches to the United States military during the war and it employed several hundred women at its Orange, New Jersey factory in painting the dials. In spite of an awareness by the company of the hazards of radium, the women were allowed (or was it encouraged?) to get their brushes to a fine point by licking them; each time they did this they ingested a small amount of radium. Grace Fryer, who had worked at the factory for just three years before leaving in 1920, had developed serious bone decay in her jaw by 1922. When an investigation was launched in 1925 by the New Jersey Consumers' League, other women working at the factory were found to have suffered similar effects. Dr E L Hoffman, who conducted the inquiry, found that 12 women had suffered from persistent infections of the jaw with, sometimes, anaemia; four had died as a result. (It was also found that the company had been warned the year before, after a secret investigation of their own, of the dangers of the procedure and of the widespread contamination in the factory.) Fryer and four other women started a lawsuit against US Radium in 1927 and, with the support of a press campaign and in spite of their declining health, won compensation in 1928.

The scientific work of Harrison S Martland, a county medical examiner, documented in some detail the medical effects and helped make the link with radiation. His publications between 1925 and 1931[60] established the presence of radon in the breath of the affected women (a clear indication of ingested radium) and established a reasonable epidemiological link with radiation. He found both the acute effects—as Fryer had experienced—and a much delayed condition; both led to necrosis and, in many cases, bone cancer. In the work reported in 1931 he examined 18 deaths from occupational disease among 800 young women all of whom had worked in the factory for less than two years. Most had severe anaemia but in five cases death was due to bone sarcoma. So, over a quarter of the deaths were from bone cancer and this was so different from what he expected (0.1% was the equivalent fraction he deduced from autopsies) that he concluded radioactivity was their cause.

Martland's work continued to the 1950s as he followed up the cases of chronic poisoning.

The story of the "radium girls" has been told by Kovarik[61], who gives a short account focussing on media interest in the case, and more completely by Mullner[62]. The understanding of the risk depended on knowing how much radium had been ingested—Martland's measurements were really just indicators—and this had to wait for the whole body counter and Robley D Evans in the mid-1930s. The results from this are taken up in Chapter 10.

The mines at Joachimsthal (now Jachymov) and Schneeberg (now on either side of the Czech Republic/German border) were a major source of silver in the 16th century and, when this was depleted and became available from the New World, of bismuth and cobalt. The discovery of uranium in the pitchblende, found in the mines by the chemist Klaporth in 1789, led to renewed life to exploit the remarkable properties of uranium salts in porcelain glazes. The extraction became more difficult and by the end of the 19th century the mines were near to closure when Mme Curie discovered radium. So valuable did the ore then become that the mines were reprieved yet again; the last was closed in the 1960s.

The miners had suffered a high mortality from a chest disease known as "mountain sickness" for a long time and it was shown in 1879, by Harting and Hesse to be a form of cancer—although they mis-diagnosed it as a lymphosarcoma. It was Arnstein, in 1913, who concluded it was a lung cancer and he found that it accounted for nearly half the deaths of miners since 1875. The further work by Rostoski et al in 1926 confirmed this: more than half the deaths looked at were due to lung cancer[63].

The association with radon was suggested in 1921 by a Schneeberg native, Margarete Uhlig. After measurements of the radioactivity in the air in the mine by Rajewsky in 1939 (levels of around 10^5 Bq/m^3 were found) this was widely accepted[64].

A material called Umbrathor, composed mainly of thorium dioxide, was introduced in Germany in the late 1920s and used as a radiographic contrast agent for the gut. It was unsuitable for injection into the circulatory system because it did not remain in solution but by 1931 Thorotrast, a different colloidal form of thorium dioxide, became available and this could be used for arteriography and venography. Thorotrast became very popular because it was such an effective contrast agent and was very well promoted by the Portuguese Professor Egaz Moniz (who received a share in the 1949 Nobel Prize for medicine for his promotion of lobotomy as a treatment for schizophrenia). Between 1928 and the mid-1950s at least 10 tonnes of thorium dioxide was distributed, mainly as Thorotrast, for use in radiography and, as Abbatt put it in 1979[65]: "... no cavity remained unknown and unexplored by Thorotrast".

He estimated that it found its way into several million patients "for practically every conceivable radiological purpose imaginable". The warnings from the work of Martland on the radium dial women were apparently unknown or ignored and, while adverse consequences were recognised after just a few years, Thorotrast use continued until it was gradually replaced by iodine-based media in the 1950s. The systematic study of risk data associated with its use began in the 1949 in Denmark and in following years in many other countries (notably Portugal 1961, Japan and Sweden 1963, Germany 1968 and USA 1972) and showed that there was a greatly elevated relative risk of liver cancer and leukaemia in patients injected with it. Relative risks for liver cancer of between about 10 and near 100 have been deduced for typical injected amounts; the risks of leukaemia are lower at 10-20. Estimates[66] have suggested that the dose to the liver following an injection of 25 ml (a typical amount) might be about 0.4 Gy/y and it seems likely that some patients would have been given injections of several times this. Some studies have shown increased bone cancer risks.

Abbatt's closing remarks are chilling:

> I believe all of us are conscious today that fashion, art and magic are still important ingredients of this particular human activity named medicine. Science, I believe, is still rarer than we pretend.

Thorium-X (Ra-224 with half-life 3.6 days)was used in treating patients before the First World War mainly for rheumatic conditions and in dermatology. Injection for rheumatism fell out of favour after two patients died of acute radiation syndrome in Germany around 1912 but the use of radioactive ointments and lacquers to treat skin conditions continued. There was a renaissance of injection for rheumatic conditions notably in France in the 1920s and in the UK in the 1940s[67].

Peteosthor was a concoction of Ra-224 chloride containing traces of platinum and the red dye eosin made up by Paul Troch, a German country doctor, in the 1940s in the belief that it could cure many ills. When Troch became head of a tuberculosis sanatorium after the Second World War he injected Peteosthor into about 2000 patients, including many children(some patients receiving as much as 100 MBq) as a cure for bone tuberculosis. Its effectiveness was never established (and the eosin and platinum were later dropped as having no value). The Troch method was also used elsewhere in Germany as a treatment for ankylosing spondylitis with rather smaller doses than those used by Troch and more than 1000 patients were treated. The potentially damaging consequences of the treatment were pointed out in 1950 by Spiess[68] but Peteosthor continued to be used in Germany until the mid-1950s.

Spiess undertook an extended study of the effects of the injections and found that among nearly 1000 patients treated for tuberculosis with the higher amounts there were 56 bone sarcomas when, from national statistics, much less than one was expected. The study and subsequent work on these German patients remain key sources of risk data for radionuclides that irradiate the bone.

Ra-224 continued to be used as a treatment for ankylosing spondylitis, particularly in Germany, until the end of the century but by then there was rather solid evidence that it was not convincing as a therapy. The risks were sufficiently high to restrict its use to clinical trials.

But it was perhaps a potion called Radithor that drew most attention—when it claimed a celebrity victim. The preparation was manufactured and sold by the Bailey Radium Laboratories through the 1920s. Each small bottle contained distilled water laced with 1 µCi(37 kBq) each of radium-226 and mesothorium (radium-228). The owner of the company was William Bailey, who was by no stretch of the imagination qualified, and the medicine was sold as "A Cure for the Living Dead": a general pick-me-up, panacea and, specifically, a cure for impotence. It went well and half a million little bottles were sold at a dollar each.

An enthusiastic user of Radithor was playboy industrialist and star amateur golfer Eben Byers. A Yale graduate, Byers won the US Amateur Open in 1906, after being runner-up twice. While returning from the annual Yale-Harvard football game on a specially chartered train in 1927, he fell out of bed and damaged his shoulder. He seems to have taken Radithor to control the pain and was soon so impressed that he was drinking three bottles a day and quaffed some 1300 bottles over the next three years.

The consequences were horrendous: tumours led his jaw to fall apart and he developed holes in his skull. He died in March 1932 and was buried in a lead-lined coffin. *The Wall Street Journal*'s headline after he died was:

> The Radium Water Worked Fine until his Jaw Came Off

The affair increased awareness of radioactive potions and probably had some effects in tighter regulation.

Establishing that there is a risk associated with radiation is one thing, judging the size of the risk is a far more demanding task.

Relative risk (RR) in this context is the risk of cancer in the presence of an agent divided by the risk with the agent absent. So a relative risk of 1.1 means that the cancer risk is increased by 10% when the agent is present. Excess relative risk (ERR) is the expression of the increased risk as a fraction of the background risk:

excess relative risk (ERR)= relative risk (RR)-1

An ERR of 1 would therefore mean that the cancer risk was doubled by the agent.

The additional risks imposed by a radiation exposure are expressed, in this context, in one of two ways: relative risk and absolute risk. Relative risk expresses the additional risk as the

8 Somatic Effects

number of times it increases the background or natural risk. So if the background risk of contracting a particular cancer at age a is r per year and the relative risk(RR) at the same age after earlier exposure to radiation is M then the modified risk at age a is M x r per year. The excess relative risk (ERR) is (M-1). Since background cancer risks, except in childhood, increase with age, a constant relative risk will mean a significant (even dramatic) increase in annual risk of cancer as the person gets older. In a relative risk or multiplicative risk model the usual assumption is that M depends on just the age of exposure and the dose.

In an absolute risk model the risk is simply expressed as the increased risk of cancer per year, A, resulting from the exposure with no presumption about the background cancer rate. In an additive risk model A is usually assumed to depend only on the age at exposure and the dose.

In both cases the cancers appear after a latent period and then continue to appear for some, maybe many, years. The difference between the two approaches is of major importance when the follow-up has not extended long enough to include all cancers. Then a projection must be made and if this is done using a relative risk model it leads, because of the growing background cancer risk with age, to generally higher predictions than an absolute risk model does.

The first tentative estimates of leukaemia risks were presented in UNSCEAR 1958 using the Hiroshima dose data then available and concluded, quoting work from the previous year by Lewis, that there might be 12 extra cases per 10,000 Gy. The Lewis analysis had suggested that the dose relationship was a proportional one but the UNSCEAR authors' re-analysis suggested that there might be a threshold.

By 1972, with the rather better dosimetry available, UNSCEAR was able to make a more confident estimate of the total impact: there would be 15-40 cases per 10,000 Gy and they settled on a number in that range, 20 cases per 10,000 Gy, for the 1977 report. However, as the new dosimetry became available and with slight adjustments to the model, the estimates moved sharply upwards: by UNSCEAR 1988 they were increased several fold to around 100 per 10,000 Gy. It became clear, at about the same time, that the dose response relationship was closer to a linear-quadratic one than a simple linear no-threshold relationship.

By the 1990s Upton[69], using the Japanese bomb data, was able to summarise the effects of several parameters on risk of leukaemia (excluding chronic lymphatic leukaemia which had not been seen). The susceptibility to leukaemia appeared to be higher in those who were children (or *in utero*) at the time of the bombs and those who were middle-aged and older. In children the relative risk was high (over 10) dropping to about 5 by the teenage years. From the teenage years the relative risk seemed roughly constant and the absolute rise reflected the increasing background risk. The relative risks for the two sexes were essentially the same but the absolute risk for men was nearly twice that of women, reflecting differences in background mortality rates for the sexes. The number of leukaemias seen as a result of *in utero* exposure was so small (just 2) that no conclusions could sensibly be drawn. However it seemed that the risk of all cancers from *in utero* exposure was very similar to the risk from exposure in early life on both a relative and absolute basis.

The A-bomb data had long suggested a short latency period of 2-5 years post-irradiation. The peak in cases occurs in the first decade and there is then a gradual decrease in relative risk getting close to unity perhaps thirty years after exposure. The ankylosing spondylitics showed a similar general pattern.

As we saw previously, the solid cancers had a longer latent period than leukaemia and began to appear in the 1960s. The excess number rose in the LSS from 79 by 1967, to 163 by 1977, 282 to 1987 and 432 in 1997[70]. It suggested that, since the LSS contained about half those survivors with significant doses and this analysis considered only the 80% of these who had been assigned doses, the total number dying from radiation-related solid cancers as a result of the bombs was about 1000. Since the number of cases continued to rise, the model used to predict the future incidence was crucial.

It proved difficult to separate any decline in the relative risk as an individual aged from the expected heightened susceptibility of people who had been young in 1945. The expectation of heightened sensitivity led to the popularity from 1985-1995 of a model that assumed that there was indeed a high susceptibility in youth but that the relative risk remained constant as the person aged. By the early 1990s it became reasonably clear that, for those exposed as children, the ERR did decrease with attained age. By the end of the century it was possible to argue that the data was slightly better represented by

8 Somatic Effects

an ERR that declined steadily with attained age with a relatively small effect from age at exposure[71].

By 1991[72] it was clear that relative risks were rather similar for the 8 solid cancer sites that could sensibly be studied. The ERR ranged from about 0.5 per Gy to around 1.5 per Gy. The uncertainties in the values were so large that, even at the end of the century, it was not absolutely certain that there was a real difference in the ERR between sites[73].

Differences in solid cancer risks between the sexes were noted from the 1980s with females[74] having generally a higher relative risk than males. The generally higher background rates of cancer in men meant that these differences were reduced and even reversed in absolute risk terms for some sites[75]. Analysis made of the Japanese data from around 1990 showed that the dose response function was linear rather than the linear-quadratic that was found for leukaemia.

Upton[76] has presented in summary the results of the assessment made in the preceding two decades and these are shown in Table 7 below. The earlier assessments used the T65D dosimetry, the last two employ DS86.

Source	Additive model	Multiplicative model
	Deaths per 10,000 persons from 1 Gy acute whole-body low-LET radiation	
BEIR I 1972	120	620
UNSCEAR 1977	250	-
BEIR III 1980	80-250	230-500
NUREG 1985	290	520
UNSCEAR 1988	400-500	700-1100
BEIR V 1989	-	885

Table 7: Excess lifetime mortality from all cancers

Upton's table rounds the values obtained at different times and by different organisations for the projected excess mortality risk from an acute dose of 1 Gray of low-LET radiation. In some cases the assumption is a linear dose effect relation in others a linear- quadratic. There are also different populations in the different assessments but the results do show some consistency within the models (to within a factor of 2) and a growing convergence of the predictions of the additive and multiplicative models.

The Japanese data derive from a single large acute exposure to radiation and it had been argued that this results in a higher risk than there would be if the same doses were protracted. This is both a consequence of the linear-quadratic response and because of the expected greater efficiency of higher dose rates in producing biological effects (deduced from animal studies). It has been argued that the risks derived from the bomb data should be reduced by a factor termed the Dose and Dose Rate Reduction Factor (DDREF) when the aim is estimation of risk from the kind of exposures encountered in routine circumstances whether from occupational exposure or natural background. It appears to have received more active attention, as a possible mitigating factor, when the bomb data suggested a leap upwards in risk at the end of the 1980s.

In its 1988 and 1993 reviews UNSCEAR concluded that, while DDREFs as high as 10 had been seen in experimental situations, the results were generally more consistent with a DDREF of around 2 ; certainly a value greater than 3 would be difficult to justify. The 1991 ICRP estimates were based on using a DDREF of 2.

In 2000 UNSCEAR gave a somewhat tentative endorsement of their earlier conclusions:

> ...difficult to arrive at a definitive conclusion on the effects of dose rate on cancer risks....However, the conclusions reached in the UNSCEAR 1993 Report... that suggested a reduction factor of less than 3 when extrapolating to low doses or low dose rates still appear to be reasonable in general.

Perhaps the last word should be from BEIR VII(Phase2) in 2006 who included the DDREF in their analysis of leukaemia as a translation from linear-quadratic to the linear form at low doses. For solid cancers they adopted a DDREF of 1.5.

8 Somatic Effects

The various projection models used, the different DDREF corrections and the different modes of presentation (BEIR have preferred to express the results as the effect of 0.1 Gy rather than 1 Gy) have made intercomparison a complex process (BEIRVII made the most recent attempt). There are still numerous differences but the central values obtained for solid cancers (and indeed leukaemias) by several authorities have converged and there is a broad consensus on the risks

The incidence of *in utero* effects was reviewed by W J Schull in 1991[77]. In the 1544 prenatally exposed individuals for whom doses could be calculated using the DS86 system, there were 30 cases of severe mental retardation. Eighteen had disproportionately small heads. The analysis of these cases revealed that the particularly vulnerable period for mental retardation was from 8 to 16 weeks after conception and there was no evidence of any increased risk after the 25th week. The 8 to 16 week risk could be fitted with a linear dose-risk line suggesting that the risk of severe mental retardation was about 43% for a foetal dose (strictly a uterine dose) of 1 Gy and there was possibly a threshold of around 0.1 Gy. For those exposed in the 16th to 25th week the risk was lower and Schull observed the possibility of a threshold at 0.5 Gy.

The results were paralleled by performance in intelligence tests and at school. They showed that the mean intelligence scores decreased with increasing dose for 7 to 25 week exposures with the greatest effects for 7-16 week foetuses. In this particularly vulnerable period the scores fell by 21-29 points per Gray. No effects were seen if exposure was before the 7th or after the 25th week.

Schull could find no other useful data for risk estimation. Diagnostic x-ray examinations in a study that gave negative results were likely to have taken place outside the sensitive period and a negative result was to be expected. Where a positive result was found in a Finnish study it seemed likely that the exposures had similarly occurred outside the sensitive period. The defects seen were then thought unlikely to be a radiation effect.

There have been many many more epidemiological studies subsequently, undertaken for a variety of reasons. Three areas have been particularly controversial and all three seemed to reach some kind of resolution—tentative to one degree or another—that was reassuring. One of them—the study of occupational exposure—was by the end of the century promising to provide the first data to rival that from the bombs for risk estimation

The Sellafield childhood leukaemia controversy—for such it remains—started when a television journalist, James Cutler, made enquiries in Sellafield about the health of workers at the plant. He was told by locals that, while this was not a concern, they had noticed that there had been a large number of leukaemias in children in Seascale, a small village a few kilometres from the plant. Cutler changed the focus of his investigation, discovered that there had indeed been more than might have been expected by chance and made the television programme *Windscale—the Nuclear Laundry* in 1983. As a result, the government set up a small advisory group under the chairmanship of Sir Douglas Black, a prominent physician, and they reported in 1984[78] that there was indeed an apparent excess but that the dose from the radioactive discharges from Sellafield had not been high enough (on the thinking on radiation risk at the time) to explain it. As a result of Black's recommendations, the Committee on Medical Aspects of Radiation in the Environment (COMARE) was set up by the UK Government and an investigation by Professor Martin Gardner of Southampton University was funded. This reported in 1987[79] and found five deaths from leukaemia and lymphoma in children born in Seascale between 1950 and 1983 compared with an expectation of about 0.5 based on national statistics. But none were found among 1500 children who had moved to Seascale[80] at some time after their birth. Subsequently Kinlen[81] showed that this striking difference disappeared if a different grouping of the diseases was used and the cancer statistics were taken over a slightly longer period.

Gardner in 1990[82] linked the cases to their fathers having been exposed to high levels of radiation, particularly in the few months before conception. This became known as the Gardner hypothesis but it appeared to be undermined by Kinlen and by Sarahan and Roberts in the Oxford survey[83]. In 1994 Doll[84] concluded that the suggested link with parental exposure was down to chance. Other clusters were claimed around Dounreay, Aldermaston and Burghfield (the UK bomb assembly facility near Aldermaston) and other nuclear installations but they always seemed inconsistent with the accepted models of radiation exposure and risk and, with the exception of Sellafield and Dounreay, could reasonably be chance occurrences. The view that radiation might be the cause of the clusters received a setback when it was found that similar clusters could be found in areas where nuclear installations had been planned and never built or were built after the clusters were found. Other data pointed to elevated childhood leukaemia levels where relatively isolated communities had seen a large influx of people, suggesting that some kind of infective agent might be involved. In their 2005 Tenth report COMARE[85] reported on the most complete survey to date of childhood cancer which included 32,000 cases between 1969 and 1993. It confirmed the clusters at Dounreay and Seascale and the elevated levels near Aldermaston and Burghfield but reaffirmed the idea that these could not be tied to radiation. Rosyth, a nuclear submarine facility in Scotland was found to have an anomalous distribution of cases but elsewhere there was no evidence for higher levels. The professional view at the end of the century was generally that, while the causes of clusters like the ones at Dounreay and Seascale were not understood, they were unlikely to be due to radiation from the facilities or from parental exposure.

However, as evidenced from the final report of CERRIE[86], the committee set up in 2001 by COMARE to review the assessment of internal exposures, there remained groups who were unconvinced and considered that the radiation risk estimates were wrong and that artificial radioactivity was the cause. Research continued to appear that kept the controversy alive—if only politically.

Several studies were carried out in other countries, notably France, the USA, Canada and Germany. The French and US results were negative. Indeed, the French study showed a slightly lower leukaemia rate near nuclear plants than elsewhere although the deficit was not statistically significant. The principal US investigation was not able to spot localised clusters like the one at Seascale but it indicated that there were no excesses of childhood leukaemias—or any other cancers—in counties with nuclear reactors. Canadian studies in the early 1990s found no significant excesses of childhood cancer within 25 km of nuclear installations in Ontario and no association with parental exposure. The study of childhood cancers in Germany in 1992 showed that, while they were generally no more frequent near nuclear reactors sites than elsewhere, increased relative risks were seen for leukaemias in the very young and lymphomas within 5 km of the facilities. As the UNSCEAR 1994 report pointed out, the control groups in this study had unusually low leukaemia incidence rates and, as in the UK, similar increases were seen in regions where a nuclear plant was simply planned.

When in 1980 the US Centre for Disease Control examined the mortality of 3200 men who had participated in the Smoky5 test in the Nevada desert in 1957 they found, between then and 1979, nine cases of leukaemia when only four would have been expected on the basis of national rates[87]. The number of cases was increased to 10 in a later report in 1983[88] but no excesses were seen for other cancers. This prompted a wider study by the National Academy of Science[89] where the Smoky test results were examined in more detail as part of a study of five series of weapon tests at both the Nevada Test Site and the Pacific Proving Ground between 1951 and 1957. This involved over 46,000 participants. No excess of leukaemias was seen in any of the other tests but the Smoky results were confirmed. Otherwise, apart from a higher number of prostate cancers than was expected in the Redwing test series (a result that might have been due to chance), there was no evidence of a general increase in cancers from any of the series. The NAS study was later found to be flawed: 4,500 people who had not been involved in the tests were included in the data while 15,000 who had participated were left out. It was withdrawn and a new epidemiological study, generally known as The Five Series study, put in place. This reported in 2000[90].

The Five Series study reviewed mortality of around 70,000 military participants in five series of tests. A similar-sized control group was used. The conclusions were rather like those of previous studies: overall similar risks of death to those of the controls with similar risk of cancer death also. What differences there were could be ascribed to chance and this included the 14% higher (not statistically significant) risk of leukaemia than that of controls. On the other hand, the authors allowed that the leukaemias were consistent with the idea of a radiation effect.

The Hardtack I tests were conducted in the Pacific in 1958 and the study led by Watanabe[91] examined the mortality of the 8500 Navy personnel involved. They had median doses of 3.9 mSv. No evidence for radiation-induced cancers was found but the power of the study was such that a small effect could not be ruled out.

8 Somatic Effects

The CROSSROADS tests took place in 1946. They were atmospheric detonations in the Bikini Atoll in the Marshall Islands. A study published in 1996[92] showed that the mortality of the roughly 40,000 US Navy personnel involved was a few percent higher than that of a control group but this was not as a result of just malignancies. In fact the mortality from malignancies was less elevated than that from all causes. The authors concluded that two factors were at play to give the higher rates: biased reporting because those suffering from a disease would be more likely than healthy people to have identified themselves for inclusion in the study and some unknown factor—other than radiation—associated with participation in the tests. Radiation could be eliminated as a cause since the pattern of malignancy simply did not square with the known risks.

The early Smoky results led to concern about participants in the British testing programme and the UK MOD commissioned a study of those involved in the tests in Australia and the Pacific in the 1950s. The results of the study of 21,000 men, published in 1988[93], showed that mortality rates for all cancers were identical with those in a control group and lower than the general rates in England and Wales. A follow-up published in 1993[94] indicated some differences between the participants and controls but all could be accounted for by chance. The small non-significant excess of leukaemia seen in participants in the first 25 years after the tests could not be linked to radiation exposure. The follow-up in 2003[95] broadly confirmed the results including the elevated level of leukaemias (other than chronic lymphatic) at 40 compared with 23 in the control group. While this might have been a chance result, the authors could not rule out a link with the tests. The concern raised in earlier analysis that there might have been a raised level of multiple myeloma was considered unfounded.

This was followed by a rather similar study of the 500 New Zealand soldiers who took part in the British tests in 1957-8. An excess of leukaemias up to 1987 was found with a relative risk compared to controls of over five (with a 90% confidence range of 1 to 41) but the temporal distribution was not thought consistent with the weapons test as a cause.

The notion that it might be possible to detect radiation–induced leukaemia in groups of workers was considered around 1960. But the analyses suggested[96] that the population size would have to be between 10^5 and 10^6 manyears for a group with an average of 0.01 Gy. Duncan and Howell[97], hardly surprisingly, found negative results when they looked at UKAEA workers—and were sharply criticised[98]. The first real step—although not necessarily a forward one—came with the Mancuso study.

After a brief preliminary study by the University of Colorado, the US AEC commissioned a feasibility study for a follow-up of workers on the Manhattan Project and associated contractor facilities. They choose Thomas F Mancuso, a physician from the University of Pittsburgh, as project director and Barkev Sanders was appointed as the consultant statistician to design and undertake the study. With health physicist Allen Brodsky, they began work in 1965. In the next five years they collected data from the main Manhattan Project facilities and actually began the analysis in 1970.

By 1971 they had found no significant differences in cancer occurrence from the rates in the general population (and Sanders went on to confirm this later) but no external publication was made; the findings were merely presented in annual progress reports. Mancuso was encouraged to publish by AEC staff but refused on the grounds that the work was incomplete and the results would be misleading. Everything changed with the findings in 1974 by Samuel Milham, a Washington State epidemiologist who claimed that Hanford workers had suffered higher rates of cancer (notably of the mouth area, lung and bone) than expected. Mancuso was asked by the AEC (which became ERDA in 1974) to produce a rebuttal but was not ready to do so. Sanders seems to have been prepared to do this and circulated a paper, without Mancuso's approval, suggesting that there was no link with radiation. The relationship between the two men deteriorated and Sanders was sacked. When the contract expired in 1977 it was renewed with Oak Ridge (ORAU) and Mancuso was dropped. However, within a few months of Sanders' departure Mancuso had recruited Dr Alice Stewart who, at age 70, was about to move from Oxford to the University of Birmingham. With the help of George Kneale, Stewart's statistician, the data were re-analysed and published in Health Physics in 1977[99]. Their conclusions were that there was a 5 to 7 percent increase in cancer deaths attributable to radiation and that the risks of radiation were about ten times higher than the current estimates.

The paper caused a storm with much professional criticism (including by Barkev Sanders) of the methods used. On the other hand it was still being quoted by sceptics of the official line on radiation risks at the end of the century[100]. Mancuso and (particularly) Stewart continued to research, addressing some of the criticisms of their original Hanford paper; both died in their nineties in the early years of this century. Mancuso's sacking was investigated in Congressional Hearings in 1978 and remains a much-quoted example of Government's unwillingness to support critical work.

The studies in the UK, after Duncan and Howell's initial exercise, began with Beral's look at mortality in the UKAEA workers between 1946 and 1979[101] with further work published in 1987[102]. This was quickly followed, in 1988, by a similar mortality study of employees at the Atomic Weapons Establishment between 1951 and 1982[103]. The mortality of workers at Sellafield was analysed by Smith and Douglas in 1986[104] and later by Douglas, Omar and Smith[105]. Otherwise, the most significant analyses of the 1990s were probably those of the National Registry for Radiation Workers in 1992 and 1999[106]. Work specifically on Capenhurst and Springfields was published in 2000.

In the USA the 1980s saw re-analysis of the Hanford workers by Kneale and Stewart and by Ethyl Gilbert from Battelle (who had reviewed the original Milham work for the AEC and received part of the reallocated contract taken from Mancuso). Through the 1990s all the major non-reactor facilities in the US were analysed in one way or another (see BEIR VII) and reports appeared of studies in Japan, Canada, Spain, Russia, China, Finland, India, France and elsewhere. An important study covering Canada, the USA and the UK usually known as the Three-country study was published in 1995[107].

Although many of the studies appeared to point to particular concerns about particular cancers and a few suggested higher cancer risks than expected, the confidence limits generally encompassed the official risk estimates. Indeed a significant number admitted the possibility that radiation reduced risk.

BEIRVII[108] compared the risk estimates from two of the more powerful worker studies, the UK NRRW and the Three-country study, with those from the atomic bomb survivors. The central estimates for leukaemia were very similar from all three but the worker studies pointed to a much lower value for all-other-cancers risk. There were still rather wide confidence intervals as can be seen from their Table 8-7 reproduced below as Table 8.

The International Agency for Research on Cancer (IARC) study of 15 countries published in 2005[109] had the largest cohort so far at over 400,000 workers. It found ERRs/Sv of, for solid cancers, 0.97 (95% CI: 0.14, 1.97) and for leukaemias, excluding chronic lymphatic (CLL), 1.93 (95%CI:<0,8.47). It thus pointed to rather higher values than the Japanese, NRRW and Three-country work for all cancers excluding leukaemia but similar ones for leukaemia—but with confidence intervals that encompassed them. At the time of writing insufficient background to the study—which necessarily drew on a diverse data source—was available for its implications to be fully appreciated.

Comparison of estimates of ERR/Gy between major nuclear industry workers combined analyses and the atomic bomb survivors with 90% confidence intervals from [110]		
Study population	All cancers but leukaemia	Leukaemia excluding CLL
Atomic bomb survivors	0.24(0.12,0.4)	2.2(0.4,4.7)
Nuclear workers:		
Three-country study	-0.07(-0.39,0.30)	2.2(0.1,5.7)
NRRW	0.09(-0.28,0.52)	2.6(-0.03,7.2)

Table 8: Risk estimates from various sources

There were of course a number of confounding factors that had to be considered in all these analyses. One obvious one was the so-called healthy worker effect: workers generally—and particularly in relatively well-organised and heavily-regulated industries like the nuclear one— are healthier than the remainder of the population. Since the early studies this has been taken into account in some way, often by choosing a well-matched control group. Since the early studies too there has been concern about the quality of dose data available for analysis; without reliability here, risk estimates make

8 Somatic Effects

little sense. Happily, this seems less of a problem than feared and where detailed reviews of dose recording have been undertaken they have not affected greatly the confidence in risk estimates[111]. What little work was done suggested that exposures to hazardous materials such as beryllium and lead did not significantly confuse the issue although one early study of employees at Portsmouth (New Hampshire, USA) naval shipyard found that radiation workers were more likely to be exposed to asbestos and welding by-products. This had possible implications for assessments, particularly, of lung cancers.

The results of analysis of data from the Mayak plutonium production facility in the Russian Federation began to appear from 2003 onwards. The cohort size (21,000) and the magnitude of the collective dose suggested that it might be possible to derive risk estimates with greater confidence—although the effects of large plutonium exposures had to be accounted for. The results were beginning to show rather lower risks for leukaemia than the bomb data at 1.0(0.5,2.0) ERR/Gy and similar ones for all-other-cancers.

Studies of the effects of internal emitters has been much more restricted than those of external dose. In their study published in 1998[112] the Oxford workers found few consistent patterns linking people monitored for radionuclide exposure. There appeared to be possible links between tritium monitoring and prostate cancer and between plutonium monitoring and cancers of all kinds. But they were unable to draw firm conclusions not least because of the effects of possible exposure to other agents like asbestos. A study of Sellafield workers[113] showed no statistically-significant association between estimated dose from plutonium intakes and leukaemia mortality. However, several studies of Mayak workers published in 2000 did link excess cancers in the lung, liver and bone to plutonium exposure [114].

Finally, the collective dose to the 200,000 front-line clean-up workers at Chernobyl (the "liquidators")had been estimated at 20,000 Sv. Unusually for occupationally exposed people it included situations where the dose was delivered in a few minutes and over a few years. The follow-ups have generally shown a doubling or trebling of leukaemia and thyroid cancer rates but studies that attempted to link risk of cancer to radiation dose had failed. The most that could be said is that the Chernobyl results were consistent with the Japanese data.

So, by around the end of the century the first estimates of risk of external exposure, with a similar standing to the Japanese bomb data, were becoming available. While individual studies had thrown up particular concerns—some of which could be seen as chance findings and others as unexplained—the overall picture from the occupational epidemiology was essentially reassuring. The confidence limits on the risk estimates were narrowing down on values that were very similar to those derived from the bombs and they related much more directly to the sort of low-dose exposure regimes encountered by workers—and indeed everyone else.

Throughout much of the century there were people inside and outside the radiological protection profession who argued that the effects of radiation were being significantly underestimated or overestimated. With some notable exceptions (for example, Martland and the radium girls and Alice Stewart and *in utero* exposure), these claims have not made much impact on the practice of protection. Much remains fertile ground for controversy: while there is no hard information about the effects of very low doses, proponents of supra- and sub-linearity can marshal arguments without many constraints. Sticking to the evolving mainstream, we have largely ignored them. However, one idea from the many is worth a brief examination because it illustrates just how large the scope for controversy is: hormesis.

The idea that a small amount of something that is dangerous in large amounts might be good for you—or even essential—has a long history. For example, it seems that Chinese and Indian physicians were using a process of inoculation in which pus from a smallpox scab was transferred to a healthy person to protect them from the disease before 1000AD. Plants seem to grow a little better if subject to some stress and, of course, the principle lies behind homeopathy, a widely-used alternative medicine.

The term hormesis itself arose in the 1940s, along with some scientific support from a rather obscure area. Southam and Ehrlich, investigating the concentrations of phenolic compounds that inhibited fungal growth in trees found that, while at high concentrations these did deter the fungi, at very low concentrations they encouraged their growth. Adapting the Greek term *hormo*= I excite, they coined the word hormesis in a 1943 paper. However, the idea of radiation hormesis owes most to Thomas D Luckey of the University of Missouri who wrote the first account[115] in 1980.

Luckey and many others have since then compiled a long list of radiation effects (he quoted over 1000 references in 1980) that, they believe, demonstrate hormesis. They include improved immunity and increased lifespan, reduced sterility and improved reproduction among experimental animals. But much later evidence has been adduced from human exposures: increased environmental radon concentrations are correlated with reduced lung cancer incidence, lung cancer was also found to be lower among workers exposed to plutonium than among controls and both leukaemia and solid cancers are found to be reduced among Japanese bomb survivors who received smaller doses.

The proponents of hormesis are able to put forward biological mechanisms to account for the effect including the possibilities that small doses stimulate the DNA repair mechanism, improve the removal of damaging free radicals and stimulate the immune system. The support for the idea did not diminish (an issue of *Health Physics* was devoted to the topic in 1987) and some of its proponents were suggesting that not only did the linear no-threshold theory need urgent revision but that a programme of controlled radiation exposure would significantly improve health. As Luckey himself put it[116] "we live with a sub-clinical deficiency of ionizing radiation".

The hormesis concept was considered by UNSCEAR in 1994 in a review of adaptive responses, by BEIR V and BEIR VII(Phase 2) in 2006. UNSCEAR concluded that "it would be premature to conclude that cellular adaptive responses could convey possible beneficial effects to the organism that would outweigh the detrimental effects of exposures to doses of low-LET radiation". But they did recommend more research. BEIR VII(Phase 2) commented rather more sharply that "the assumption that any stimulatory hormetic effects from low doses of ionizing radiation will have a significant health benefit to humans that exceeds potential detrimental effects from the radiation exposure is unwarranted at this time".

Radiation hormesis had by the end of the century evolved a substantial scientific following in areas where it is difficult to prove a negative not least because of the potential confounding factors that abound. However, it seemed unlikely, in these precautionary days, to become an orthodoxy.

Notes Chapter 8

1 (IXRPC, 1928)
2 (Mutscheller A, 1925)
3 (Cantril S T and Parker H M, 1945)
4 (Warren S et al, 1949)
5 (Taylor L S, 1984)
6 (Neary G J, 1952)
7 (Court Brown W M and Doll R, 1957)
8 (UNSCEAR, 1958)
9 (UNSCEAR, 1962)
10 (ICRP, 1964)
11 (ICRP, 1959)
12 (Hedge A R, 1959)
13 (Boveri T, 1914)
14 (Tyzzer E E, 1916)
15 (Edler L and Kopp-Schneider A, 2005)
16 (Bauer K H, 1928)
17 (Muller H J, 1927)
18 (McCombs R S and McCombs R P, 1930)
19 (Nordling C O, 1953)
20 (Armitage P and Doll R, 1954)
21 (Doll R, 2004)
22 (Armitage P and Doll R, 1957)
23 (Platt R, 1955)
24 (Nowell P C, 1976)
25 (Moolgavkar S H, 1983, 1986)
26 (Knudson A G, 1971)
27 (Temin H M and Mizutani S, 1970)
28 (Baltimore D, 1970)
29 (Huebner R J and Todaro G J, 1969)
30 (Bishop M B, 1996)
31 (Alberts B, Johnson A et al, 2002)
32 (Doll R, 1995)
33 (Colwell H A and Russ S, 1934)

34 (March H C, 1944)
35 (Ulrich H, 1946)
36 (March H C, 1950, 1961)
37 (Dublin L I and Spiegelman M, 1948)
38 (Lewis E B, 1957, 1963)
39 (MRC, 1956)
40 (Folley J H, Borges W and Yamawaki T, 1952; Moloney W C and Kastenbaum M A, 1955)
41 (MRC, 1956)
42 (Schull W J, 1995)
43 (Beebe G W, 1979)
44 (Schull W J, 1995)
45 (Plummer G, 1952)
46 (MRC, 1956)
47 (Schull W J, 1995)
48 (Folley J H, Borges W and Yamawaki T, 1952)
49 (Moloney W C and Kastenbaum M A, 1955)
50 (Glasstone S, 1950)
51 (Kato H and Schull W J, 1982)
52 (Socolow R L, Hashizume E, Neriishi S and Niitani R, 1963)
53 (Glasstone S, 1950)
54 (Court Brown W M and Doll R, 1958)
55 (Smith P G and Doll R G, 1981)
56 (Berrington, Weiss H A and Doll R Darby S C, 2001)
57 (Stewart A, Webb J, Giles D and Hewitt D, 1956)
58 (Harvey E B, Boice Jr J D, Honeyman M and Flannery J T, 1985)
59 (Doll R, 1995)
60 (Martland H S, Conlon P and Knef J P, 1925; Martland H S, 1929, 1931)
61 In (Neuzil M and Kovarik W, 1996)
62 (Mullner R, 1999)
63 See (Doll R, 1995)
64 ibid
65 (Abbatt J, 1979)
66 (UNSCEAR, 2000)
67 (Schales F, 1978)
68 (Spiess H, 2002, 2005)
69 (Upton A C, 1991)
70 (Preston D L, Shimizu Y, Pierce D A Suyama A and Mabuchi K, 2003)
71 (Pierce D A, 2002)

72 (Upton A C, 1991)
73 (Preston D L, Shimizu Y, Pierce D A Suyama A and Mabuchi K, 2003)
74 (Preston D L,Kato H, Kopecky K J and Fujita S, 1987)
75 (Upton A C, 1991)
76 ibid
77 (Schull W J, 1991)
78 (Black D, 1984)
79 (Gardner M J, Hall A J, Downes S et al, 1987b)
80 (Gardner M J, Hall A J, Downes S et al, 1987a)
81 (Kinlen L J, 1993)
82 (Gardner M J, Snee M P, Hall A J et al, 1990)
83 (Sorahan T and Roberts P J, 1993)
84 (Doll R, 1994)
85 (COMARE, 2005)
86 (Goodhead D (Chairman), 2004)
87 (Caldwell G G, Kelley D B and Heath C W Jr, 1980)
88 (Caldwell G G, Kelley D B, Zack M et al, 1983)
89 (Robinette C D, Jablon S and Preston D L, 1985)
90 (Thaul S, 2000)
91 (Watanabe K K, Kang H K and Dalager N A, 1995)
92 (Johnson J C, Thaul S, Page W F, Crawford H, 1996)
93 (Darby S C,Kendall G etal, 1988)
94 (Darby S C, Kendall G M et al, 1993)
95 (Muirhead C R et al, 2003)
96 (Hems G, 1966)
97 (Duncan K P and Howell R W, 1970)
98 (Sanders B S, 1970)
99 (Mancuso T F, Stewart A M and Kneale G W, 1977)
100 (Alvarez R, 2006)
101 (Beral V, Inskip H et al, 1985)
102 (Inskip H, Beral V, Fraser P, Booth M, Coleman D and Brown A, 1987)
103 (Beral V, Fraser P et al, 1988)
104 (Smith PG and Douglas AJ, 1986)
105 (Douglas A J, Omar R Z and Smith P G, 1994)
106 (Kendall G M, Muirhead C R , MacGibbon B H et al, 1992; Muirhead C R, Goodill A A, Haylock R G et al, 1999)
107 (Cardis E , Gilbert E S , Capenter L et al, 1995)
108 (BEIR Committee, 2006)
109 (Cardis E, Vrijheid M, Blettner M et a, 2005)

110(BEIR Committee, 2006)
111 ibid
112 (Carpenter L M, Higgins C D, Douglas A J et al, 1998)
113(Omar R Z, Barber J A and Smith P G, 1999) 114(BEIR Committee, 2006)
115(Luckey T D, 1980)
116(Luckey T D, 1999)
---

9 Measuring External Radiation

Some of the early measurement systems for measuring x-ray and gamma-ray intensities and exposures were described in Chapter 1. Further developments led to the instruments and systems that form the basis of radiation protection of workers and members of the public from external sources of radiation. The story is largely one of detectors: new ones, based on novel physical processes, that have either proved practical and reliable or been dropped and old ones, known from the earliest days of radiation. Indeed many of the devices regularly used today rely on physics that was well understood a century ago.

Of course, the electronics that turns a detector into a usable instrument have developed enormously, first with the valve ("tube" in the USA) and then the transistor and integrated circuit. The increasing complexity and functionality have been built around the same basic devices so, on the whole, no attempt has been made here to trace in detail the developments in commercial instruments in the last few decades.

The fact that radiation could cause ionisation was known early on. Electroscopes discharged under the influence of x-rays and Becquerel found the same effect near radioactive substances only weeks after his initial discovery. However, it was to need decades of development of radiation detectors and of the electronics to go with them before practical robust instruments were readily available.

9 Measuring External Radiation

When an ionising particle passes through a gas it produces electrons and positive ions. A typical counter set-up is a thin cylindrical cathode surrounding a central anode. If an electric field is applied across the gas then the electrons migrate towards the positive electrode and the positive ions the other way. At low voltages the probability is that the ions will recombine before they reach the electrodes but, as the voltage increases, it is more and more likely that they will be collected at the electrodes. When all the ions produced are collected there is said to be saturation and increasing the voltage will, for a significant increase in voltage, not increase the collected charge because there are no more electrons and ions to collect. This is the region where ionisation chambers work. The chambers can be used in either a current mode and a pulse one. In the current mode the chamber current is related to the intensity of radiation while in pulse mode the tiny charge released in individual events is proportional to the energy deposited by individual particles or photons.

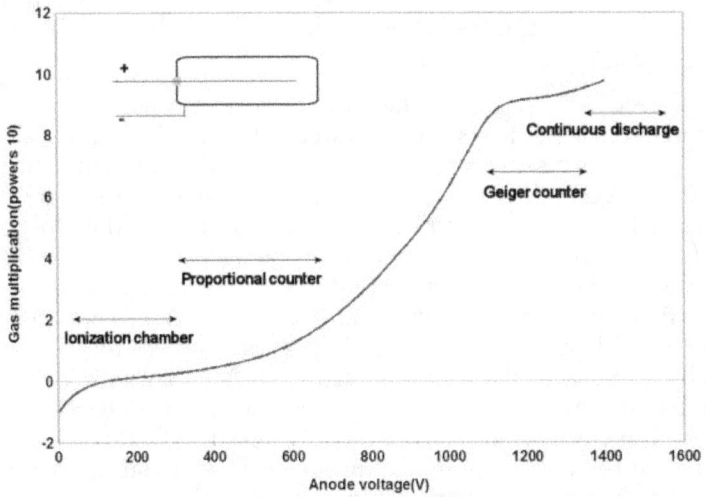

Figure 10: Schematic counter tube performance

If the voltage is increased substantially then a new process will start to occur. As electrons reach the high electric field near the anode they may acquire enough energy to themselves cause ionisation on collision and introduce another ion pair. The process may then be repeated in a kind of avalanche so that the initial electrons are multiplied many times. This is gas amplification and the amount of amplification increases with the voltage applied. Devices that operate in this region are called proportional counters because the output charge from them for an event is proportional to the initial amount of ionisation. The useful gas amplification ranges to perhaps 10^6. Proportional counters are used in pulse mode with the significant advantage that they give a relatively large signal pulse which is proportional to the energy deposited in the counter.

If the voltage is increased still further then the proportionality relation begins to degrade until a regime is reached where the output is much larger than from a proportional counter but is independent of the size of the initial ionisation. At this point the electron avalanche spreads through the tube because the uv photons produced in ion recombination cause further ionisation. This is the region where Geiger counters work. They operate only in pulse mode and, while the output is the same whatever the energy of the ionising particles, the output pulse is so large that relatively unsophisticated electronics can be used. If we take the voltage higher still a continuous discharge takes place making the tube useless for counting.

The electroscope used by the earliest workers was an essentially qualitative device. If measurements were to be made with ionisation chambers, a very sensitive device was required to measure the tiny electric charges released. Quadrant electrometers were established devices of high accuracy and sensitivity (with perhaps the Dolezalek quadrant electrometer of around 1896 being a pinnacle) but they depended on the charges generated deflecting plates suspended from a fine filament. While accurate and sensitive, they were delicate laboratory instruments. The string electrometer that became available in a practical form just after 1910 was rather more robust. It depended on measuring, with a microscope, the deflection of a tensioned, metal-coated quartz fibre in an electric field when charge was introduced onto it. It was a non-linear instrument but it was to be, in one variant or another (and there were versions down to Lindemann, Lutz-Edelmann and Wulf among others), a key component of many of the devices developed up to the outbreak of the Second World War.

The first practical ionisation chambers that could be used for routine measurements on patients were probably those made around

1920 by William Duane[1] but they were quickly followed by those produced by Glasser and Fricke[2] and the ionometer of I Solomon[3]. It was the Glasser and Fricke design that was the core of the first practical instrument when, in 1928, John Victoreen founded the Victoreen Instrument Company in Cleveland, Ohio to manufacture the Fricke-Glasser x-ray dosimeter, an instrument calibrated in R-units that would "eliminate the possibility of x-ray burns"[4]. This was the condenser r-meter in which a small ionisation chamber was connected to a string electrometer that could be viewed with a built-in microscope. It was intended for measurements in x-ray beams to control dose delivery to patients and was calibrated in roentgens per minute.

In 1930 Victoreen began to manufacture an r-meter based on the Glasser-Seitz dosimeter (Glasser, Seitz and others at the Cleveland Clinic had designed a new dosimeter in 1929). In this version the chamber (usually made using Bakelite) could be charged, removed from the electrometer during exposure and returned for reading. Known as the Model 70, it was provided with a selection of chambers with ranges from 0—0.25 R to 0— 100R. A rugged and reliable instrument (similar in appearance the earlier instrument), it was in use for more than 50 years.

Figure 11: The Victoreen Model 70 [ORAU]

Lauriston S Taylor at the National Bureau of Standards developed a portable survey meter in 1929 after he suffered an accidental exposure[5] as he sat for several minutes in the full beam of an x-ray set after someone left out a lead shielding panel. Investigation showed that the whole-body exposure was around 200R which left Taylor relieved: it was "less that that commonly used in therapy at the time". He later said: "Since I was not aware that one should become nauseated at this exposure, I was not nauseated either". However, it decided him to make a portable instrument. This was made with a spherical spun-aluminium ionisation chamber connected to a version of the Lutz-Edelmann string electrometer then in use at his laboratory. The three batteries for the ionisation chamber, total potential about 135 V, were housed in a wooden box with the chamber and electrometer on the top. The system was "relatively rugged" and maintained its calibration over a long period. Three chambers were made and these could be interchanged to vary sensitivity. With the largest, 75 mm diameter, chamber the full scale-range was 2 mR.

The Lauritsen electroscope[6] was developed in 1937. The electrometer here was a gold-covered quartz fibre where the electrostatic forces were balanced by tension in the charged fibre. The deflection was measured against a graduated scale with a microscope. Charging was done with a friction charger, batteries or, a little later, a valve rectifier. It was produced by the Fred C Henson Company of Pasadena and a Model 2, with the electroscope enclosed in a case with a built-in power supply and light source, appeared shortly afterwards. It made a rugged, portable and sensitive survey meter and it was reportedly still being used in the 1970s.

There were clearly possibilities to use radio valves ("tubes" in the USA) to replace the measuring electrometer and a number of circuits were designed in the late 1920s so that ordinary radio valves could be used to measure very small currents. However, while some were apparently successful, the electrometer remained the instrument for such measurements. The problem was that the available valves were simply not designed for low-current dc measurement.

The first custom-designed tube, the General Electric "FP-54 pliotron" described in 1930 by Metcalf and Thompson[7], was capable of amplifying currents as small as 10^{-17} A. To measure such small currents it is essential that the grid input resistance of the valve is extremely high because it is the high resistance that results in measurable voltages when the minute currents pass through it. In ordinary valves of the time this resistance was perhaps 10^8 Ω. By running the filament at low power, keeping the anode voltage below 8 V and introducing an extra grid, Metcalf and Thompson managed to increase this resistance to about 10^{16} Ω. A number of practical circuits using the valve, really adaptations of existing circuits, were described by DuBridge in 1930[8]. In particular he showed how a second valve could be used to amplify the tiny current enough to register on a conventional microammeter and he speculated how further improvements could be made. While he did encounter a few problems—the valves needed to warm up for at least 30 min before reliable readings could be taken—the way to produce practical, rugged ionisation chamber instruments based on valves was now clear. The FP-54 was the only electrometer valve available until the end of the 1930s and it, and its 6 V storage battery, were too big to incorporate into a portable ion chamber survey meter. This had to wait for a miniature electrometer valve, the VX-41, developed by a Dr Ewing, a professor of physics at Northwestern University, for use in hearing aids[9]. A number of these were obtained by the US health physicist Dale Trout and they formed the basis for the first commercial ion chamber survey meter. The prototype of this was put together by Trout and John Victoreen with materials to hand: for example, the counting tube was a cardboard mailing tube coated with Aquadag to make it conducting. A version of the survey instrument, probably the model 241 dating from the early 1940s, is in the Oak Ridge Associated Universities(ORAU) museum[10].

An example of some early work in the UK using an electrometer valve circuit for measurements of voltage on small ionisation chambers is that of Farmer[11] reported in 1942. Farmer sought a simpler system than the usual one—then based on the Wulf or Lindemann string electrometer. He used a Marconi ET1 electrometer valve in a two valve circuit so that the electrometer valve could be used at low anode voltage. The capacity of the chambers were around 1 pF while that of the valve was about 3 pF so simply connecting then to the grid would drop the voltage by a factor of four and introduce uncertainty. Farmer solved this problem by shielding the cable from the chamber and connecting the shield to the cathode of the electrometer valve.

Ion chambers have almost always been used in the direct current or integrating mode but pulse operation caused by cosmic rays was recorded by Lindholm in 1928 and the gridded ion chamber devised by Frisch in 1944 removed the problem of position dependence of pulse size, to produce a high-resolution alpha device used to detect alpha-emitters in gases. Pulse-type ion chambers have also been used as fission chambers for neutron detection. They were made sensitive to neutrons by coating the cathode with a fissile material—usually $^{235}U_3O_8$. When a neutron is absorbed by the uranium and fission occurs, the fission fragments cause large pulses that can be fairly readily detected.

Gas amplification was discovered in about 1900 by J H Townsend[12] but practical use came from Ernest Rutherford and Hans Geiger[13] in 1907. They were using cylindrical gas-filled tubes with central electrodes and discovered that alpha particles gave much larger electrometer readings than had previously been seen. Geiger and Klemperer[14] exploited the proportional region in 1928 and could differentiate between alpha and beta by pulse size. This characteristic, with the benefits of gas amplification and the relatively short dead-times, was to make them valuable instruments throughout the rest of the century.

Three broad types of proportional counter evolved:

- Gas flow counters with and without windows
- Air proportional devices for alpha counting
- Sealed proportional counters for special purposes e.g. BF3 and He-3 neutron detectors

Argon has been preferred as a fill gas but hydrocarbons have been used and the tissue equivalent gas 64% methane, 32% carbon dioxide and 3% nitrogen called P-10, developed by John Simpson in the 1940s, has been widely employed for gas flow proportional counters. Air-filled counters have been used for alpha counting because they allow the use of thin windows without gas flow.

Taylor[15] describes the use by Simpson[16] of large area air proportional counters with thin nylon film windows in an installed alpha-contamination hand monitor[17]. For such uses they had "no competitors".

9 Measuring External Radiation

The devices first used by Rutherford and Geiger at Manchester were refined by Geiger in 1912 and 1913 after his return to Germany as a point counter in spherical geometry. The counter was used in laboratories and was the detector employed in 1925 to confirm the Compton effect. Hans Geiger continued to refine the device and in 1928, working with Walther Muller, he developed counters[18] with much larger sensitive volumes and using various fill gases that were very close to the Geiger-Muller(G-M) tubes we know today.

One problem with early G-M tubes was the long dead-time—the time after one event before another could be detected. This occurs because the avalanche initiated in the tube by a particle continues until the potential gradients near the anode are reduced below a critical value by the accumulation of positive ions. This brings the avalanche to an end and makes the counter ready to respond to the next particle. This "quenching" action was initially achieved with a high "quenching" resistance but this resulted in dead times of perhaps 10 ms, limiting greatly the use of the device. In the 1930s active valve circuits were designed[19] that removed the potential from the tube briefly after a pulse and gave shorter dead-time values, down to perhaps 100 μs these but added to complexity. In 1937 Trost discovered organic quenching[20]. This meant adding a small quantity of an organic vapour such as ethanol that would quickly terminate avalanches and make the tube self-quenching with a dead time of around 200 μs. Tubes that relied on organic quenching had a limited life because a tiny fraction of the organic quenching agent is destroyed in terminating the discharge but it was found by S H Liebson, while working on his PhD thesis in 1947 [21], that an added halogen-compound vapour would have a similar quenching effect. Since the halogen compound normally recombines after quenching, such tubes had a much longer life. Halogen-quenched tubes quickly caught on once they became commercially available around 1950 and remain the norm.

Practical tubes were made from the 1930's in a variety of shapes and sizes. The early ones were made of glass but by the late 1940's they were being routinely made from metal with, if required for alpha or beta detection, thin windows made from mica.

The first commercial instrument, using a glass-walled G-M tube and earphones, was probably developed by Victoreen in the 1930s[22]:

a pair of high-impedance headphones were used to hear the events in the tube. The refinements of the ratemeter—combined with headphones in the ORNL "Walkie Talkie" instrument shown in Figure 12—and counting circuits were to come in the next decade. The simplicity resulting from the large pulse coming from the tube meant that the G-M tube was rapidly established as a tool for surveys and contamination measurement.

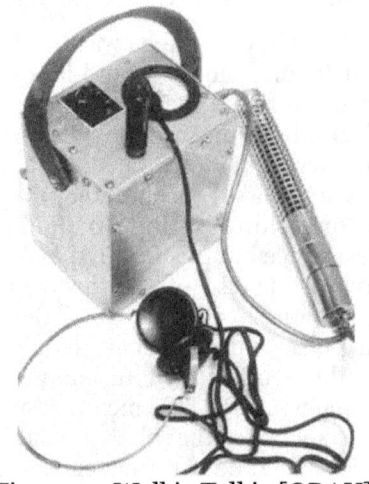

Figure 12: Walkie Talkie [ORAU]

By the late 1940s several companies were producing survey meters not too different from those in use today—although usually rather heavier—with shutters to exclude shorter-range radiation. The large civil defence and uranium prospecting markets after the war gave rise to a large number of basic but robust instruments.

The versatility of the G-M detector is illustrated by Taylor[23] who describes a doorpost monitor to detect gamma-emitters developed at Harwell before 1950, based on two 1.5 m long G-M counters.

The spinthariscope was a popular scientific toy in the early days of radioactivity. It relied on the fact that when an alpha-particle struck a zinc sulphide screen a tiny flash of light was generated. In darkness the flashes could be seen through a low-power microscope as minute scintillations. This was the scintillation detector with the human eye as the register of the scintillations (and the eye could detect a flash of as little as 30 photons) used in some of the most important experiments of early nuclear physics. Birks[24] recounts that, in one famous lab, students interested in pursuing nuclear physics were tested in a dark room; if they failed the eye test they were directed towards less physically exacting areas of physics. The last major experiment to use a human detector was that of Cockcroft and Walton in 1932 when they studied the disintegration of lithium by protons. Of course, while it depended on the trained eye in the darkened room it was not a practical instrument for radiation measurement so the scintillation counter had to wait until a device that could reliably measure very low light levels became available during the Second World War.

The photomultiplier tube (PMT) depends on a simple effect: if an electron is accelerated by an electric field and made to impact on a suitable surface, several secondary electrons are ejected. The electrons can then in turn be accelerated and the process repeated for perhaps a dozen stages so that the initial electron is multiplied many millions of times to give an easily-measurable electrical pulse from just a small number of initial electrons. If the initial electrons derive from a photocathode—a material that emits electrons when struck by light photons—we have a very sensitive detector of light. Following the work on direct detection of ionising particles by Zoltan Bay[25], the practical device that we recognise today originates from the work of J A Rajchman and R L Snyder[26] in 1940.

The first publication of the use of a photomultiplier and zinc sulphide screen is probably that of Blau and Dreyfus[27] for source measurements in 1945. In these the brightness of the ZnS screen caused by an alpha source was measured as proportional to the current in the photomultiplier tube. It was reckoned that, with suitable calibration, an accuracy of about 5% in measuring source strength was possible. However, the first significant use in counting may well have been in 1944 by Curran and Baker[28], in which individual alpha-particles were detected using an RCA 1P21 PMT and an oscilloscope. This was not published (and then as only a rather drastic summary) until 1948 so the modern scintillation detector is generally considered to have come into being in 1947 through the work of Coltman and Marshall[29].

The search for scintillators other than zinc sulphide quickly brought results. Kallmann[30] in 1947 detected beta and gamma scintillations from naphthalene and the high efficiency was confirmed by Deutsch[31] who also confirmed the very short pulse length and found solid benzene to have similar behaviour. In 1948 Bell[32] showed that anthracene was a better scintillator still and detected fast neutrons. Also in 1948 Hofstadter made the important discovery of the thallium-activated alkali halide scintillator NaI(Tl)[33]. This material, in which radiation energy absorbed in the sodium iodide is transferred to the thallium luminescence centres, has several important properties: iodine has a relatively high atomic number, it is efficient at converting gamma energy into light, the pulse of light produced is quite short (<1μs) and well-matched to photomultipler photocathodes and it can be made as large single crystals. The high atomic number means a high probability of full absorption of gamma rays within a smallish crystal and thus the generation of a characteristic pulse-height spectrum that made spectrometry possible. One of the disadvantages of the organic scintillators is the likelihood of Compton scattering and loss of the scattered gamma ray (and hence some of the initial energy) from the system. This leads to very degraded spectra virtually useless for gamma spectrometry.

In parallel with the improvements in scintillators there were improvements in PMT design. The original devices (the 1P21 for example) had a curved side window but a design with a flat end window, the RCA 5819 of 1949, gave much better light collection. The development of the Venetian Blind type tube by EMI[34] in the UK and improvements in photocathodes were both significant.

There was a brief interest in photosensitive Geiger counters as alternatives to the PMT in the late 1940s because of their potential simplicity and of the modest electronics needed to run them[35]. However they proved to be less stable and had much larger deadtimes than PMTs and they were relatively quickly dropped.

By the late 1950s methods had been found by the Harshaw Company for producing large single crystals of NaI(Tl) and optically coupling them efficiently to photomultipliers with a photocathode

9 Measuring External Radiation

response well matched to the emission spectrum of the scintillator. The efficiency of the resulting detector for gamma rays has not been surpassed for general use although many other materials such as CsI(Tl) and CsI(Na) are available. It found uses in survey instruments, analytical systems and, taking advantage of the ability to produce very large crystals, the whole-body monitor and the gamma camera.

The liquid scintillator arose from the need to count particles of very low energy. The principal driver was the potential for the diagnostic use of the artificial radionuclides that became available in the 1940s. In 1950 Kallmann and Furst in New York[36] and Reynolds, Harrison and alvini[37] at Princeton found that efficient scintillating solutions could be produced by dissolving substances such as anthracene in solvents like toluene. These teams worked with external gamma sources but in 1951 Raben and Bloembergen showed that, where active materials could be dissolved in the scintillating liquid, it was possible to detect very weak beta-emitters like C-14, S- 35 and H-3 with reasonable efficiency [38]. This efficiency came from the fact that all the particle energy was dissipated in the scintillating medium—there were no walls to pass through. At Los Alamos there were further major steps forward with the systematic study of systems for internal scintillation counting[39]. All the groups began using coincidence counting to reduce the effects of photomultiplier dark noise[40]—the technique reduced the background noise by a factor of 1000. The first commercial system was produced by the Packard Instrument Company in 1953 and, once an automated system was developed shortly afterwards, the liquid scintillation counter established itself as a key tool in the biological sciences for the detection of C-14 and H-3 for much of the rest of the century.

The discovery of liquid scintillators led quickly to plastic scintillators when Schorr and Torney experimented with scintillator material dissolved in plastic solids [41] in 1950. These were rapidly developed to produce relatively inexpensive scintillators that could have many shapes and could be produced in large sizes. They suffered from poor energy discrimination for gammas but had very short pulse lengths (of the order nanoseconds) and found wide use in experimental work. Health physics applications grew, taking advantage of their relatively low cost and ability to be made in large sizes and various shapes. They (notably PVT) found applications in, for example, portal monitors where increased computing power made a useful degree of spectrum analysis possible.

The invention of the transistor and the investment in semiconductor research it led to had two distinct consequences for radiation protection: it made electronics more compact and less fragile and it introduced a new detection device. The invention of the first point-contact transistor by Bardeen and Brattain in 1947 provoked experiments in the growth of single crystals of germanium and this made possible the junction transistor, conceived shortly after the point-contact device by William Shockley. It was 1951 before the first junction transistor was working convincingly but from there progress was astounding: the preparation of large single crystals of germanium and then silicon quickly led to a burgeoning technology that produced a transistor radio in 1954, a computer in 1955 and a simple integrated circuit in 1958. The valve was quickly doomed for most purposes, especially for instruments with a premium on portability and robustness.

The possibility of using a semiconductor crystal rather than a gas-filled chamber as a detector was first raised by the work of Jaffe in 1932[42] ; the work over the subsequent decade or so was summarised by Van Hardeen[43]. However, the first detectors based on modern principles derive from K G McKay's work published in 1951[44] and by 1960 these had improved enough for diffused-junction silicon detectors to give respectable pulse height spectra for alphas. The development of lithium-drifted germanium (Ge(Li)) at Chalk River, by neutralising impurities with lithium ions, provided a large sensitive volume and, in 1963, a detector with a gamma-ray energy resolution about ten times better than that available from NaI(Tl). This exploited the small ionisation energy of germanium in crystal form to produce a large number of ions from an event. However, the germanium had to be kept at liquid nitrogen temperature[45]. Such detectors have remained the mainstay of high-resolution gamma-ray spectrometry (with dramatic increases in efficiency) and there have been improvements such as the use of high-purity germanium (HPGE) and the development of electrical cooling that have made them more usable. Room temperature detectors, notably CdZnTe (CZT), promised even more convenience with useful sensitive volumes and resolutions significantly better than NaI(Tl) if not as good as germanium.

9 Measuring External Radiation

Although valve amplifiers had been constructed since the invention of the triode valve in 1907, stable amplifiers suitable for interfacing with radiation detectors had to wait for the work of Henri Greinacher in 1924[46] and, particularly, the pulse amplifier of J R Dunning in 1934[47]. The further development of amplifiers after this is rather too technical a diversion here but other electronic circuits were anyway at least as important.

Until 1932 detection systems were very limited in value because there was no way, other than by mechanical register, of counting the pulses that came from them. Count rates had therefore to be very low. However, in 1932 C E Wynn-Williams[48] designed a circuit that would scale down the count rate, using a series of valve circuits as binary counters or flip-flops (the flip-flop or bistable circuit using two triode valves had been conceived much earlier by Eccles and Jordan[49].

With each of these flip-flops feeding its output to the next one in the series, the scaler could count in binary. A series of six flip-flops could thus count up to 63 with the final one outputting a signal at the 64th input pulse. Lights could be used to record the binary count accumulated by the 6 flip-flops and each 64th pulse could be used to drive a mechanical counter. This was a scale-of-64 counter; higher scales were possible by simple extension. Since the flip-flops could operate very quickly, much higher count rates could thus be accommodated. This was the basis for a useful tool although binary scaling such as 64 made the device inconvenient to read. The development of a scale-of-10 or decade counter by Victor Regener in 1946[50] much simplified the arithmetic and these replaced the binary scalers over the next 10 or so years.

A step forward in simplicity came with the Dekatron tube in the early 1950s. These gas discharge tubes used a set of electrodes designed so that pulses would transfer the glow from one electrode to the adjacent one. With a circle of ten electrodes the glow could be made to hop round in a circle, so counting to ten, and then for each tenth pulse to be transferred to the next tube in the sequence. Decades could then be counted with relatively simple external circuitry and the tubes were widely used until perhaps the 1980s.

As an alternative to the counting circuit Robley Evans in 1939[51] designed the ratemeter: a device subsequently more used in health physics survey equipment than counters. Originally produced in response to the unreliability of mechanical counters, the ratemeter proved a simple, robust and reliable alternative for many applications. It worked on the principle that, if each pulse from a detector was made to place a fixed charge onto a capacitor, the average leakage current through a resistor placed across the capacitor would be proportional to the pulse rate. The statistical variations in the rate of arrival of pulses would cause fluctuations but, with a suitable choice of capacitance and resistance, these could be limited—at the expense of the system not responding promptly to rate variations.

Another circuit of considerable importance was the Schmitt trigger[52]. This circuit gives an output only when the input pulses have an amplitude greater than a set (but adjustable) threshold value. It formed the basis for pulse height discrimination and pulse height analysis. The spectrum of input pulse heights (and thus particle energies with suitable detectors) can be found by varying the threshold voltage of the Schmitt trigger and making a series of counts. This takes some time and a more efficient device, the kicksorter or multichannel analyser, counts the pulses in a series of small, adjacent voltage ranges or channels to produce a pulse height spectrum. The first practical kicksorter of 12 channels is credited to Otto Frisch in Liverpool, England about 1944 but modern devices derive more from the work of D H Wilkinson at AERE, Harwell and are based on a linear ramp and Analogue to Digital Converter. With this kind of technology 100 channel devices were in use by 1951 and by the mid-1950s 2048 channels kicksorters were developed.

If routine measurements are to be made of alpha surface contamination then thin window chambers are essential. Since thin-window Geiger counters were not available in the 1940s the measurements were made with thin-window unpressurised ionisation and air proportional counters. These were of low sensitivity and unreliable—especially the proportional counters—and certainly demanded a stable power supply. The development of a stabilising circuit by Overbeck at Oak Ridge led to "Zeus"—a battery-powered ionisation chamber monitor for surface alpha, beta and gamma contamination—and the later more sensitive "Zeuto".

At Hanford a proportional counter was used but, to achieve voltage stability, this was AC powered and the weight meant that it had to be trolley-mounted. It had both a scaler and an audio output; the popping sound from the latter giving it the name "Cart Poppy". The Cart Poppy was still in use at Hanford in the 1960s. Proportional counters were also used at Berkeley and Los Alamos, initially as the 100kg "Pee Wee" but the scintillation "Poppy" was available for alpha contamination measurement at Hanford. This was based on a ZnS(Ag) scintillator and a speaker that "popped" when each particle was detected. A ratemeter was added later.

A version of the proportional counter has found wide use as a thermal neutron detector. The BF3 counter is a cylindrical counter made of aluminium, brass or copper with a central wire anode and filled with boron trifluoride gas at a pressure between half and one bar. The boron reacts with a thermal neutron to give an alpha particle.

$$B\text{-}10 + n > Li\text{-}7 + \alpha + 2.3 \text{ MeV}$$

The alpha particle and the recoil Li-7 ion then produce ionisation and the BF3 doubles as the detector gas with the tube operating in the proportional counter region. In modern tubes the boron is usually enriched in B-10 for higher efficiency. The boron reaction occurs only at thermal energies but the device can be made to detect fast neutrons by slowing them down with a moderator. The BF3 counter was used widely by the end of the 1930s[53].

An alternative filling for neutron detection is helium where the reaction

$$He\text{-}3 + n = H\text{-}3 + p + 0.76 \text{ MeV}$$

Although more expensive that the BF3 it is more sensitive and more stable. Helium-filled G-M tubes were used in the 1930s for neutron detection[54].

The fast neutron detector "Chang and Eng" (Figure 13) was developed for the Manhattan Project at Chicago by K Z Morgan and others. The following is from an interview with Morgan in 1995[55].

> Dr. Gamerstfelder and I, along with some help from Parker, developed what we called a "chang and eng." (This instrument consisted of) two small cylinders; one was filled with nonhydrogenous gas, like argon, and the other with gas like hydrogen or methane (CH_4). As you know, neutrons don't produce ionization along their path because they have no charge, and their only ability to cause ionization is when they strike one of the nuclear components—that is, a proton or a neutron or a collection of nuclear particles.
>
> So, with two chambers—one filled with gas containing hydrogen, the other with no hydrogen—and having them under pressure to give a large cross-section, we measured the differential output of these two chambers. I could measure (accurately) the neutron contribution from fast neutrons. Now, these chambers were very effective and very quantitative in their evaluation, so I used them later in some experiments in Oak Ridge.

The fill gases were normally argon and butane at something over atmospheric pressure and a Lindemann electrometer was used to measure the difference between the two ionisation currents. This was balanced out in a gamma field by adjusting the pressures of the gases. Then, in a neutron or mixed field, the current difference represented the fast neutron flux. This measurement was converted into reps. The device weighed about 35 kg; its name came from the conjoined twins Chang and Eng Bunker who were taken to the USA from Thailand (then Siam) in 1829 and who were the original "Siamese twins".

The BF3 and He detectors described respond to thermal neutrons and their sensitivity drops very rapidly with energy. This can be countered to some degree by moderating the neutrons before they reach the detector. The "Long Counter" described by Hanson and McKibben[56] in 1947 consists of a cylindrical paraffin moderator (diameter 20 cm and length 36 cm) surrounding a 34 cm long BF3 counter. A thick boron-loaded (or lined) paraffin shield is arranged around this so only neutrons arriving at the front face, along the detector axis, are detected. The response was found to be independent of energy, within a factor of two, over an energy range from about 20 keV to 20 MeV. The designs have been improved and the response flattened and they are widely used a secondary standards for neutron fluence measurement. The directional response and size (mass about 45 kg) has limited their use in routine protection.

Figure 13: The Chang and Eng neutron dosemeter
[ORAU]

The devices most widely used for routine neutron surveys are so-called rem counters where a detector sensitive to thermal neutrons is placed at the centre of a moderating sphere or cylinder which contains, like the long counter, a thermal neutron shield. They give, like the long counter, a response independent of energy but are (or aim to be) non-directional and more compact. The rem counter devised by Andersson and Braun described in 1963 and 1964[57] is perhaps the most widely used model still. A central BF3 counter is placed axially in a cylindrical polyethylene moderator, this is surrounded with a cadmium or boron-loaded layer with perforations, acting as a thermal neutron shield, and there is then a further cylindrical layer of polyethylene. The outside dimensions of the cylinder are about 22 cm diameter and 25 cm long. Fast neutrons will mainly be slowed down within the central moderator so they are not attenuated by the shield. By adjustment of the shield perforations a response can be achieved which is proportional to dose equivalent over a wide range of neutron energies.

A detector based on the same principle having a 25.4 cm sphere with 4x4mm LiI(Eu) scintillator at the centre was devised by Hankins in 1962[58] and a device constructed by Ladu at about the same time used a hollow paraffin sphere with a central BF3[59].

The Leake type counter, designed by John Leake of AERE Harwell in the late 1960's[60], had a central spherical He-3 counter surrounded by a polyethylene layer, then a perforated natural cadmium shell and then an outer polyethylene layer. The principles of operation are the same as the Andersson and Braun device. The diameter was 21cm and the design remained in use with little alteration to the end of the century. The paper by Rogers[61] compares the different types.

The recognition that it was necessary to measure the doses received by individual workers demanded new instruments that had new and demanding constraints: they had to be small and light as well as being reliable and reasonably accurate. A number of solutions evolved.

One of the best-known personal dosimeters was the Victoreen Minometer, a charger-reader developed in the late 1930s and made for several decades. Shown in Figure 14, the ionisation chamber was about the size of a pen and could be plugged into the side of the charger-reader to be read with a string electrometer. This was an indirect reading dosimeter—it had to be plugged into the reader to see the dose.

9 Measuring External Radiation

Figure 14: The Victoreen Minometer
[ORAU]

Direct reading dosimeters had appeared earlier: the first was made by C C Lauritsen in about 1932 (Figure 15). Here, while the external charger was needed, the electroscope, about the size of a pen, had a built-in microscope. It could be worn in the pocket and checked at any time. These dosimeters do not seem to have been put into production and been widely available until the early 1950s.

Figure 15: The Lauritsen Dosimeter [ORAU]

After this, a number of companies (including Tracerlab, Landswerk, Victoreen, Cambridge Instruments and Bendix) made direct reading pocket dosimeters and these gradually replaced the indirect-reading type. Sensitivities of around 0-200 mR were available but higher range models reading to several R were also produced. The simplicity, robustness and low cost of pocket dosimeters meant they continued in widespread use (under the acronym QFD for quartz fibre dosimeter) until near the end of the century.

An alternative to the electroscope was the ionisation chamber with electronic circuitry attached—a miniature version of a hand-held instrument. An early example is that of P R Bell from around 1950. This had an electrometer valve and an alarm was triggered by high dose rates[62]. Such systems could be made more compact as semiconductor technology improved and would replace the QFD—but only after a long time.

In the first decades of the 20th century a number of radiologists advocated the use of films to check for excessive exposures. For example, in 1922 Rollins, the x-ray pioneering dentist, used film to test x-ray shielding and in 1907 Rome Wagner[63] recommended x-ray workers carry a film to check for excessive radiation exposure and no doubt a number did. If they did, it may not have helped them much because the exposures required to blacken the film were large enough to present a severe long-term health risk. There were anyway at least two problems that needed to be solved before films could form the basis for a reliable dosimetry system. The first was to establish the relationship between exposure to x-rays and blackening and it was not until about 1920 that this was done by German and British workers to give some correlation between the two.

The second problem with film is that its response is—unlike tissue —very energy dependent. The energy dependence arises from the high atomic numbers of the silver and bromine which can lead to responses to the same exposure being 50 times greater at around 100 keV than to gamma rays of 1MeV or more.

The credit for the first use of films for routine monitoring of exposures is generally given to George Pfahler, a Philadelphia radiologist, who recommended the use of film packets for x-ray and radium workers in 1922[64]. These dental film packs were to be carried in the breast pocket and developed every two weeks. In 1926 Edith Quimby, a New York medical physicist, was the first to devise a practical film badge, based on strips of x-ray film, which addressed the problem of energy dependence by the use of compensating metal filters and she set up a monitoring programme within her laboratory[65]. The replacement of the strips with small readily available dental x-ray packs produced a system that would develop and be the mainstay of personnel monitoring for many years.

Following the definition of the rontgen Heinrich Franke working in Germany in 1928 linked the degree of blackening (the optical density) of film to its exposure in rontgens[66]. This was extended by Bouwers and van der Tuuk in Holland in 1930[67] to doses well below the then-current daily exposure limit. They also described a multi-filter film badge. However, although the technical basis for a film badge dosimetry system was in place by perhaps 1930 there remained practical difficulties arising from the need to apply energy compensation and for consistency in processing and calibration. One of the first documented large-scale film-badge programmes was in the UK.

The UK National Physical Laboratory had been responsible for the voluntary programme of inspection of hospital x-ray equipment that followed the 1921 protection recommendations[68]. The Laboratory had started regular film badge monitoring of the staff of the Radiology Division in 1937 using dental film[69] after work that established the value of photographic film as a dosimeter for both x-rays and gamma rays[70] Prompted by the arrival of American
diagnostic x-ray equipment for use in emergency hospitals, the Laboratory was asked by the Ministry of Health to set up a film badge service and this began, using dental x-ray films, in November 1942. Industrial users of radiation were added the following year and, by the end of 1945, 2000 hospital staff in 500 hospitals and 1000 workers in 160 industrial workplaces were supplied by the service. The industrial users were engaged in either radiography or radium luminising and the films issued to   luminisers were covered at one
end with 1 mm lead filter to give an indication of beta exposures. The films were used to trigger more detailed surveys and these revealed any defects and poor working conditions[71]. By 1947 the service was issuing about 10,000 films annually and by 1950 20,000. By 1956 the Laboratory was issuing 50,000 film badges a year and the service was transferred to the new Radiological Protection Service.

Whether the NPL was first or not, the most significant step in the development of the film badge came with the widespread use of filtered film badges at the Met Lab at Chicago, part of the Manhattan Project, from 1943. Each badge had an open window and 1 mm cadmium filter to compensate for the higher response of the film (DuPont dental x-ray film 502) to lower energy photons [72]. The design was patented (US Patent 2 483 991) and was in use at ORNL until the early 1950s when it was replaced by a badge that combined a more complex dosimeter with an identification badge.

The badge worn by Ernest Wollan, the head of radiation protection at the Met Lab in Chicago is shown in Figure 16. Figure 17 is an exploded view of its contents.

Figure 16: Ernest Wollan's film badge [ORAU]

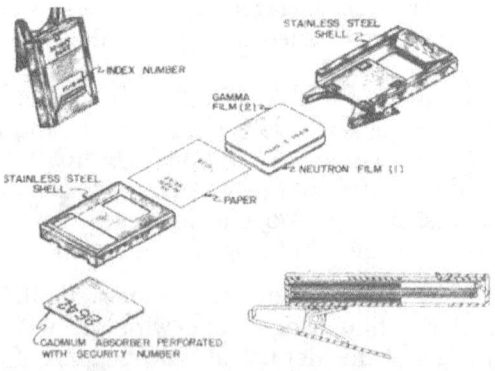

Figure 17: Components of an early film badge [ORAU]

9 Measuring External Radiation

The film (DuPont 553) used in the replacement was a pack of three emulsions[73] and there were now three circular filters of lead, cadmium and copper. The cadmium and copper filters could be used to estimate any accidental exposures to neutrons but there was also an NTA film (see later) for fast neutron measurement. The further development of the ORNL badge resulted in the Model 2 in 1960 (the 1958 Model 1 was quickly replaced).

In the Model 2 the same film pack was used but a more elaborate system of filters was used to compensate for different gamma energies and to allow estimation of beta exposures. At the back of the badge were two filters, one plastic and the other a plastic lead sandwich. At the front there were four: a thin plastic window, a thicker plastic filter, a plastic/aluminium sandwich and then a 5-layer laminate of plastic, cadmium, gold, cadmium and plastic. The badge incorporated an Eastman Type A emulsion for routine neutron exposure. For accidental exposures there was a S-32 pellet, an indium foil and a gold foil. There was also provision for a silver phosphate glass dosimeter and a chemical dosimeter[74].

There are variations on this in the ORAU museum. From the late 1940s comes an example with just cadmium filtering and the Tracerlab stamped-tin badge with a single film pack (DuPont 552) with filtration by lead and cadmium sheets wrapped around the ends of the film pack. In the early 1950s Gardray produced an unfiltered badge with two DuPont films ranges 2-35 R (Type 555) and 90-700 R (Type 834). From the mid-1950s there is a plastic badge with three filters (Cd, Al, Pb) front and back by Nuclear Chicago. At Hanford in the 1960s a badge was being developed with iron and tantalum filters that incorporated two radiophotoluminescent (RPL) "fluorods" for accident dosimetry as well as activation foils and a sulphur rod for criticality accident dosimetry[75].

So the general trend in the USA was from the simple film badge of the MetLab with just one filter and one film in 1943 to double or triple films for x-rays and gamma-rays with several filters by the mid-1950s. By 1960 the ORNL badge had a sophisticated set of filters that took the photon element of the film badge about as far as it could reasonably go and permitted estimates to be made of beta exposure. By the mid-1950s badges also incorporated several means for neutron measurement. The ORNL badge included an NTA emulsion and cadmium and copper foils; the Model 2 had more still.

The development in the UK followed a similar path. Dolphin et al[76] describe the film badge in use at Harwell in the late 1950s. This had an exposed area and two 1 mm thick filters of tin and cadmium. The tin filter compensated for the higher sensitivity of the film to lower energy photons. With the tin filter the sensitivity varied by no more than +/- 20% between 0.15 and 1.0 MeV but there was hardly any response below 70 keV. The cadmium filter was used for the detection of slow neutrons: slow neutrons captured by the cadmium result in gamma radiation that was detected by the emulsion underneath. Since the photon attenuation of the tin and cadmium are similar, extra blackening under the cadmium is due to neutrons. An unshielded window in the film badge allowed low-energy gammas and betas to strike the film directly but it was not possible to distinguish between the two radiations. A plastic film badge that would allow this was under development.

Dresel and Wachsmann[77] in Germany devised more complicated film badges in the late 1950s and 1960s. One used 3 copper filters of different thicknesses (0.05, 0.5 and 1.2 mm) and a lead one of 1 mm. This allowed the film to be used for the energy range from 5 keV to several MeV. Two emulsions in the holder had different sensitivity ranges: one covered 0.05 to 20 r and the other up to several hundred r. Another contained a track film and Sn/Cd filter pairs for neutron dosimetry [78].

An alternative to filtration for energy compensation, briefly considered in the 1950s, was to sandwich the film between layers of organic scintillator so that it picked up the emitted light. With careful matching, the increasing response of the scintillator with energy could compensate for the falling response of the film to ionising radiation. Although it could be made to give good compensation, there were problems with dose rate dependence and the technique was dropped.

By 1960 the double emulsion Kodak RM film had become available. One side could be used to measure doses in the range 0.02 to 100r while to other could cover up to 1000r so that it could record any serious accidental exposures that might saturate the other emulsion.

The construction of a practical, energy-compensated film badge was achieved substantially by the 1960's—although minor improvements continued to appear for many years. However, equally important was the achievement of a standardised film processing system that was reproducible and could derive the information available from the filtration. This built on the enormous experience of processing x-ray plates and films to handle the many thousands of films regularly produced in the Manhattan Project dosimetry programme.

Film has its limitations. While, once developed, it is a permanent record of exposure, it fades between exposure and processing. Convenient in its highly refined badge, it requires extremely careful quality management of the processing. It could not readily be made to give a quick result although Frame[79] has described a system developed by Land of the Polaroid Corporation in the mid-1950s where a self developing film could be developed on the spot by a worker. He crushed the self-developing film pack and placed it in his mouth so the processing took place at a well-defined temperature. After the specified time the pack was removed and the blackening could be measured. It never caught on.

The use of photographic emulsions for fast neutron dosimetry was pioneered by Cheka in the late 1940s and early 1950s using nuclear track emulsions produced by Kodak. Of these the most useful was the Personal Monitoring Film Type A (usually known as NTA) and this was widely used. This type of emulsion has a higher concentration of silver halide, smaller grains and is much thicker (around 30 µm) than a conventional x-ray emulsion. The particles detected are principally the recoil protons produced in the gelatine which leave tracks that can be counted using a microscope at magnifications between about 500 and 1000. The counting was generally performed manually and it was a slow and exacting task which meant that a technician could count no more than about 50 films a day.

When strongly ionising particles pass through some insulating materials they produce a more-or-less permanent track of damage. It is sometimes possible to make this track, when at the surface, visible through a microscope as a so-called etch pit by chemically etching the material and this can be enhanced with electrochemical etching.

The technique was discovered by Young in 1958[80] when he revealed fission tracks in LiF after chemically etching. Silk and Barnes[81] discovered similar tracks in mica the following year but the next significant step, from a dosimetry viewpoint, was the work of Walker, Price and Fleischer in 1963[82] showing the use of mica, glass and polycarbonate as possible materials. Over the next ten years phosphate glasses were trialled, with thin fissionable films and doped with fissionable materials, as neutron detectors. However the discovery of Becker in 1966 that recoil and (n,α) reactions could be used for neutron dosimetry brought a new direction in the research. Cellulose nitrate was widely used in the form first of plates and then Kodak-Pathe films CA 80-15 and LR115. Boron-doped films were developed for thermal and intermediate neutrons and systems for automatic counting of the etch pits were developed. However, the significant advance came from a new detecting material[83].

The plastic CR-39 was developed in the early part of the Second World War when the Pittsburgh Plate Glass Company set up a project to investigate clear resins. Plexiglass and Lucite had already been developed by other companies. PPG (as it later became known) placed the project with a subsidiary, the Columbia Southern Chemical Company, and called it "Columbia Resins". By May 1940 an allyl diglycol carbonate monomer was discovered and used to make a resin which was trademarked under its project batch name—CR-39®. A number of uses, particularly in aviation, were found for the compound during the war, taking advantage of its property of combining well with other materials.

Post-war CR-39—sometimes known under its chemical name polyallyl diglycol carbonate (PADC)—gradually established itself as a material for making plastic lenses (this accounted for more than 80% of output) and was found to be a suitable material for track etch work in 1978[84]. It was recognised as a potential material for personal, higher-energy neutron dosimetry[85] because of its ability to register recoil protons, its sensitivity and its wide energy range from around 0.1 MeV to 15 MeV. It also had a very smooth surface and the clear etch pits that could be produced made it very suitable for automatic counting. It remained the most used material to the end of the century.

Etching regimes have developed. Etching at elevated temperatures was found to give quicker results and an electrochemical   process   improved   the sensitivity. Methods for automatic recording of etch pits included the spark-through method of Cross and Tommasino (1972) and techniques similar to the recording of biological colonies. A different approach has been the measurement   of radiation dose (through pit number density) by
measuring the light scattered by the etched foil. This was originated

in the late 1960s, perhaps by Schultz[86]. The possibilities for development as a practical system for routine neutron dosimetry, based on CR-39, are illustrated by the LITES system proposed for LANL[87].

The bubble detector works in a similar way to the bubble chamber of high energy physics. A liquid with a normal boiling point below room temperature is kept under pressure. When the pressure is released the liquid becomes super-heated and bubbles form along the path of ionisation of any charged particles that pass through it. Under normal circumstances this period of sensitivity would be quite short but where superheated droplets are dispersed in a suitable elastic solid it can be much longer. This was discovered by Apfel in 1979 and used in the bubble detector for neutrons described by Ing and Birnboim[88] in 1984. In this detector small droplets of freon were dispersed in a transparent acrylamide polymer that prevented them from vaporising and kept them in position. The passage of a charged particle then causes bubbles that expand to a size that can be readily counted optically to give a dosimeter that is sensitive and easily read. The detectors initially suffered from temperature dependence and shock sensitivity but these problems were generally solved and by the end of the century they were available as routine neutron dosimeters that could be supported with automated reading.

When some solids are exposed to radiation the free electrons produced may be trapped at imperfections or impurities and retain some of the energy they received from the radiation. When warmed so the thermal energy is enough to release them from the traps, the electrons return to their original state with the release of energy and a proportion of this energy may appear as visible or ultra-violet light in a transient glow. The first proper record of thermoluminescence (as the glow is called) is from an address to the Royal Society by Robert Boyle in 1663. He, for a reason never established, took a diamond to bed with him and found that, when it touched his naked skin for a time, it glowed. The stored energy in this case came from background radiation and had been accumulating over many thousands of years.

Many natural substances show the effect and in the early years of radiation Wiedemann and Schmidt found thermoluminescence caused by cathode rays. They also investigated systematically the activation effect on thermoluminescence by impurities in systems like $CaSO_4$:Mn. The first glow curves (the varying light output as the material is warmed up) observed were probably those seen by Przibram in Vienna in 1922-3. The understanding of the underlying processes grew as the understanding of solid state physics—and luminescence phenomena in particular—grew in the first half of the century. The specific shapes of the glow curves seen was explained theoretically by Randall and Wilkins[89] in 1945 for first order processes and extended in 1948 by Garlick and Gibson.

The most widespread thermoluminescence dosimeter (TLD) material appeared when, as a result of the work in the late 1940s and early 1950s by Professor Farrington Daniels and his team at the
University of Wisconsin, lithium fluoride was discovered.

A brief account of at least some of the work has been published[90] by Louis Heckelsberg who, from 1948, worked with Daniels studying the glow curves of alkali halide crystals after excitation with x-rays. This work led to Daniels proposing a thermoluminescence dosimeter based on NaCl that could measure the high doses from bomb detonation ("up 1000 Roentgen units") with a very simple system using an inflammable pellet to heat the crystal and relying on estimation of the glow intensity by eye. It was soon recognised that LiF would be the basis for a much more sensitive dosimeter ("half a Roentgen unit could be detected readily"). Heckelsberg then persuaded Daniels to abandon the flammable pellet idea and developed an electrical heater system based on cementing the LiF chip onto an aluminium disc and placing this on a heater made from a car cigarette lighter. It was more compact than the experimental rig he had been using and several hundred measurements were made with it in early 1951. Heckelsberg was awarded his PhD in June 1951.

Daniels and his team then appear to have examined pressed pellets of LiF made by Harshaw Chemical Co but abandoned the material for Al_3O_3 because of lack of interest (and the complicating low temperature peak). The lack of interest in TLD was demonstrated when Daniels was told by Dr. Ernst Pohle, Chairman of the Radiology Department at the University, that there was no need for a new dosimeter since the Victoreen r-meter was adequate for all their needs.

Daniels stopped work on thermoluminescence dosimetry altogether in 1956 and in 1960 gave John Cameron, a medical physicist at the university, his remaining LiF so that he could continue the work. This Cameron did and found the material more satisfactory than the radioluminescent glass rods he had been using. In 1961, when his stock was nearly exhausted Cameron ordered new supplies and discovered that Harshaw had improved their refining process so that the (unknown) activating impurities essential for thermoluminescence had been removed. The Wisconsin team then attempted to grow their own doped LiF crystals, using measurements on impurities in their old LiF powder as a guide. While they found (through a lucky accident) the value of titanium, they soon turned to cooperation with Harshaw who developed a TLD material with titanium and magnesium activators that Harshaw patented as TLD-100. It became the most widely used low Z TLD material.

Natural lithium contains 92.5% Li-7 but the other isotope, Li-6, gives tritium and alpha-particles when neutrons interact with it. TLD chips are made with enrichment in both the isotopes. TLD-600 contains 96% ^{6}LiF while TLD-700 contains 99.9% of ^{7}LiF. TLD-600 is thus sensitive to betas, gammas and neutrons while TLD-700 to just betas and gammas. Used in pairs, the difference in their responses gives a measurement of neutron doses. They are frequently used in an albedo arrangement.

Lithium fluoride has a number of undesirable characteristics: it is non-linear at very high doses and it requires a fairly complex annealing procedure. Other low Z materials have been investigated. Lithium borate activated with manganese was investigated by Schulman et al 1967. Like LiF, it is approximately tissue equivalent and has a wider, more uniform, energy response than LiF but is more sensitive to thermal neutrons. Beryllium oxide, similarly activated, was also studied about the same time by several workers.

Other higher Z materials have been investigated and used as dosimeters. For example, $CaSO_4$:Mn had been used since the 1930s as a uv dosimeter in rocket experiments and was taken up in early medical TLD studies in the mid-1950s. Other activators were tried including dysprosium (Dy) and thulium (Tm) and some of the problems of instability that had been seen were solved[91]. Similarly, the response of calcium fluoride to x-rays had been studied as early as 1928 by Wick and Slattery and CaF_2:Mn was introduced as a dosimetry material in the late 1950s. A system based on sealed CaF_2:Mn was described by Schulmann et al in 1963. The material TLD-200, CaF: Dy, is much more sensitive than LiF but, with the higher Z constituents, requires energy compensation before it can be used in personal dosimetry systems. It has been used in several environmental dosimetry systems.

The materials are but a part of the story of thermoluminescence. The development of practical read-out systems has been at least as important. The very early work was done with a modified soldering iron as the heater and Lucas[92] has described how the early dosimeters, developed largely for military use, used the thermoluminescence material encapsulated in a glass bulb in contact with an electrical heating element. The system developed at MBLE (Manufacture Belge de Lampes Electriques)in Brussels was widely used   in   European   armies. The Harshaw system, CP-1112/PD, remained in use by the US Navy to the end of the century.

Lucas divides the development of systems into three phases. In the first, in the middle 1960s, the heating of the sample was controlled only through the current passing through a heating coil. By the early 1970s readers were appearing with feedback for better temperature control but the mechanical arrangements were relatively simple for acceptable reliability. Problems associated with the warming of the read-out PMT were solved, optics were improved and the thermoluminescence material was, in some systems, sealed in Teflon film. Later in the 1970s systems became available that could read chips automatically and these became the basis for replacing film badges with TLDs in a number of organisations. In these systems the thermoluminescence material was generally encapsulated in Teflon and in some the heating was by hot gas. For some the mechanical aspects of reliably handling the dosemeters was the greatest challenge but, as they evolved further, the readout was frequently controlled by microprocessors and the data fed directly to microcomputers.

Systems developed by large organisations were perhaps more influential in establishing TLD as an alternative to film badges than the commercial systems. In the UK for example the service developed by the NRPB, and used by many organisations in the UK, employed LiF/Teflon cards and was almost completely automatic from receipt of a exposed card in the mail to production of a dose record. Such systems swept away the film badge for many routine applications.

Some materials luminesce when exposed to ultraviolet light but others only do so after previous exposure to radiation. The mechanism is that radiation causes the creation of new permanent luminescence centres in the material and these are excited to luminesce by the uv light. Since they are permanent centres they can be stimulated to luminesce any number of times. The effect was known from about 1912 and studied by Przibram (who called it radiophotoluminescence or RPL) in the 1920s and 1930s mainly using minerals and synthesised inorganic compounds with rare earth impurities.

Much later—around 1950—it was found that dilute solutions of silver salts in some alkali halides and certain glasses gave a strong orange emission after exposure to radiation, with an intensity proportional to dose, when excited with 365 nm Hg uv light. Studies of such glass systems led to the first mass-produced system based on a metaphosphate glass with composition 44% $Al(PO_3)_3$, 23%$Ba(PO_3)_2$, 23%KPO_3 and 8% $AgPO_3$ and some 4 million of these dosimeters were made for the US Navy in a form with a perforated lead filter to compensate for over-response to low energy photons. They were re-readable—because the radiation induced colour centres are permanent—and stable. The sensitivity was not adequate for routine purposes (minimum detectable dose about 0.1 Gy) but the system was robust and simple and suitable as a casualty dosimeter for accident or wartime dosimetry. Similar systems were developed in the UK, Sweden and elsewhere for military use. The US system was just being phased out at the end of the century.

The reliability and the possibility of re-reading of RPL dosimeters led to efforts to improve the sensitivity and by 1961 glasses had been developed which could read down to 0.1 mGy and were suitable for routine monitoring. However, by then the investment in film badge technology was so great that RPL was not adopted.

Optically-stimulated luminescence (OSL) is rather similar to thermoluminescence except the trapped charges are released not by heat but by exposure to light. The only material currently used for dosimeters based on this principle is aluminium oxide Al_2O_3. The Al_2O_3 is grown so that defects are produced and these form the trapping sites for the electrons and holes released by the radiation. When the electrons and holes are remobilised by irradiation with light, they recombine at defect centres with characteristic light emission.

The light emission—proportional to dose—is detected by a PMT. There is clearly the possibility for confusion between the stimulating light and the luminescence. This is avoided by either pulsing the stimulating light and measuring between the pulses or measuring light only outside the emission band of the stimulating light. The Landauer Luxel system is the only one known to use OSL.

Several other detectors have been considered as personal dosimeters and while some have seemed to have potential for a while, they have fallen by the wayside. Thermally Stimulated Electron Emission (TSEE)—based detectors might be taken as an example.

When some materials are irradiated the electrons released by the radiation become trapped at defect centres at the surface of the materials. If the material is heated or exposed to light then these electrons (called exoelectrons) may be released and can be counted.

A proportionality between the electron emission and radiation dose in calcium sulphate was first noted in 1957 by Kramer but early results as a dosimeter were disappointing because of non-linearity in response, difficulties of reproducibility and the rapid fading that occurred. Some improvements in linearity and reproducibility were achieved during the late 1960s by the incorporation of carbon into the materials and by the time of Becker's review in 1973 the method seemed to hold considerable promise. Its advantages over TLD at that time included its sensitivity, the wide range of materials potentially available, and their particular suitability, because the detecting layer is thin, to short-range radiations.

Methods for detecting the exoelectrons included the use of gas flow G-M counters, proportional counters and ionisation chambers with the TSEE material being heated within the gas. In some approaches the exoelectron collection was improved with an applied electrical potential. Materials used included $CaSO_4$, ceramic BeO and $SrSO_4$. The heating regimes were similar to those of TLD but some workers used light stimulation to release the exoelectrons with the advantage that the surface could be selectively interrogated. When Becker wrote in 1973 he could find only one commercial system. He noted also that there were unsolved problems: memory effects were experienced with some materials and there was non-linearity with higher dose rates and heating levels.

9 Measuring External Radiation

In fact one of the more promising developments was the electronic personal dosimeter using a semiconductor detector and with quite sophisticated electronics built in. Such systems were beginning to be developed and accepted as statutory dosemeters as the century closed.

Arguably not much has happened in basic measurement equipment since the 1980s[93]. This might no be too surprising, simply reflecting the fact that all detectors rely on the same physical process as the one that causes damage in humans: the deposition of energy and ionisation. There have thus been no striking new detectors and electronics have probably not significantly increased the functionality of individual instruments. However, there have been important advances in ergonomic design, robustness and, particularly, reliability and these have been achieved with tight cost control. While these may not have the glamour of some of the earlier discovery and development work, they have given the health physicist a stable of instruments that meets the evolving needs of the profession.

Notes Chapter 9

1. (Duane W, 1922, 1923)
2. (Fricke H and Glasser O, 1925; Fricke R and Glasser O, 1925)
3. (Solomon I, 1926)
4. (Kathren R L, 1980)
5. (Taylor L S, 1967)
6. (Lauritsen C C and Lauritsen T, 1937)
7. (Metcalf G F and Thompson B J, 1930)
8. (DuBridge L A, 1931)
9. (Kathren R L, 1980)
10. (Frame P W, 2004)
11. (Farmer F T, 1942)
12. (Townsend J S, 1901)
13. (Rutherford E and Geiger H, 1908)
14. (Geiger H and Klemperer O, 1928)
15. (Taylor L S, 1967)
16. (Simpson J A, 1948)
17. (Taylor D, 1950)
18. (Geiger H and Muller W, 1928, 1929)
19. (Neher H V and Harper W W, 1936; Neher H V and Pickering W H, 1938)
20. (Trost A, 1937)
21. See (Present R D, 1947)
22. (Kathren R L, 1980)
23. (Taylor D, 1950)
24. (Birks J B, 1964)
25. (Bay Z, 1938)
26. (Rajchman J A and Snyder R L, 1940)
27. (Blau M and Dreyfus B, 1945)
28. (Curran S C and Baker W, 1948)
29. (Coltman J W and Marshall F H, 1947)
30. (Broser I and Kallmann H, 1947; Kallmann H, 1949)
31. (Deutsch M, 1948)
32. (Bell P R, 1948)
33. (Hofstadter R, 1948, 1949)
34. (Sommer A and Turk W E, 1950)
35. (Mandeville C E and Scherb M V, 1950)
36. (Kallmann H and Furst M, 1950, 1951, 1952)
37. (Reynolds G T, Harrison F B and Salvini G, 1950)
38. (Raben M S and Bloembergen N, 1951)
39. (Newton Hayes F, Hiebert R D and Schuch R L, 1952)

40 (Hiebert R D and Watts R J, 1953)
41 (Schorr M G and Torney F L, 1950)
42 (G Jaffe, 1932)
43 (Van Hardeen P J, 1945)
44 (McKay K G, 1951)
45 (McKenzie J M, 1979; Frame P W, 2004)
46 (Greinacher H, 1924)
47 (Dunning J R, 1934)
48 (Wynn-Williams C E, 1932)
49 (Eccles W H and Jordan F W, 1919)
50 (Regener V H, 1946)
51 (Evans R D and Alder R L, 1939)
52 (Schmitt O E, 1938)
53 (Korff S A and Danforth W E, 1939; Korff S A, 1946; Kathren R L, 1980)
54 (Kathren R L, 1980)
55 (Morgan K Z, 1995)
56 (Hanson A O and McKibben J L, 1947)
57 (Andersson I O and Braun J, 1963, 1964)
58 (Hankins D E, 1962)
59 (Ladu M, Pelliccioni M and Rotondi E, 1963; Ladu M, Pelliccioni M and Rotondi E, 1965)
60 (Leake J W, 1967, 1968)
61 (Rogers D W O, 1979)
62 See (Taylor D, 1950)
63 See (Frame P W, 2004)
64 (Pfahler G E, 1922)
65 (Quimby E H, 1926)
66 (Franke H, 1928)
67 (Bouwers A and van der Tuuk J H, 1930)
68 (Glasser O(ed), 1933)
69 (Smith E E, 1975)
70 (Bell G E, 1936a, 1936b)
71 (Smith E E, 1975)
72 (Pardue, 1944)
73 (Lehman R L, 1961)
74 (Gupton E D, Davies D M and Hart J C, 1961)
75 (Fitzgerald J J, 1969)
76 (Dolphin G W, Megaw W J and Rundo J, 1962)
77 ibid
78 (Becker K, 1973)

79 (Frame P W, 2004)
80 (Young D A, 1958)
81 (Silk E C H and Barnes R S, 1959)
82 (Walker R M, Price P B and Fleischer R L, 1963)
83 (Tsuruta T, no date)
84 (Cartwright B G, Shirk E K and Price P B, 1978)
85 (Benton E V and Oswald R, 1980)
86 (Schultz W W, 1968)
87 (Moore M E, 2002)
88 (Ing H and Birnboim H C, 1984)
89 (Randall J T and Wilkins M A F, 1945)
90 (Heckelsberg L F, 1980)
91 (Becker K, 1973)
92 (Lucas A C, 1993)
93 (Bronson F, 2006)

10 Measuring Internal Activity

The measurement of internal doses needs three things: an understanding of how much of a radionuclide has entered the body, how it has spread itself over time around the organs and what dose thus results. The questions of metabolism and dose were discussed in Chapter 6. Now we look at the methods for assessing how much radioactivity has entered or remains in the body after inhalation—the most significant route for those people with higher intakes. Detecting what goes in is the key function of air sampling; body monitoring and excretion monitoring aim to measure what is there.

The two essential stages in air sampling are collecting a sample and then finding how much activity in contains by counting it. Sampling was established for industrial hazards before the Manhattan Project was set up and samples were collected by various techniques including impingers, where the particles were driven against a surface by air flow and stuck to it, and electrostatic precipitators, where electric charge was used to attract particles. The standard impingers of the time were the koniometer, where the particles were driven onto a plate coated with glycerine, and the Greenburg-Smith impinger, developed in the 1920s by the US Bureau of Mines where the surface was a wetted glass slide. According to Kathren[1], electrostatic precipitators were initially the preferred collection methods in the Manhattan Project because dust was simply collected onto a dry surface and was easier to process for counting. Indeed, given that the collection was onto the inside of a precipitator tube, it was quite possible to make an immediate rough measurement by inserting a counter into the tube. More usually though the samples were taken to a laboratory for analysis.

Samples were also collected by evacuated flask ("can" or "pig") which could be opened to collect a sample (particulate or radioactive gas) and then taken off for analysis but the most significant development was probably the use of filter papers to collect samples —particularly for beta-gamma monitoring. While this technique had been used before in industrial hygiene it had not lent itself to chemical analysis as well as the other available methods: the filter paper was a confusing factor. However, it was near perfect for radioactive samples which could easily be counted by placing a detector in front of the paper. With a simple system based on a vacuum-cleaner motor the health physicists had a robust and reliable device at their disposal and such systems were widely used for routine and spot sampling. By 1944 a continuous air monitor was developed for beta-gamma activity with recording of count rate on a chart recorder.

Electrostatic collection methods remained popular for alpha-particles (largely because of concern about the attenuation caused by the filter paper) but, as Kathren points out, the equipment was heavy and bulky and there was "inconvenient and disconcerting arcing that occurred during the collection procedure, particularly in humid atmospheres". By the late 1940s and 1950s systems based on glass fibre or cellulose acetate filter papers dominated the scene, even for alpha contaminants. The basic designs developed then were to survive through the century: air was drawn through a filter paper by an air pump which was often, at least in the early years, one intended for a commercial vacuum cleaner. Static air samplers (SAS), which required the papers to be collected and analysed elsewhere, and monitors (SAM), with a detector which could sound an alarm if the activity on the paper exceeded a pre-determined level, were both used. They became more compact and reliable and began to be linked into computerised systems that allowed large areas to be monitored efficiently. But SASs and SAMs were essentially the same devices in 1999 as those of the 1940s and 1950s.

Measuring radon and its daughters has always been a distinct and specialised area and has challenged physicists and engineers for decades. Early measurements were made by collecting samples in evacuated vessels and then using an electroscope or ionisation

10 Measuring Internal Activity

chamber. One system used two identical chambers: one of them was sealed and the air to be tested was passed through the other. The difference current was measured with a string electrometer. While such devices were quite sensitive they did require skill to operate reliably and should perhaps be regarded as laboratory instruments. Practical field measurement began to appear in the 1950s when the Lucas cell[2] came onto the scene. The cell was a small chamber (200 ml) coated internally with a zinc sulphide phosphor. It was evacuated and then filled with air through a filter (which removed radon daughters). The scintillations then produced by the trapped radon and its (new) daughters were detected by a photomultiplier tube and gave a direct measure of the radon concentration in the air.

The two-filter method, developed in the late 1960s, worked on a similar principle but was much more sensitive and easier to use. Here air was drawn through a one metre long tube with a filter at each end. The filter at the input end removed all the radon daughters so that only radon itself entered; the daughters that were generated while the air was in transit were collected on the outlet filter and the remaining radon exited. Counting of the outlet filter gave a direct measure of radon concentration.

Of course, it was not the radon itself that really caused the hazard, it was the sequence of alpha particles emitted by its daughters. This was expressed through the Working Level concept; the significant quantity was the potential alpha energy carried by the radon and its daughters in any given volume of air. Measuring this, rather than radon concentration itself, would be much more meaningful and this thought led to the development of a number of working level meters.

The first of them was devised by Kusnetz[3] in 1956. It was rather simple in concept: air was drawn through a filter paper in a standard way for a standard time then, after a standard delay of 40 or 90 minutes, a gross alpha count was made on the paper for two minutes. The gross count was then proportional to the Working Level in the sampled air. The method was widely used but had the disadvantage that it took around an hour to obtain a result and this led to the search for the "instantaneous" working level meter. A number of these were devised based on alpha spectroscopy, essentially measuring the concentrations of the alpha-emitting daughters, but others used gross counts after short delays and some depended on a combination of beta and alpha measurements [4].

TAMING THE RAYS

With the discovery of the significance of radon as a public health hazard, there was an interest in passive (or at least inexpensive) detectors that could be used in mass surveys. These really took up ideas developed in the 1970s with the search for a radon personal dosimeter. They were generally based on track counting either on film or some track etch medium.

One of the other challenges that did bring real innovation was that of monitoring plutonium in air. While it was easy enough to collect samples, counting them was confused by the presence of alpha-emitting radon daughters. The samples could, of course, be taken away and counted after allowing the radon daughters to decay —to give rather delayed warning of airborne contamination. However continuous monitoring was a more serious problem because the higher energy alphas (6.0 and 7.8 MeV) from the daughters interfered with the 5.1 MeV plutonium alphas. The problem was partially solved in the 1960s as semiconductor detectors capable of alpha-spectrometry began to become available and could be incorporated into simple pulse height analysis circuits that provided some discrimination between plutonium and the interfering radionuclides. Degradation of the energies of the daughter alphas by the filter paper and by the dust accumulated on it made this less effective. However, as more sophisticated electronics became available, more complex algorithms for compensation were developed and effective systems that would alarm at quite low levels were developed. They did though suffer from one shortcoming that had little to do with sensitivity: in glove box facilities, where most plutonium work was done, the most common source of exposure was a failure in either a glove or in the bags used to transfer material. The activity was released right into the breathing zone of the worker and would be very much more diluted before it reached an air monitor; given the nature of air currents it might miss it altogether.

This was a extreme example of a problem that faced (and faces) air sampling: how to obtain a sample that represents what a worker is (or, at least, has been) breathing—to give an equivalent of the personal dosimeter used for external radiation. Several such devices were produced for radon monitoring in mines (with varying degrees of success) from the 1970s and Personal Air Samplers (PASs) probably began to be used in radioactive facilities from about the same time. In these air was drawn through a filter paper in a holder worn on the worker's lapel (intending to collect a sample from the breathing zone) by a small pump usually worn on a belt at the waist. The paper was then removed and counted at the end of each shift (although some versions eventually incorporated an alarm capability). The variability of results when compared with biological measurement (particularly nose blows or swipes) made it clear that it could not be regarded as an analogue of the personal dosimeter when the contamination came from point sources nearby. PASs therefore became regarded in these circumstances as devices which would trigger further investigation. The assessment of internal doses proved a much more challenging problem. Air sampling had to be combined with *in vivo* and excretion measurements to get results that were satisfactory.

The first reference to detection of radioactivity in a living human[5] was that of Martland in 1925 as part of his work on radium poisoning of dial painters[6]. He put a Wulf electroscope in front of a patient's chest and found that the leakage current was increased because of the "penetrative radiations coming from the... patient". Measurements of the amount of radium first came from Schlundt et al in 1929[7] who placed a fibre electroscope behind the back of a subject and timed the drift of the fibres. The subjects had between 4 and 126 µg of radium in their bodies; the person with the highest amount served as a calibration because, at their death shortly after the measurement, the radium content of their body was measured. The measurement with the electroscope was not an easy one since there were large variations in background from day to day but Schlundt thought that the limit of detection (as we would say now) was 2 or 3 µg. He was aware of complications that might come from the presence of Ra-228 (mesothorium, a popular alternative for dial paint to Ra-226 for a while) but was unable to correct for it. Flinn, one of Schlundt's co-workers, pointed out[8] the importance of taking a background reading with an unexposed subject in place of the exposed one and recommended corrugated cardboard as a thermal shield between subject and electroscope.

After much lobbying by supporters of the radium dial painters (Martland included) and with pressure from the journalist Walter Lippmann, the US Surgeon General Hugh S Cumming was persuaded to hold a conference on the health effects of radium in December 1928. A Commission on radium hazards was then set up by the US Public Health Service and the Service published four papers in 1933 under the general title of *Health Aspects of Radium Dial Painting*.

One of these, by Ives et al[9], was on *in vivo* measurement and used a Wulf electroscope placed as close as possible to the centre of gravity of the seated subject to obtain a sensitivity rather better than that Schlundt et al. Great trouble was taken to avoid contamination of clothing and subjects were asked to bathe and shampoo their hair before measurements and were not allowed into the factory on the day of examination so they inhaled no radon. The system was calibrated by injecting eleven radon sources (seeds) into a cadaver to reflect expected radium distribution in the skeleton and then arranging the cadaver in the measurement position. The limit of detection of the system was about 0.4 µg.

Sensitivity greatly improved when, in 1937, Robley Evans used a Geiger-Muller tube in place of the electroscope with his newly-devised count-rate meter: the device was more than ten times more sensitive. Evans had the subject lie bent in a 1 metre arc centred on the detector both facing and looking away from it. The system was calibrated using standard radium sources with and without the subject in position to allow for attenuation in the body and Evans obtained, according to Rundo[10], "the first reliable estimates of radium content".

The method depends on measuring the gamma radiation produced by a short-lived daughter of Ra-226: Ra-226 itself is an extremely weak gamma-emitter while Bi-214 emits copious gammas with energies to more than 2 MeV. This produced an uncertainty because the immediate daughter of Ra-226 is radon (Ra-222) and its half-life (3.8d) is long enough for a proportion to find its way into the bloodstream and be exhaled. As a result about one half of the gamma-emitters are lost before they decay and can be counted. It is possible to measure the radon content of the breath and correct the result of the gamma measurement but we will restrict ourselves to gamma measurement results and here Evans managed to achieve a limit of detection of about 0.2 µg of radium for which the daughters remained in the body. This corresponded to about 0.4 µg of total radium once allowance had been made for escape of radon—about four times the tolerance level accepted at the time.

The detection efficiency was improved again by Hess (who had won the 1936 Nobel Prize and devised the variant of the Wulf electroscope used by many of the earlier workers) and McNiff[11] in 1947. They used a large cylindrical ionization chamber (20 cm diameter and 41 cm high) with thick brass walls filled with nitrogen at atmospheric pressure. The ionization current was measured with a Lindemann electrometer (in fact used as a null instrument) and the chamber was placed in front of and behind the subject. The set-up was calibrated with sources placed in a water phantom of the thorax —for some reason they thought that radium would be in the thorax rather than spread around the skeleton—and they claimed to be able to detect a "tolerance dose of radium" in an hour. One reason for the improvement was the use of a shadow-shield behind the subject's chair to reduce background.

Rolf Sievert could report much improved sensitivity in 1951[12] y using larger detectors and by reducing the background. The detectors were 10 ionization chambers 0.22 m in diameter and 1.25 m long filled with nitrogen at high pressure (2.45 MPa) and they were arranged in a ring 0.5 m in diameter with space for the subject to lie inside. The ionisation current in the array was measured with an electrometer tube. The equipment was placed below ground, partially shielded by concrete and tanks filled with water (taken from the River Thames because local Swedish water was considered to have too high a radium content). Although the shielding was not complete, Sievert found that "under favourable conditions the standard error could be reduced to +/- 0.002 µg Ra equivalent with 6-8 observations for persons with no radioactive contamination" and Rundo interpreted this as a limit of detection for a single observation equivalent to about 0.015 µg of radium. Sievert went on to build an underground laboratory in the side of a mountain near Stockholm shielded in all directions by at least one metre of water with twelve even larger detectors (now 2m long and 0.4 m in diameter). He used this for measurements of a range of radionuclides but most importantly for measuring naturally-occurring K-40, establishing it as the most important of the contributors to natural background before radon was recognised.

Other workers built similar equipment to that of Sievert in the early 1950s with detection limits of a few nanograms of radium. They used high pressure ionisation chambers with compensating chambers so that residual natural background could be "backed off" in differential measurements using a vibrating reed electrometer. But the most significant development of the time was when Reines et al[13] built a system around a large liquid scintillation detector. These detectors were being constructed at Los Alamos for neutrino research but proved to be good detectors of gamma rays from bulk samples and were being used to check materials intended for counter construction to eliminate the more active ones. They made a detector with a hole in the centre and placed the materials in it for counting. Before long one of the team, Robert Schuch, suggested making a detector large enough to hold a human. The subject crouched inside a 0.5 m diameter hole in an annular scintillator 0.76 m high with a diameter of 0.76 m surrounded by a thick lead shield except for an opening to allow the subject in. The scintillator was viewed by 45 photomultipliers. With it it was possible to detect the potassium content of volunteers who had been "trussed up and lowered into the 18 inch hole" (of the development team only Wright Langham was small enough to fit in)[14] and simple calibration suggested that a few nanograms of radium might be detected. Encouraged by the results, part of the team went on to build the Los Alamos Human Counter on the same principle specifically for the measurement of radioactivity in humans. HUMCO I was arranged so that the subject could be slid in and out horizontally and was much larger: the volume of scintillator nearly tripled to 530 litres and the number of photomultiplier tubes went up to 108. The average potassium content could be determined with a statistical accuracy of about 20 g —about 14% of typical body content and the device was used—with its successor HUMCO II —for body radioactivity studies often assessing the uptake that had resulted from weapon test fallout.

Such large detectors could offer the possibility of high detection capability but they lacked the energy resolution of the NaI(Tl) scintillators that were, by the mid-1950s, being produced as large crystals. It was the use of these that was pioneered at Argonne National Laboratory by Marinelli et al[15] who used a single crystal 57 mm long by 38 mm diameter,and placed the subject in both the metre arc configuration and in a chair[16]. The counter, subject and operator were all enclosed in a substantial shield with walls of 200 mm steel or 100 mm lead. Rundo[17] constructed a counter at Harwell using four NaI(Tl) crystals 110 mm diameter and 51 mm long, suspending two of them above the supine subject and two below. Massive lead shielding was used and he had a multichannel analyser available. Potassium could be measured to within about 7 g while, counting the gammas above the 1.46 MeV of K-40, a limit of detection for radium was quoted at 8 ng. Miller et al[18] at Argonne obtained slightly better results with a single crystal 200 mm diameter and 100 mm long suspended above a subject in a tilted- back chair.

Large and multiple NaI(Tl) detectors established themselves as the preferred detector for general body monitoring. Massive shields became the norm made of steel, lead, concrete, water or in one case, in Brazil, sugar (the "Sugar Bowl" constructed in Rio de Janeiro in 1967 to investigate suspected high natural thorium series levels showed that sugar was a good enough shield until turned to syrup by the high humidity[19]). The configuration was generally, for single detector set-ups, the metre arc of Evans or the Argonne tilting chair or, for multiple detectors, an array of these above and below a supine subject. Electronics improved greatly and, for radium determination, there were tries at using coincidence counting for Bi-214 and indeed counting Ra-226 directly (and therefore avoiding the uncertainties associated with radon loss) using its weak 186 keV gamma-ray. Neither were successful in the late 1950s. An attempt to use the lower energy gammas (around 0.3 MeV) from Pb-214 rather than those with energies above the K-40 1.46 MeV photons did result in an improvement in detection—but not a large one.

When the IAEA put together a catalogue of whole body monitors in 1964 it listed 120; by 1970 the number had grown to 230. The main uses of these counters through the 1960s were in studies of natural internal radioactivity and of the uptake of fallout radionuclides. Apart from measurement of radium (which had driven their development for several decades) there had been limited application to other occupational internal exposures (although they had been used after accidents). As Ramsden and Foster point out in their short 1984 review[20] this was partly down to the problems of data management from early kicksorters (which made them more research devices than operational ones) but more to a lack of demand: it was not until after ICRP Publication 10 in 1968 that there were guidelines on general occupational internal exposure. As improved kicksorters were developed and computers with the capability to run spectral analysis software became more generally available, whole body monitors became accepted as part of routine health physics monitoring. As a consequence, simpler and more transportable monitoring systems became available commercially.

As the century closed there promised to be a revolution in whole body monitoring with the availability of large-volume high-purity germanium detectors. These had sensitive volumes approaching those of largish NaI(Tl) crystals with much better energy resolution. However, they were much more expensive than the scintillator, difficult to calibrate in multiple arrays and had to be maintained at liquid nitrogen temperature. So, while they were finding specialised uses, sodium iodide detectors remained the workhorse.

The determination of plutonium-239 in the lung by external counting of the 20 keV x-rays emitted in a small fraction of the decays has been an aim since the 1970s. Proportional counters, large diameter but thin NaI(Tl) scintillators, phoswich detectors and, later, semiconductor detectors have all been tried to obtain the necessary sensitivity. As Ramsden and Foster pointed out in 1984

> ...the fundamental problems remain that even if we could achieve zero background and zero interference from other activities present in the body, the low specific emission of the L x-rays and the attenuation of these x-rays in tissue means that, in any reasonable counting period we could not measure sub- ALI levels for pure Pu-239 a few days after the intake.

This problem was nowhere near as acute when mixed oxide fuel of high burnup were involved. For these situations the lung counter was available in several commercial forms and at least provided vital information on excess exposures that could confirm that there was no long-term buildup of activity.

Not surprisingly the first studies of excretion of radioactive materials involved radium. Rundo[21] could quote an example from as early as 1915 of work which measured the excretion following two intravenous injections of 100 μg of radium (probably in chloride form). Seil, Viol and Gordon[22] found that after injection or ingestion there was much less radium in the urine than in the faeces and that, up to the ninth day after administration, something over half the radium remained in the body. They also measured the radon in the breath. Seil and his co-workers came from the research laboratories of the Standard Chemical Company, a business set-up in the USA specifically to mine and process radium. The research lab concentrated on discovering the beneficial effects of radium administration[23] and Seil presumably had no qualms about being one of the guinea pigs and receiving two doses of 50 µCi of radium by mouth, seven days apart. The fate of the other subject, a young man who received two injections of 100 µCi, is not known but Seil, 33 years old at the time of the experiment, lived to the age of 69 when he died of diverticulosis of the colon. His exhumed body was found to contain 0.15 µCi.

In his 1933 review Evans[24] summarised what was then known of radium metabolism as:

- 90% was excreted by faeces: 10% by urine
- if by mouth 2 to 35% remained after five days; if intravenous injection 55-65%
- by the 10th day excretion was less that 1% per day
- after several years the excretion was down to 0.002-0.005% per day
- 2-40% of radon generated was lost through breathing
- the biological half-life was about 45 years.

Much of the data came from Seil's work but Evans also had available the work of Schlundt and Failla[25], one of four papers published by Herman Schlundt's group (Schlundt was a member of the Cumming's radium commission and a long-time radium researcher) at the University of Missouri between 1929 and 1933[26]. These papers reported measurements on individuals who had been administered radium for a variety of reasons.

With Nerancy and Morris, Schlundt[27] studied 32 patients at the Elgin State Hospital (a mental institution in Elgin, Illinois) who had been injected with radium as a treatment for schizophrenia (then known as *demetia praecox*). The work added to the body of knowledge at the time but it was when it was reviewed at Argonne in the mid-1950s and 19 of the patients still at Elgin were monitored again that it allowed a much-extended profile of radium retention to be constructed. Records of radium administration and measurements of radium contents of ampoules dating from the 1930s, similar to those that had been supplied to Elgin, allowed Norris[28] to construct rather simple expressions for retention and excretion at t days after the administration:

Percentage retained $54 \times t^{-0.52}$

Percentage excreted each day $28 \times t^{-1.52}$

It was a retention function that was to stand up well: for example later work by ICRP and others[29] was to show that while it rather underestimated the re-evaluated long-term retention, it did so by no more than about 50% at 30 years. Indeed—although there was now information about longer-term retention—there was not much change from the conclusions of Evans in 1933 and even from Seil's of nearly twenty years before that.

While this meant that the metabolism of radium was, by the 1950s, fairly well understood from the point of view of retention and excretion characteristics, it had long been the case that the best indicator of exposure was radon in the breath and the most reliable measurement techniques was whole body counting. For some of the other nuclides, that overtook radium in importance once the Manhattan Project got under way, excretion monitoring was to be much more important. For two it was to be vital. Tritium intakes could be assessed in no other way but its metabolism was relatively easily understood and the measurement problems tractable—making urine analysis a reliable technique. Plutonium inhalation was, however, another matter. It proved very difficult to measure by lung counting—especially if this was delayed by a few days from intake—and, in insoluble form, excretion was largely through the faeces. Faecal monitoring was thus a routine part of life for many nuclear workers.

The excretion monitoring methods demanded more-or-less sophisticated sample preparation and counting techniques in the laboratory and for some nuclides could provide rather definitive assessments of exposures on their own. However, for some nuclides

10 Measuring Internal Activity

—and plutonium was perhaps the outstanding one—assessment and (more importantly) control of doses required a combination of techniques in a well-constructed monitoring regime. Typically this involved a potential problem being detected by an air monitoring system (either by routine measurements or from an alarm) and an assessment of personal air samplers and perhaps nose blows or swipes. If these indicated possible intakes—and maybe if they did not—excretion sampling would be started and whole body or lung counting begun. The results would be assessed using the metabolic models of the time—often to work back to an estimated intake for comparison with local standards. If the intake did prove to be significant then the full arsenal of measurement techniques, with complex computer analysis using detailed (and perhaps personalised) metabolic data, would be needed for dose assessment. Although the instruments and models improved constantly, the need to combine the various more-or-less reliable inputs remained. The assessment of internal doses was, even at the end of the century, a much more significant challenge than what was required for external exposures.

Notes Chapter 10

1. (Kathren R L, 1980)
2. (Lucas H F, 1957)
3. (Kusnetz H L, 1956)
4. (Stannard J N, 1988)
5. (Rundo J, 1958)
6. (Martland H S, Conlon P and Knef J P, 1925)
7. (Schlundt H, 1929)
8. (Flinn F B, 1929)
9. (Ives J E, Knowles F L and Britten R H, 1933)
10. (Rundo J, 1993)
11. (Hess V F and McNiff W T, 1947)
12. (Sievert R M, 1951)
13. (Reines F, Schuch R L, Cowan C L, Harrison F B, Anderson E C and Hayes F N, 1953)
14. (Unknown, 1997)
15. (Marinelli L D, Miller C E, Gustafson P F and Rowlands R E, 1955)
16. (Marinelli L D, Miller C E, Rowland R E and Rose J E, 1955)

17 (Rundo J, 1958)
18 (Miller C E, 1958)
19 (Pachoa A S, Nogueira de Oliviera C A et al, 1993)
20 (Ramsden D and Foster P P, 1984)
21 (Rundo J, 1993)
22 (Seil H A, Viol C H and Gordon M A, 1915)
23 (Rowland R E, 1994)
24 (Evans R D, 1933)
25 (Schlundt H, Nerancy J T and Morris J P, 1933)
26 (Schlundt H and Barker H H et al, 1929; Barker H H and Schlundt H, 1930; Schlundt H, Nerancy J T and Morris J P, 1933)
27 (Schlundt H, Nerancy J T and Morris J P, 1933)
28 (Norris W P, Speckman T W and Gustafson P F, 1955)
29 (Rowland R E, 1994)

11 The Evolution of Organisations

Numerous bodies have been involved in researching radiological issues, formulating standards, applying them or, like radiological protection societies, supporting their members in some activity or other. It would be tedious and probably impracticable to cover all of them. However, it would be wrong to ignore them: they have driven, moulded and mediated the advances in protection. So, while below it has been possible to give a reasonable if short account of the evolution of international bodies with an interest in protection, the descriptions at the national level are limited (and even then are not complete) to the UK and the United States. The justification for this is that these countries have played a leading (but not unique) role in developments internationally and, for that reason, are an essential part of the story to date.

The first International Congress of Radiology (ICR) was held in London at the invitation of British radiologists in 1925 and congresses have subsequently taken place every two or three years other than during the Second World War. Until 1953 they were organised on an essentially *ad hoc* basis by the host country but, in 1953, the International Society of Radiology was formed with organisation of the congresses as its main function.

The International X-ray Units and Measurement Committee (which was to become the International Commission on Radiation Units and Measurements, ICRU) was set up at the first ICR in 1925 to develop internationally-accepted definitions of quantities and units for use by radiologists. The first recommendations were approved by the second ICR in 1928 and included the definition of the Roentgen as well as guidance on measurement. By the next Congress in 1931 (when the name International Committee for Radiological Units was adopted) it was clear that this definition was an adequate basis for measurement on lower-energy x-rays but it was not possible to extend it to higher energy x-rays and gamma-rays until 1937.

When the Committee was reconstituted after the Second World War it was set up as the ICRU with thirteen members and slowly evolved into the current organisation. Notably, in the early 1950s (by which time it was, following ICRP, a Commission rather than a Committee) it moved towards clear formulation of quantities based on physical principles rather than on measurement and in 1953 approved the rad as the unit of dose. At this time it also started to develop a sub-committee structure and began to hold symposia on specialised topics in units and measurement—an important step in maintaining links with advances in the field and identifying potential areas of ICRU interest. By the early 1960s the Commission had managed to draw a large number of people into its work: over 120 people from 18 countries were involved in the four committees that had been set up.

The ICRU continued and expanded its work (the Radiological in its titled changed to Radiation in 1968 to express its wish to extend its range) and by 2000 it had published 63 reports on all aspects of measurement and units, some in collaboration with ICRP. A non-profit organisation, it has been supported by a range of international organisations, national bodies and businesses[1]

The International Commission on Radiological Protection (ICRP) originated[2] at the 1928 ICR meeting as the International X-ray and Radium Protection Committee (the IXRPC, an acronym not used at the time). It had been realised that such a body was needed at ICR1 in 1925 but it took three years and the commitment of the British workers Kaye and Melville and the American L S Taylor at the Second Congress to bring it into existence; the similar British and US committees had been established for some time. They, with the support of Grossman from Germany and Sievert from Sweden, prepared and presented their plans to the Congress. These were approved and the committee was set up with the five initiators as the

11 The Evolution of Organisations

members, Kaye as Chairman, and a conviction that the committee should remain small and be composed of workers in the field.

The IXRPC then met for the first time and considered proposals prepared by Kaye and based on the earlier British recommendations. These were agreed and the first international recommendations were published by the end of the year in Britain[3]. They were also issued as an NBS circular.

At the second meeting, three years later in Paris, two more members joined the Committee: Solomon from France and Ceresole from Italy. There was discussion of the need for tolerance doses for workers but there was little information on which they could be based and there were therefore no specific recommendations. However, in 1934 meeting in Zurich, the IXRPC for the first time made a recommendation on permissible exposure to radiation and set the level at 0.2 r/day.

Until 1937 the Committee was principally concerned with specifying the barriers that would give adequate protection. The 1934 permissible dose made this less arbitrary and allowed some correlation to be established between required barriers and consequent dose levels. By 1937 the relationship between barrier thickness and exposure could be put on a much sounder quantitative basis. But, after this the Committee was not to meet for thirteen years and, when they did it was under the chairmanship of Sir Ernest Rock Carling (both Kaye and Melville had died), there were 10 members and its name had been changed to the International Commission on Radiological Protection. The links with ICR remained: membership of ICRP is reviewed on a four-year cycle at the time of ICR and 3 to 5 members are replaced.

In 1950 the Commission faced a new world. The Manhattan Project had produced its fruits from a massive organisation and the British were well-advanced with their bomb and civil programmes. There had been information exchange between UK, Canada and US in the form of Tripartite conferences. The US National Committee (later Council) on Radiation Protection and Measurement (NCRP) had assembled an enormous amount of information on radiation protection since its formation in 1946 and some of this was discussed with the Canadians and British in the Tripartite Conferences at Chalk River, Harriman House and Harwell. The impetus given by the Manhattan Project led the NCRP also to look more closely at x-ray protection. The ICRP was thus presented with much new data and itself was a different organisation from the IXRPC. It now had a set of rules for the selection of members and a more formal definition of its scope of work. To cope with the new role and the information that poured in, it set up a number of subcommittees. The recommendations it agreed in 1950 were those proposed by the NCRP and these covered both external exposure—the NCRP's 0.3 r/week—and internal exposure (using previously unpublished data from NCRP).

By 1962 the ICRP had evolved into a main Commission (now with 12 members and a Chairman) with four standing Committees: Committee 1 on Radiation Effects, Committee 2 on Doses from Radiation Exposure (at first called the Committee on Derived Limits), Committee 3 on Protection in Medicine, and Committee 4 on the Application of the Commission's Recommendations. Committee 5 on Protection of the Environment was added in 2005. On the recommendation of the Committees, the Main Commission sets up Task Groups to address specific issues and draw in expertise from outside ICRP. The standing Committees themselves set up Working Parties, usually composed of Committee members, for specific tasks. In 2000 there were 17 Task Groups in action.

General recommendations appeared in 1959 (retrospectively, called Publication 1), 1964 (Publication 6 revised as 9 in 1966), 1977 (Publication 26) and—after a number of documents clarifying and extending Publication 26—1991 (Publication 60). As the century's end approached the Commission was consulting on new general recommendations, expecting to publish them in 2008. Meanwhile a stream of specific recommendations and supporting technical information flowed from them and, in the year 2000, the Commission issued its 82nd Publication (on protecting the public in situations of prolonged exposure).

The ICRP is set up as a British charity located in Sweden where its Scientific Secretary, its only paid official, is based. It is funded by voluntary contributions from international bodies (IRPA and IAEA) and governmental bodies in 14 (in 2000) countries. In 2000 its income was around $300,000. Since 1959 it has produced its own publications and since 1977 this has been through the Annals of the ICRP, a dedicated journal published initially by Pergamon and more recently by Elsevier[4].

11 The Evolution of Organisations

The United Nations Scientific Committee on the Effects of Atomic Radiation (UNSCEAR) was created[5] by the General Assembly at its 10th session in December 1955 as a response to the world-wide concern about the effects of fallout from nuclear weapons testing and perhaps finally triggered by the disastrous Castle Bravo explosion. It is generally presumed to have been created as a means of avoiding a UN resolution calling for an immediate halt to all nuclear weapons testing. Its terms of reference—with the key ones below—were set out in resolution 913(X):

The General Assembly,

Recognizing the importance of, and the widespread attention being given to, problems relating to the effects of ionizing radiation upon man and his environment,

Believing that the widest distribution should be given to all available scientific data on the short-term and long-term effects upon man and his environment of ionizing radiation, including radiation levels and radioactive "fall-out",

Noting that studies of this problem are being conducted in various countries,

Believing that the peoples of the world should be more fully informed on this subject

established the Committee:

(a) To receive and assemble in an appropriate and useful form the following radiological information furnished by States

Members of the United Nations or members of the specialized agencies:

(i) Reports on observed levels of ionizing radiation and radioactivity in the environment;

(ii) Reports on scientific observations and experiments relevant to the effects of ionizing radiation upon man and his environment already under way or later undertaken by national scientific bodies or by authorities of national Governments

With its original 15 members, UNSCEAR met for the first time in March 1956. The membership was enlarged in 1973 and again in 1986 when China became the 21st state to be offered membership. The Committee appoints a Chairman, Vice-Chairman and, since 1968, a Rapporteur to serve usually for two Sessions (two years). A short report of a few pages, largely formal but with mention of work areas, is prepared each year and is discussed by the so-called Fourth Committee (the Special Political Committee) of the UN. This Committee then prepares a Resolution for the General Assembly commending UNSCEAR and recommending that it continues its work. Since 1974 UNSCEAR has been located in Vienna where it is part of the United Nations Environment Programme (UNEP). Its work is supported by a small Secretariat led by a full-time Scientific Secretary.

UNSCEAR's first scientific report was made in 1958. Since then the Committee has, roughly every three years, prepared major reports on the sources of exposures and the current understanding of radiation effects. At nearly 250 pages the first report was substantial but by 2000 the report had reached almost 1200 pages. The Contents of the 1982 report indicates their breadth:

A: Dose assessment models
B: Exposures to natural radiation sources
C: Technologically modified exposures to natural radiation
D: Exposures to radon and thoron and their decay products
E: Exposures resulting from nuclear explosions
F: Exposures resulting from nuclear power production
G: Medical exposures
H: Occupational exposures

11 The Evolution of Organisations

I: Genetic effects of radiation
J: Non-stochastic effects of irradiation
K: Radiation-induced life shortening
L: Biological effects of radiation in combination with other physical, chemical or biological agents

Other issues have been addressed from time to time: epidemiology, developmental effects, DNA repair and mutagenesis for example. Those who have studied the reports will know just how detailed and well-argued they can be when required. As such they have a major influence on the other international bodies.

The Nuclear Energy Agency was set up in February 1958 as the European Nuclear Energy Agency (ENEA) of the Organisation for European Economic Cooperation (OEEC)—the body set up to administer aid from the USA and Canada to European countries after the war under the Marshall Plan[6]. The OEEC's role changed in 1961 when it became the OECD, the Organisation for Economic Co-operation and Development, a forum where economic and social challenges of industrial development could be discussed by its (currently 30) members. In 1972 the ENEA became the NEA when Japan joined as the first non-European member. By 2000 twenty eight of the OECD members states were members.

The mission of the NEA is:

> - to assist its member countries in maintaining and further developing, through international cooperation, the scientific, technological and legal bases required for a safe, environmentally friendly and economical use of nuclear energy for peaceful purposes, as well as

> -to provide authoritative assessments and to forge common understandings on key issues as input to government decisions on nuclear energy policy and to broader OECD policy analyses in areas such as energy and sustainable development.

The Agency has focussed on issues that directly relate to nuclear power: nuclear safety, radioactive waste management, nuclear science, economics of the nuclear fuel cycle and nuclear law and liability. It also supports developments in nuclear science and the NEA database is an international repository of computer codes and nuclear data. It has taken an active interest in radiological protection through its Committee on Radiation Protection and Public Health (CRPPH), set up in 1957 slightly in advance of the formal establishment of the ENEA itself. The committee is largely composed of representatives of national regulatory authorities but it has sponsored a wide range of supporting activities. Its Expert Reports have covered issues ranging from the management of tritium and I-129, the biological behaviour of transuranics, radon dosimetry and gastrointestinal absorption of important radionuclides. Projects have included its role in the assessment of sea dumping through its CRESP programme, compilation of data on occupational exposure (ISOE), the extensive studies associated with Chernobyl and the INEX programme on international emergency exercises, which it initiated post-Chernobyl[7].

The International Labour Organisation (ILO) was created in 1919 at the peace conference that followed First World War. The first International Labour Conference with representation from government, workers and employers in all Member States, took place later that year and adopted six basic Conventions on workers' rights. Such Conventions have something of the nature of treaties and have to be ratified and implemented by Member States. The executive arm, made up of the Governing Body (appointed by the Conference) and the Secretariat, the International Labour Office, was set up in Geneva in 1920. The Conference, the Governing Body and the International Labour Office make up the ILO. The International
Labour Office grew so that, by the 1990s there were nearly 2000 officials working in Geneva and some 40 field offices around the world. The organisation relies heavily on international experts.

The very broad aims were expressed in 1919 in the Constitution and these were clarified in 1944 with the so-called Declaration of Philadelphia. The ILO's mandate is extensive, covering all aspects of

11 The Evolution of Organisations

conditions of employment and including health and safety. It has had a limited role in radiation protection.

The earliest mention of ionising radiation is in the 1934 Convention 42 on compensation payments which added x-rays and radium to the hazards of lead, mercury and anthrax that should be covered but the first document concerned with health physics was the 1949 Model Code of Safety Regulations for Industrial Establishments which included a section on radiation protection. The radiation provisions were revised and issued in 1959 before being included as Part II of the Manual of Industrial Radiation Protection in the early 1960s.

There have been numerous Conventions on working conditions but the one specifically related to radiation protection was Convention 115. It and the associated Recommendations 114 were adopted by the ILO Conference in 1960. The Convention, to be ratified by member states, contained 11 non-procedural Articles that, in very basic terms, laid down a legislative framework for radiation protection: the need for dose limits and medical examinations for example. The Recommendations amplified these, proposing for instance that the recommendations of the ICRP should be followed. They were included as Part I of the Manual in 1963.

Part III was a general guide with basic data for radiation protection. Three other parts were published over the next few years.

The Manual was almost the ILO's final independent excursion into radiation protection. After its publication, the ILO joined with other international agencies in co-sponsoring the IAEA's Basic Safety Series and several guides, notably on mining and milling of uranium. However, the meeting to revise the Manual, held in Geneva in 1986, did result in "Radiation Protection of Workers (ionising radiations): An ILO Code of Practice"in 1987 and related guidelines were issued in 1989.

In January 1946 the UN set up its Atomic Energy Commission (UNAEC) to "consider problems arising from the discovery of atomic energy" at the first session of the General Assembly. In June Bernard Baruch—the wealthy financier, advisor to Democratic presidents and benefactor of the party who had been appointed the US representative on the Commission—presented to the UNAEC a plan to create an International Atomic Development Authority (IADA) which would have control of all potentially dangerous nuclear activities, would not be subject to national vetos and would have power of sanction to enforce its decisions. In the US many thought the Baruch Plan unrealistic (and Baruch the wrong choice as Commissioner) while the USSR, having begun its own weapons programme, proposed nuclear disarmament before such a system of control was introduced. A Soviet proposal for inspection was rejected by the USA and UK. In the USA the McMahon Act was passed in July 1946 with the aim of the USA maintaining a monopoly of nuclear technology and therefore thoroughly undermining Baruch's plan. By 1949 the UNAEC was wound up and in the same year the USSR exploded its first atomic bomb. The British atomic bomb followed in 1952.

In early 1953 Dwight D Eisenhower followed Harry S Truman as US President and Stalin died. The USSR exploded its first H-bomb in August and by the end of the year Eisenhower had formulated the idea for a new agency and presented it in his "Atoms for Peace" speech to the UN General Assembly in December 1953. It was primarily seen as an agency that would control nuclear material by being a banker: it would take weapons materials from the weapons states and redistribute it (with necessary safeguards) for peaceful use. In this way controlled disarmament would take place. Negotiations with the USSR took several years (they were still pressing for complete nuclear disarmament), during which time the Statute of the agency was drafted first by eight and then by twelve countries (not including the USSR). It was finally accepted by the Soviet Union, with some changes, and the first General Conference of the International Atomic Energy Agency (IAEA) took place in October 1957 in Vienna, where the Agency was to be established.

It was apparent by the early 1960s that the nuclear banker idea was not likely to be a major activity, not least because the USA made bilateral agreements on technology transfer rather than trusting the IAEA as an independent intermediary. However, the possibility of the Agency having a international safeguards and inspection role was stronger (it was to become central to the Non-Proliferation Treaty that came into force in 1970) and there was a growing flow of technical assistance to countries who wished to develop civil nuclear power. As part of this flow of assistance came an Agency initiative for health and safety standards.

11 The Evolution of Organisations

It had been foreseen that IAEA would appoint safety inspectors (who would operate separately from safeguards inspectors) but this was never done and the idea was dropped in 1976. Instead, the IAEA would sponsor advisory "safety missions". From 1960 the Agency had issued recommendations on safety (including their important regulations for safe transport) and in June 1962 the first Basic Safety Standards for Radiation Protection, derived from ICRP recommendations, were adopted. These were revised in 1967 and again in 1981 and 1982. The 1982 standard was revised in collaboration with several other international agencies, following ICRP's recommendations of 1990, and approved in 1994.

A new structure for publications, the IAEA Safety Series, was introduced in 1989:

- Safety Fundamentals stating basic objectives, concepts and principles
- Safety Standards with requirements for particular operations sponsored or undertaken by the Agency(it includes the transport regulations)
- Safety Guides giving recommendations on observing the Standards Safety Practices with examples of methods for meeting Standards and Guides

This was later replaced by the Safety Standards Series with Fundamentals, Requirements and Guides. Altogether the Agency has published several hundred guidance documents in these series covering many aspects of nuclear safety, radiation protection, medical applications and environmental matters.

Following the Three Mile Island accident it was recognised that there needed to be a straightforward classification of the severity of nuclear accidents. The International Nuclear Event Scale (INES) was launched in 1990 and classifies events at reactors on a scale from 1 (minor) to 7 (most severe). By 1995 some 59 Member States were reporting such events classified as 2 or above to the Agency for world-wide distribution.

Technical assistance was being delivered to about 100 countries by 1995—with a major effort in assisting Central and Eastern

European states bring their nuclear activities up to international standards. By 1995 the Agency had a total budget of over $300M, 2000 staff, over 1000 associated staff and was providing upwards of 3000 experts for various activities from a wide variety of external organisations[8].

The European Atomic Energy Community (EURATOM) was set up by a treaty signed on the same day in 1957 that the European Economic Community was created by the Treaty of Rome. Initially a distinct organisation, although composed of EEC members, it was integrated with the EEC in 1967—although it remains a distinct legal entity. Originally having its own Assembly, Council and Commission the functions of these transferred to EEC institutions with the merger. Its aims are listed in Article 2 of the EURATOM Treaty, which remains the mechanism for EU actions on radiation safety:

For the attainment of its aims the Community shall, in accordance with the provisions set out in this Treaty:

(a) develop research and ensure the dissemination of technical knowledge,

(b) establish, and ensure the application of, uniform safety standards to protect the health of workers and of the general public,

(c) facilitate investment and ensure, particularly by encouraging business enterprise, the construction of the basic facilities required for the development of nuclear energy within the Community,

(d) ensure a regular and equitable supply of ores and nuclear fuels to all users in the Community,

(e) guarantee, by appropriate measures of control, that nuclear materials are not diverted for purposes other than those for which they are intended,

(f) exercise the property rights conferred upon it in respect of special fissionable materials,

(g) ensure extensive markets and access to the best technical means by the creation of a common market for specialised materials

11 The Evolution of Organisations

and equipment, by the free movement of capital for nuclear investment, and by freedom of employment for specialists within the Community,

(h) establish with other countries and with international organisations any contacts likely to promote progress in the peaceful uses of nuclear energy.

The funds for research have been made available through a series of Framework Programmes which include radiation protection as well as power generation, fusion research and waste management. Since the mid-1990s the EU has also supported work on the safety and efficiency of Eastern European reactor systems.

Under Chapter III (Health Protection) of the treaty, EURATOM (and hence the EU) is empowered to publish Basic Standards of protection for ionising radiation and this has been done through Directives which Member States are required to implement in their national legislation. There are also requirements to provide information: Article 37 for example requires the Commission to be notified on any activities likely to spread contamination across Community borders.

The first Directive was issued in 1959 with subsequent amendments, largely to keep it in line with ICRP thinking. The 1996 Directive(96/29/EURATOM) incorporated the principles of ICRP Publication 60 and was to be implemented by Member States by 2000. The 1990 Directive on protection of outside workers (90/641/EURATOM) extended protection to workers when employed in countries where the employer is not legally registered. The 1989 Framework Directive was aimed at the protection of all employees and did have some implications for those working with radiation. There were other Directives on protection of patients, provision of information in emergencies and radioactive wastes. EU Regulations have more force than Directives since they immediately come into force without any national action. They have not been widely used in radiation protection; the few examples are restricted to the control of the movement of contaminated foodstuffs following the Chernobyl accident

The Röntgen Society (originally to be called the X-ray Society) was founded in London in 1897. It had a physical rather than medical bias and this led medical radiologists to set up the Electrotherapeutic Society in 1902. This became the Electrotherapeutic Faculty of the Royal Society of Medicine in 1906. This body combined with the medical members of the Röntgen Society in 1921 to become the British Association for the Advancement of Radiology and Physiotherapy (BARP) subsequently, in 1927, renamed the British Institute of Radiology. In 1927 the BIR took the remains of the Röntgen Society under its wing[9].

The first journal devoted to x-rays (*Archives of Clinical Skiagraphy*) was founded in London in early 1896 with Sydney Rowland, who had been responsible for the publication of the series of articles in the *British Medical Journal,* as the first editor. Renamed *Archives of Skiagraphy* for just one issue in July 1897, it soon became the *Archives of Radiology*, a title it retained with minor changes to 1928 when it merged with the *Journal of the Röntgen Society* to become the *British Journal of Radiology.*

The Röntgen Society had met to discuss protection measures in June 1915. Professor Sidney Russ from the Middlesex Hospital presented a paper on protective devices and, after the subsequent discussion, it was agreed that some "strict rules" for operators should adopted and that the Society should take the lead in framing these. The Society published its first recommendations on protection in November of that year. They were quite basic but clarified some issues about which there had been doubt. In brief:

Recommendations for the Protection of X-ray Operators:

- X-rays of all degrees of hardness are harmful and the effects are cumulative
- An experienced and qualified medical practitioner should direct all treatment
- X-ray tubes should be shielded leaving a beam just large enough for the work
- Gloves and aprons should be worn

11 The Evolution of Organisations

- Cubicles should be used for x-ray treatment with the operator working from a protected space
- Thick lead glass shields should be used in work with screens
- The hand should never be used to test hardness

There was also a warning that commercial shields were often ineffective and so should be tested for opacity

While a number of x-ray pioneers had already died it was the death of Ironside Bruce in March 1921 that triggered the next step. Aware that this tragic event, clearly due to radiation, could undermine faith in the safety of x-rays and deprive medicine of a technology with outstanding potential, Robert Knox wrote a letter to the Times on 21 March hoping to calm public fears. In it he announced the setting up of a standing committee on x-ray safety. This was the X-ray and Radium Protection Committee (XRPC) under the chairmanship of Sir Humphrey Rolleston. Members were: Knox and Sir Archibald Reid (from BARP), Stanley Melville and S Gilbert Scott (from the Royal Society of Medicine), Cuthbert Andrews and G Harrison Orton (from the Röntgen Society), Russ (representing the Institute of Physics), J C Mottram (from the Radium Institute) and G W C Kaye (National Physical Laboratory).

Russ, having done so much to stimulate its formation, was joint secretary, initially with Melville. It was a position he was to hold until the Committee was disbanded in 1952.

The Committee's Preliminary report was ready in June and was published in the July issue of the *Journal of the Röntgen Society* [10]. Further reports followed:

- Revised Report No1 and No2 December 1923
- Third Report   May 1927
- Fourth Report   1934
- Fifth Report   1938

•Sixth Report 1943

•Seventh Report 1948

The third report is of particular significance since it was the one presented to, and accepted with minor modification by, the first meeting of the International X-ray and Radium Protection Committee in 1928.

The Medical Research Council (MRC) originated in arrangements for funding of national research on tuberculosis when, in 1913, a Medical Research Committee was set up to manage the funding with an overseeing Advisory Council. The Committee became the Council by Royal Charter in 1920.

In the immediate post-war period it was recognised that a great deal more needed to be known about the effects of radiation for both effective medical use and general protection and the Council set up a number of research units specifically to look at this. The Radiobiological Research Unit was set up in 1947 with J F Loutit as director to study the effects of radiation on living cells. In 1995 it split into RAGSU—the Radiation and Genome Stability Unit—and the Mammalian Genetics Unit. The Experimental Radiopathology Unit was set up at the Hammersmith Hospital, London in 1953 with a somewhat similar remit but with a much greater emphasis on radiotherapy. The Group for Research on the General Effects of Radiation (later the Clinical Effects of Radiation Research Unit) was established at the Western General Hospital, Edinburgh in 1956. The Unit's original aim was to study the incidence of leukaemia in ankylosing spondylitis patients who had been treated with x-rays but the remit became much wider and it was re-named the MRC Clinical and Population Cytogenetics Unit in 1967 and later the Human Genetics Unit. In 1959 the Environmental Radiation Research Unit, was established in Leeds under the direction of F W Spiers, to assess the dose received from environmental sources and consider its biological significance. It was disbanded in 1972 on Professor Spiers's retirement.

The MRC Medical Statistics Unit at the London School of Hygiene and Tropical Medicine in London has been a leading establishment in the application of epidemiological techniques to public health and

11 The Evolution of Organisations

this continued, under various names and different funding, to the end of the century. Ionising radiation research started in 1955 following concern about the health impact of hydrogen bomb test that caused widespread radioactive fallout. The study, described in Chapter 8, of more than 13,000 ankylosing spondylitis sufferers treated with radiation resulted and established a link between radiation and cancer mortality. Then the scientists largely responsible for the work—Richard Doll (later Sir Richard), best remembered for his work that established the connection between smoking and cancer, and Michael Court Brown—left the unit with several staff and, with Doll as Regius Professor of Medicine, continued the work at the Cancer Epidemiology and Clinical Trials Unit at Oxford. With a change of name to the ICRF Cancer Epidemiology Unit in 1988, it undertook a wide range of public health studies with a limited, but important, element on radiation effects. Doll also conducted major studies on mortality of radiologists and the effects of radon.

Apart from its research activities, both within dedicated units and funded elsewhere, the MRC became responsible, with the Ministry of Health, for the Radiological Protection Service from its inception in 1953 when it took over the film badge service previously run by NPL. Located in Sutton, Surrey, England the RPS maintained a number of regional centres and provided a monitoring and advisory service for radiation protection. It formed part of NRPB when it was set up in 1971.

A number of publications by the MRC have had a significant impact but two reports prepared in the late 1950s are perhaps outstanding and have already been referred to several times. After debates in Parliament, the MRC was asked to prepare a detailed statement on the health effects of radiation, including genetic effects, The Committee set up in 1955 produced a detailed report in 1956[11] and an update in 1960[12]. These closely-argued, self-contained and lucid papers remained accessible and authoritative sources for many years. The 1956 report (*The Hazards to Man of Nuclear and Allied Radiations*) introduced the principle of justification, generally endorsed the recommendations of the ICRP, suggested that bomb fallout could potentially lead to ill-effects and recommended caution in the use of radiation, including in medicine. By 1960 and the second report more information had become available: internationally, two reports from the US committee on the Biological

Effects of Atomic Radiation had been published and UNSCEAR had issued its first report. There were also better data on fallout and its impact and on the population and on medical exposures.

The United Kingdom Atomic Energy Authority was set up in 1954 as the vehicle for weapons research and production and the development of nuclear power. It designed and built the Calder Hall power station—arguably the first proper commercial one—at Windscale in the 1950s along with several material testing reactors and the Dounreay Fast Reactor (DFR). With principal sites at Harwell, Windscale, Risley, Capenhurst, Dounreay, Springfields and Winfrith, UKAEA led much of the subsequent development work on British gas-cooled graphite-moderated reactors reactor systems(as well as the ill-fated Steam Generating Heavy Water Reactor at Winfrith). The Magnox type—a direct descendent of the Calder Hall station—gave way to the more efficient Advanced Gas-Cooled Reactor, for which a prototype was built at Windscale. The Authority continued to operate the Dounreay site which included the Prototype Fast Reactor and its associated reprocessing plant until these closed at the end of the century but other activities were hived off in the 1970s. First fuel fabrication and the Windscale (later Sellafield) commercial reprocessing were split off to British Nuclear Fuels Ltd (BNFL) while the Capenhurst centrifuge enrichment plant went to URENCO in 1971. Then the weapons development and manufacture work was separated in 1973 to be managed by the Atomic Weapons Research Establishment(AWRE) with sites at Aldermaston and Burghfield in Berkshire with outposts in Cardiff and elsewhere. In 1983 the isotope production activity was removed when Amersham International was set up and privatised.

The Authority had been permitted to do some commercial work after 1965 and this grew but in the 1980s and 1990s, with the decline in interest in nuclear power from government and the completion of much of the essential development work (and specifically the lack of funding for work on fast reactors), the UKAEA was forced to shrink and seek new opportunities. In the late 1980s it was set on a new financial basis, a Trading Fund, to encourage commercial work. It then underwent a split in the mid-1990s into a part which managed the existing legacy nuclear facilities and prepared to decommission them (which stayed as the UKAEA), a facilities management company and a new entity, AEA Technology, created from the assets

11 The Evolution of Organisations

that seemed to have commercial potential. AEA Technology was fully privatised in 1996 and rapidly expanded its portfolio of non-nuclear interests. But by the early years of the next century it had shrunk by selling assets to be a company with predominantly environmental interests.

British Nuclear Fuels developed the activities at Springfields, where fuel is manufactured and at Sellafield, building plants to reprocess oxide fuel, to become a global player in reprocessing. It was fully privatised in 1984 but remained under government ownership.

After Calder Hall and with the exception of prototypes all the commercial thermal nuclear generation in England and Wales was installed for the Central Electricity Generating Board, a nationalised conglomerate formed from the multitude of electricity companies that existed until the 1940s. In Scotland, with its different energy organisation, four plants were built for the South of Scotland Electricity Board. The privatisation process for CEGB began in 1990 when generation was split between two companies: PowerGen (which took all the former CEGB nuclear capacity) and National Power. The nuclear part of PowerGen was soon separated out when it became clear that it would be a major impediment to full privatisation and it was set up as Nuclear Electric. In a more or less parallel way Scottish Nuclear was set up with the three functioning nuclear stations in Scotland.

The evolution of the industry, the background of public and political concern that seemed inseparable from nuclear power and the increasingly complex safety-related demands on operators challenged the regulatory structure and process. The Health and Safety Executive (HSE) is the UK body responsible for enforcing health and safety legislation as it applies to people at work, guided by the Health and Safety Commission. It is also responsible for monitoring several aspects of the impact that activities at work might have on members of the public. Before the HSE was created in 1974 workplaces were subject to inspections by a variety of ill-related bodies; the creation of the Executive aimed to bring a coordinated and consistent approach. The key legislation is that of the Health and Safety at Work Act 1974 with its associated Regulations—but there is other applying to specific areas.

Within the UK, operators of large nuclear establishments have since the mid-1960s been subject to the Nuclear Installations Act (with its subsequent modifications) and must obtain a licence before they can operate. These impose additional requirements on them, beyond those of the H&SW Act ones, and these are regulated by the Nuclear Installations Inspectorate (NII) of the HSE. A key element of all such activities is the production of a series of safety cases which must be approved by the NII before activities are allowed to progress and part of those will be a demonstration that there are adequate arrangements for radiological protection. Because of the way it was set-up, for most of its life the UKAEA was not subject to the Nuclear Installation Acts and was therefore not required to have a licence to operate. Until this was changed in the early 1990s, the Authority's major nuclear facilities were assessed by an internal body, the Safety and Reliability Directorate, formed from the Authority Health and Safety Branch which had been set up after the Windscale accident in 1957.

Where operations affect the environment through for example waste discharge or disposal then regulatory responsibilities lie with different government bodies: the Environment Agency in England and Wales, the Environment and Heritage Service in Northern Ireland and in Scotland the Scottish Environment Protection Agency. Each applies the various relevant Radioactive Substances Acts.

The Ministry of Agriculture, Fisheries and Food (MAFF) became involved in radioactivity monitoring in 1947 when the Ministry of Supply was planning the reprocessing plant at Sellafield. The Fisheries Division of the Ministry of Agriculture and Food (MAF) were warned that effluents would be discharged into the Irish Sea from about 1950. There was little idea what the impact of this would be on fish (or on the humans who might eat them) and MAF planned to undertake the necessary research to find out. Staff were recruited into the "Fisheries Research Laboratory" (an innocuous name to conceal the sensitive nature of the work) and work began on brown trout—which were quickly found to be rather more resistant to external radiation than people are. The consequences of radioactivity taken up by the fish—on the fish and on consumers—were rather slower to appear. However, MAF did begin to get information on this and on fish movements (through a tagging programme) so that, by 1952, when the discharges started there was a basis for calculating discharge limits.

11 The Evolution of Organisations

All these functions were refined but MAF's role changed significantly when they were given an enforcing role over the newly-formed UKAEA in 1954. This introduced formal discharge authorisations for the first time. The associated inspection activity was run in parallel with the research activity and there was continuing study of the fish exposure pathway. For a while a new pathway—the consumption of the seaweed *Porphyra* (the ingredient of laverbread)—was very significant but this became less so as collection of the seaweed declined in the 1970s. MAFF (as it became in 1956 when Food was brought in) was also deeply involved in the debate about the sea dumping that took place in the English Channel to 1963 and at Atlantic sites up to the 1980s. The growth of nuclear power and the various public enquiries that took place brought an increase in staff: the 5 people of 1947 was 42 by 1965.

The formation of the Environment Agency (EA) in 1996 removed the enforcement role of MAFF, although the EA were required to consult them on specific issues. In the following year the Directorate of Fisheries Research (as it was then called) became an Executive Agency of MAFF as the Centre for Environment, Fisheries and Aquaculture Science (CEFAS).

The focus for general radiological protection advice and guidance of best practice for Governments and much of industry was the National Radiological Protection Board. Set up by Act of Parliament in 1970, the Board combined the MRC's Radiological Protection Service and the part of UKAEA's Authority Health and Safety Branch concerned with radiological protection matters, AHSB(RP). With its strong expertise in all areas of radiological protection including medical, operational, policy and modelling it had a major influence on the development of radiological protection in the UK and internationally through ICRP. It issued regular and authoritative updates on individual exposures in the UK, reflecting an early involvement in measuring public radon exposures and an active interest in the reduction of unnecessary medical exposures. There was also strong expertise in practical and theoretical optimisation
and in environmental modelling, including the assessment of radiation dose from potential major reactor accidents. The headquarters were at Chilton in Oxfordshire with regional offices in Leeds and Glasgow. In 2005 the Board was merged into the Health Protection Agency[13].

TAMING THE RAYS

The American Roentgen Ray Society (ARRS) was founded in 1900 and was still very active at the end of the century. Its journal, the *American Journal of Roentgenology*, continues to be published today. However, schism occurred early on when its east-coast orientation led to mid-western radiologists setting up the Western Roentgen Society (WRS) at a meeting in Chicago in 1915[14]. By 1918 the society had its own journal—the *Journal of Roentgenology*. The name of the WRS was changed to reflect growing national and international ambitions in 1919 to the Radiological Society of North America (RSNA) and the name of the journal was changed, to avoid confusion with the ARRS publication, to the *Journal of Radiology*. The journal became *Radiology* in 1923 after a dispute over ownership, involving litigation, which led to the old journal being closed down. The RSNA survived difficult financial circumstances in the years before the Second World War and the proliferation of radiological societies that followed it, triggered by technological developments and a perception by some that the society—like the ARRS—was identified with diagnostic rather than therapeutic applications of radiation.

For many years the journals were major outlets for work on radiation protection as well as radiology and the societies were both very active at the end of the century. They were closely—and antagonistically—involved in some of the earliest international activity that led to the formation of the precursor on the National Council on Radiation Protection and Measurement.

Representatives of three organisations were invited to attend the Stockholm ICR in July 1928: the National Bureau of Standards (which inevitably sent the person responsible for the invitation, Lauriston Taylor), the American Roentgen Ray Society and the Radiological Society of North America. While the NBS, there as an observer according to Taylor, had no recommendations to offer, the two professional bodies each offered different ones and, since each claimed to be the national authority, a compromise agreement could not be reached between them. As a result, the first internationally agreed protection recommendations were (as we have seen) those presented by the British delegation. In parallel with the ICR, Kaye, Melville and Taylor were organising the IXRPC—the predecessor of ICRP—and agreed that this should be kept small and that the representatives should be drawn, where possible, from national laboratories. Kaye, the first Chairman, recommended that a single

11 The Evolution of Organisations

committee should be formed in each country to put together national recommendations for international consideration.

Lauriston Taylor then took on the task of organising such a committee in the USA and by early 1929, with the agreement of the two professional societies, the Advisory Committee on X-ray and Radium Protection was formed. Taylor was chairman and there were representatives (2 each) from the two societies, one representative from the American Medical Association and two from x-ray equipment manufacturers—one was W D Coolidge.

The Committee first met in September 1929 and by May 1931 the first recommendations on x-ray protection were published as NBS Handbook 15. A Handbook on radium protection, prepared by a suitably extended committee, was published as NBS Handbook 18 in March 1934. Handbook 15 was revised and re-issued as Handbook 20 in 1936 when it contained the first numeric limit (the tolerance dose) of 0.1 r/week. The radium handbook was revised and published in 1938 as Handbook 23.

The Advisory Committee did not meet during the war and, when it did in 1946, it was clear that there had been dramatic changes in the needs for radiation protection. Its membership was expanded to include representatives of the Manhattan Project and the Public Health Service but it was soon apparent that a major reorganisation was required to deal with the many new problems. The outcome was the setting up of a Main Committee, a small Executive Committee and seven subcommittees. Taylor remained as Chairman of the Main Committee and became chair of the Executive Committee. To reflect the expanded remit, the committee was renamed the National Committee on Radiation Protection. By 1958 the Main Committee had nearly 40 members, the Executive Committee 9 members and 18 subcommittees had been established (including 4 dedicated to measurement issues). By then, fifteen sets of recommendations had been published as NBS documents[15].

Other US organisations had emerged in the 1940s and 1950s. The Atomic Energy Commission (AEC) was formed in 1946 and took over the Manhattan Project in 1947. In 1955 it asked the National Academy of Sciences and the National Research Council to study the effects of low levels of radiation. The NAS/NRC set up the Biological Effects of Atomic Radiation Committee (BEAR) which issued its first report in 1956 and raised concerns about the possible genetic effects of radiation. The AEC's first published standards, 10CFR Part 20 in 1957, which included weekly dose limits and permissible concentrations in air and water, relied heavily on NCRP recommendations.

By the late 1950s radiation was a significant political issue—largely through the increasing awareness of test fallout and the possible damage it might cause. During the hearings of the Congressional oversight committee on the AEC between 1957 and 1959 it became apparent that standards were being formulated by NCRP, a self-appointed group of scientists, rather than by a central Government body. Each Federal agency was free to choose which standards it adopted.

To achieve some coherence in the face of growing public concern the Federal Radiation Council (FRC) was created by presidential Executive Order in 1959 to advise the President on radiation matters affecting health and give guidance to all Federal agencies on radiation protection standards. This was a very high level body composed of the Secretaries of departments including Agriculture, Health, Defense and the Chairman of the AEC. Although advisory it was clearly more powerful than NCRP not least because Federal agencies were required to report to it annually on their progress in implementing its guidance. The NCRP remained the key technical body: the FRC was required by law to consult its Chairman (as well as the President of NAS) on matters of radiation safety.

The FRC issued its first Federal recommendations (as Radiation Protection Guides or RPGs) in 1960 on occupational and population radiation exposures. This was followed by guidance on a number of radiation protection issues including, in 1969, guidance on exposure to radon and its daughters in uranium mines—introducing the term "Working Level".

Three Guides were promulgated under Presidential signature:

- Radiation Protection Guidance for Federal Agencies, Federal Radiation Council 25 FR 4402 May 18, 1960

11 The Evolution of Organisations

- Radiation Protection Guidance for Federal Agencies, Federal Radiation Council 25 FR 9057 September 26, 1961
- Underground Mining of Uranium Ore, Federal Radiation Council 34 FR 576 January 15, 1969 35 FR 245 December 18, 1970

Eight reports were issued as background information to the Presidential guidance or as technical guidance on other matters:

- Background Material for the Development of Radiation Protection Standards (1960)—for occupationally exposed and public
- Background Material for the Development of Radiation Protection Standards (1961)—internal emitters
- Health Implications of Fallout from Nuclear Weapons Testing Through 1961 (1962)
- Estimates and Evaluation of Fallout in the United States from Nuclear Weapons Testing Conducted Through 1962 (1963)
- Background Material for the Development of Radiation Protection Standards (1964)
- Revised Fallout Estimates for 1964-1965 and Verification of the 1963 Predictions (1964)
- Background Material for the Development of Radiation Protection Standards Protective Action Guides for Strontium-89, Strontium-90, and Caesium-137 (1965)
- Guidance for the Control of Radiation Hazards in Uranium Mining(1967)

By the end of the 1950s the position of the NBS in sponsoring NCRP and publishing its reports as Bureau Handbooks was giving concern —many of the biological and medical topics were well outside NBS's strict competence. NCRP had anyway strengthened links with Congress in the debate on the effects of bomb fallout and sought a

Federal charter. The charter was enacted in 1964 as Public Law 88-376 and the NCRP became a non-profit corporation: the National Council on Radiation Protection and Measurements. The four specific roles mentioned in the charter can be summarised as:

- Collect and disseminate information and recommendations on protection and measurement
- Act as a means for cooperation of organisations with an interest in radiation protection
- Develop concepts of radiation protection and measurement
- Cooperate with ICRP, ICRU and other bodies with interests in radiation protection

It continued with these aims to the end of the century. By then it had published some 130 Reports, 15 Commentaries and 8 Statements and the 80 members of the Council were active on almost 20 subgroups. Its founder in 1931, Lauriston S Taylor, survived into the 21st century, the Honorary President of NCRP until his death in 2004 at age 102. He is commemorated with a series of annual lectures that started in 1977.

While the NCRP remained active after more than 70 years, the FRC performed one of its last significant acts in the mid-1970s when it commissioned a study from the NAS reviewing the current state of knowledge of the effects of ionising radiation on humans including the response to low doses. By the time the NAS/NRC Advisory Committee on the Biological Effects of Ionizing Radiation (successor to BEAR), published its first report—BEIR I—in 1972 the FRC had gone. It disappeared as part of a Federal reorganisation (Reorganisation Plan No 3) undertaken by President Nixon in late 1970 and its role was taken over by a new Federal Agency, the Environmental Protection Agency (EPA).

The EPA's general federal responsibility for standard setting was expressed through the publication of Federal Guidance Reports. Those requiring the signature of the President are concerned with principles and basic standards and four were published by the end of the century. These covered underground mining of uranium ore

11 The Evolution of Organisations

(1971), diagnostic use of x-rays (1978), occupational exposure (1987) and exposure of the general public (1994).

Five Guidance Reports not requiring Presidential approval (sometimes called Technical reports) have also been published. These provide current scientific information to support Federal and State programmes and cover: diagnostic x-rays (No9, 1976), radioactivity concentration guides (No.10, 1984), limits on intake, air concentration and dose conversion factors (No.11, 1988), external exposure to environmental activity (No.12, 1993) and cancer risk coefficients for environmental exposures (No 13, 1999).

While EPA had broad enforcement authority for other pollutants, its role in the radiation field was more constrained. It was given an enforcement role outside the boundaries of nuclear facilities but within the plant itself this remained the responsibility of the AEC. In fact EPA did develop a more intrusive role through environmental statutes and the development of standards for managing environmental issues but the interface with AEC remained difficult and ill-defined in these areas.

The Atomic Energy Commission's early years[16] were dominated by the Cold War and the organisation concentrated on weapons development, production and testing. By the end of the 1940s wartime production facilities had been overhauled and a substantial arsenal of fission weapons had been built up. But the Soviet A-bomb test of 1949 brought a massive drive—one of the largest peacetime programmes in US history—towards the development of thermonuclear weapons. The result was the Mike test at Enewetak atoll in the Pacific in the autumn of 1952. By the following year there had been a further 12 thermonuclear tests and the immense resources of the AEC were devoted almost entirely to military work.

However, while the military emphasis continued, by the mid-1950s the AEC was devoting more effort to peaceful uses. This was encouraged by the Atomic Energy Act of 1954 which both facilitated international exchanges on nuclear power and enabled the involvement of private companies in the evolution of atomic power within the US. As well as requiring the AEC to promote the development of nuclear power it gave it the responsibility to protect public health and safety. A two-stage licensing process for commercial development was introduced: a construction permit based on the reasonable assurance that the planned reactor would be safe and a license to authorise fuel loading and operation. The passing of the Price-Anderson Act in 1957, which provided for a government-supported insurance safety net, removed a barrier to commercial exploitation (and incidentally required that more information be made publicly available). So, by 1962 six large power reactors were operating—two built entirely with private money—and in the next six years the AEC issued a further 38 construction projects in what became known as the "great bandwagon market"[17]. The fixed-price turnkey contract for the Oyster Creek power plant in New Jersey offered in 1963 by General Electric to Jersey Central Power and Light was a factor in removing the doubts of utilities about the risks of the new technology. The AEC itself began to focus more on civil power as the testing of nuclear weapons came to an end—the Limited Test Ban Treaty was signed in 1963—and the Johnson administration allowed the private ownership of Special Nuclear Materials (plutonium and enriched uranium) in 1964.

By the early 1970s the heightened awareness of the hazards of radiation resulting from the debate on fallout and the perception that the AEC was not, as a promoter of nuclear power, to be trusted in its regulation was leading to much questioning of its role. The Commission had also not emerged well from three issues:

- There had been a growing controversy over the effectiveness of Emergency Core Cooling Systems(ECCS) and the way this had been handled by the AEC. Public hearings during 1972 and 1973 about interim rules for ECCS resulted in "acrimonious testimony"[18] and damaging headlines.

- The Commission emerged badly from the Calvert Cliffs Nuclear Power Plant case in 1971 when the federal Appeal Court for the District of Columbia found that the AEC had failed to implement fully the National Environmental Policy Act of 1970 in approving discharges form the plant

- An announcement was made that an abandoned salt-mine near Lyons in Kansas would be used for high-level waste disposal before a proper geological assessment had been made. This was challenged and, when it was made, the assessment showed the site to be unsuitable.

11 The Evolution of Organisations

The AEC was abolished in 1974, by the Energy Reorganisation Act and replaced by two organisations: the Energy Research and Development Administration (ERDA) and the Nuclear Regulatory Commission (NRC). ERDA took over research and development responsibilities for both weapons and the peaceful uses while NRC became an independent regulatory agency responsible for licensing and monitoring nuclear fuel cycle operations. ERDA was subsequently absorbed into the Department of Energy (DOE) when this was created in 1977[19]

The NRC began operations in January 1975 and within the year was embroiled in two quite different reactor safety issues. First, in March a technician investigating air leaks with a lighted candle at Browns Ferry Nuclear Power Plant in Alabama ignited cable insulation. The fire lasted several hours and nearly disabled the safety systems of one of the reactors. Second, the Reactor Safety Study, commissioned by the AEC in 1972, was released in October 1975 (although there had been a draft available in 1974). This study, conducted by Norman Rasmussen, applied risk and reliability analysis techniques to the possibility of serious reactor accidents. The calculated risks from nuclear power were then compared with risks from natural events and other technologies and it was concluded that the risks from nuclear reactors were, relatively, very small. There was concern within and outside of NRC about the completeness of the assessments and whether they supported Rasmussen's conclusions. After years of review, in 1979, the Commission withdrew its endorsement of it.

However, 1979 was to be remembered for something else. On 28 March 1979, a pressure relief valve on Unit2 of the Three Mile Island NPP stuck open and allowed large quantities of reactor cooling water to escape. The operators failed to identify what had happened and made matters worse by initially disabling the emergency core cooling systems. The core of the reactor became exposed causing it severe damage and partial meltdown. However the containment was effective and only a small quantity of radioactive material escaped. The other good news was that the emergency core cooling systems had worked effectively once they were allowed to.

The impact of the accident was enormous. The operators were perceived as incompetent in handling the accident itself and the NRC was faulted for never having envisaged such an accident. The arrangements for keeping the media informed were deficient and the emergency plans were clearly inadequate.

NRC responded—and TMI influenced its activities for the next fifteen years—in several ways including:

- A much greater importance was attached to "human factors" and hence the training and licensing of operators
- There was much closer scrutiny of operating experience of utilities
- Emergency planning was upgraded.

The public concern was such that NRC was forced to suspend operating licenses for reactors in the pipeline for a year. While licenses began to be issued again after this pause (and over 40 were granted by the end of the 1980s) the concerns that nuclear power was an inherently dangerous business controlled only by a complex technology managed by very fallible people persisted. As the veteran reporter Walter Cronkite told US viewers at the time of the accident: "the danger faced by man for tampering with natural forces, a theme from the myths of Prometheus to the story of Frankenstein, moved closer to fact from fancy"[20].

Public concern was stimulated again on 26 April 1986 by the Chernobyl accident—which produced levels of I-131 around TMI three times higher than those after the TMI accident. Against this background the NRC's 1990 proposal to identify radiation levels they considered "Below Regulatory Concern" brought such an outcry that the idea was shelved.

By this time the NRC had a new focus. New plant was not being built and interest turned to the 100 or so plants in operation. It was recognised that many of the incidents that occurred were directly down to inadequate maintenance and the NRC gave high priority to improving this situation. A rule on funding and planning decommissioning was published in 1988 and a long-running issue of plant life extension was resolved, in NRC terms, by regulation in the early 1990s. In 1998 the Calvert Cliffs plant became the first for which extension was sought.

The NRC has, in its history, consistently been accused of over-regulation by the industry and a specific target became its reliance on "prescriptive" regulation. By the 1990s the risk-based methods of Rasmussen had had little impact on the licensing process and the NRC had relied on the defence-in-depth philosophy of engineered barriers. The demonstration of compliance with risk-based goals by probabilistic risk assessment (PRA) was seen by the industry as means of ensuring adequate safety while giving the licensee more freedom. The Commission endorsed the general principle but, by the end of the century, there was little agreement on how PRA should fit into the regulatory process.

A number of other agencies are engaged in regulation:

Occupational Safety and Health Administration (OSHA)

US Department of Transportation Office of Hazardous Materials Safety

US Food & Drug Administration (FDA)

Mine Safety & Health Administration (MSHA)

Individual states have extensive programs for managing radiation safety sometimes in partnership with the NRC or OSHA.

The US Department of Energy (DOE) remained self-regulating and was responsible for the oversight of its contractors. Their standards were specified in DOE Orders—such as Radiation Protection of the Public and the Environment, DOE Order 5400.5—and through the Code of Federal Regulations where 10 CFR Part 835--Occupational Radiation Protection—applied. Special arrangements applied to the Joint US Department of Energy & US Department of Defense Naval Nuclear Propulsion Program.

The National Academy of Sciences (NAS) was set up by President Abraham Lincoln in 1863, at the height of the Civil War. Its remit was to "investigate, examine, experiment and report on any subject of science or art when called upon by government". Since 1916 it had done this through the National Research Council set up to stimulate research, promote cooperation in science and gather and collate scientific and technical information for government. The NAS created the National Academy of Engineering in 1964 and the Institute of Medicine in 1970. Together, as the National Academies, they have brought together a formidable array of scientist, engineers and physicians on a voluntary basis to address an enormous range of issues.

The role of the NAS in the study of the aftermath of the Japanese bombs has been described earlier. A related activity was to lead to an authoritative role that was to continue for at least 50 years when, in 1955, the NAS President Detlev Bronk set up the BEAR Committees to review what was known of the biological effects of radiation. Six committees were formed : Pathologic Effects of Atomic Radiation (with various subcommittees); Meteorological Aspects of the Effects of Atomic Radiation; Effects of Atomic Radiation on Agriculture and Food Supplies; Disposal and Dispersal of Radioactive Wastes; and Oceanography and Fisheries. A number of reports, including two summary reports, the first released in 1956 and second released in 1960, were published by the various BEAR committees and subcommittees before the study was terminated in 1964.

The work was continued by the Committee on the Biological Effects of Ionizing Radiation (BEIR), a major source of information in the US and world-wide on the effects of low levels of radiation. It issued a number of reports:

> BEIR Committee (The Advisory Committee on the Biological Effects of Ionizing Radiation). (1972). BEIR-I: The effects on populations of exposure to low levels of ionizing radiation. Division of Medical Sciences, the National Academy of Sciences, National Research Council, Washington, D.C. (Generally known as the BEIR Report.)

11 The Evolution of Organisations

BEIR Committee. (1979). BEIR-II: The effects on populations of exposure to low levels of ionizing radiation. Draft Report. Division of Medical Sciences, Assembly of Life Sciences, National Research Council, National Academy of Sciences. (Generally known as the BEIR-II Report.)

BEIR Committee. (1980). BEIR-III: The effects on populations of exposure to low levels of ionizing radiation. Final Report. Division of Medical Sciences, Assembly of Life Sciences, National Research Council, National Academy of Sciences. (Generally known as the BEIR-III Report, Final.)

BEIR Committee. (1988). BEIR-IV: Health risks of radon and other internally deposited alpha-emitters. National Research Council. Committee on the Biological Effects of Ionizing Radiations. National Academy Press, Washington, D.C.

BEIR Committee. (1990). BEIR-V: Health effects of exposures to low levels of ionizing radiation. National Research Council. Committee on the Biological Effects of Ionizing Radiations. National Academy Press, Washington, D.C

BEIR Committee. (1999). BEIR-VI: Health effects of exposures to radon. National Research Council. Committee on the Biological Effects of Ionizing Radiations. National Academy Press, Washington, D.C

Preparation of BEIR-VII was in progress at the end of the century.

The Health Physics Society came into existence in the USA in 1955 [21] at the prompting of people who had been engaged in the Manhattan Project and particularly through the encouragement of Elda E Anderson, the person running the post-war training programme at

Oak Ridge. One of her former students, Saul Harris wrote to the journal *Nucleonics* in 1952 proposing a health physics society but the idea met with opposition from nuclear engineers, hygienists and medical radiologists. The engineers thought health physics was an integral part of the jobs, the industrial hygienists already had a section for radiation safety and medical radiologists already had (at least) two societies with similar interests. However, interest within the health physics community grew and a conference on health physics was organised at Ohio State University in June 1955. On the first day of the three-day meeting, which featured eight scientific sessions, the attendees voted overwhelmingly to form a society and agreed that the membership dues would be $2.00 per year. In three months the society had over 200 members.

The name "Health Physics Society" was settled at the first annual conference the following year and Karl Z Morgan—who had been the principal driving force behind the creation of the Society—was installed as President. The HPS journal *Health Physics* began publication in 1958 with Morgan as Editor-in-Chief, a position he was to hold for twenty years.

A concern about qualifications for health physicists had been raised by Lauriston Taylor at the Ohio meeting and this was addressed in 1959 with the creation of the American Board of Health Physics in 1959 to administer professional qualifications. (Much later, in 1976, it was to lead to the setting up, with HPS and ABHP support, of the National Registry of Radiation Protection Technologists as a basis for qualification for technicians.) By 1959 there were more than 1000 members (including over 100 overseas) and this was to double in the next two years.

The members voted to set up an international society and, in 1964, exploratory contacts were made with health physicists in 41 countries, leading to the creation of the International Radiation Protection Association (IRPA), with first President K Z Morgan, by the end of the year. Overseas members, as a result, drifted away since many of the benefits of HPS membership came through IRPA—including *Health Physics* at preferential rates. This coincided with some operational health physicists, especially those at power reactors, deciding that the society was not servicing the needs of the operational health physics community but becoming too absorbed in research. These two factors led to a 20% decrease in membership, falling from 3100 to about 2500. There were some breakaway moves

11 The Evolution of Organisations

but they were averted by changes in committee appointments by the 1968-9 President Wright Langham to reflect the wider interest and this stabilised the membership. At the same time he addressed another problem— the perceived dominance of the journal and even the Society by Oak Ridge—by removing the voting rights of the Editor-in-Chief.

The Society continued to grow—although there was another dip in membership in the early 1970s—and began to exert a much more sustained and coordinated influence over public affairs. The senior officers became more adept at lobbying government for a professional health physics view of the growing issues of nuclear safety and the Society took a vigorous stand against what was seen as poor science. The maverick member Dr Ernest J Sternglass was formally denounced in 1971 for his allegation that radiation from nuclear power plants was causing infant deaths. The need for developing a Society policy on many public issues led to the preparation of an agreed stance in position papers. The first of these, on the safety of food irradiation, was issued in 1988.

By the end of the century the membership was over 6000 and the Society was grappling with a new problem: the age structure of the health physics community and the decline in investment in the nuclear industry was expected to lead to a dramatic decrease in membership in the early years of the 21st century.

While there were a number of bodies with an interest in radiation protection already in existence in the United Kingdom in the early 1960s there was no society dedicated to the subject[22]. The preceding years had been eventful: the Windscale fire and the scrutiny that resulted, two nuclear power stations at full power, bomb testing in full swing. There was a perceived need for a society to speak for the profession and 66 people met in London in May 1963 to create The Society for Radiological Protection, as a section of the US HPS. Probably stimulated by the interest around the creation of IRPA in 1964, the SRP broke away from HPS in December 1965, by which time its membership had doubled. From the beginning there was a programme of several scientific meetings a year and by 1966 the society was running its first international symposium in Bournemouth. By the year 2000 there had been six such symposia, latterly in partnership with other European societies, with each attracting several hundred delegates.

As for HPS, an early and continuing concern was the education and certification of health physicists and its first Certificate of Competence for operational health physicists was issued in 1979 and by the late 1990s the Society was prominent in developing a unified scheme endorsed by all the other interested professional bodies. The Society's interest in training extended to the development of an National Vocational Qualification in radiation protection and this received accreditation in 1997. As this was happening the membership continued to grow from around 500 in 1978 to over 1000 by the end of the century.

The Society's journal, *Journal of the Society for Radiological Protection*, was first published in 1981 on a kitchen table-top. By the end of the decade it had become a respected quarterly international journal published by the Institute of Physics Publishing under the title of the *Journal of Radiological Protection*.

A problem arose on the creation of IRPA because only one UK society could be an affiliate and there were eight in the UK with some claim. The solution was to create the British Radiation Protection Association (BRadPA) as an umbrella organisation representing the interests of all eight societies. From the beginning there were complaints that, with one society (however large or small) having one vote, the arrangement was unfair and there were suspicions that BRadPA was, at times, trying to develop an independent life. It was finally wound up on 31 December 1997 and SRP took over as the UK affiliated society. An international committee, representing all the societies, was created.

Notes Chapter 11

1. (Taylor L S, 1958b, 1989, 1990; Jennings W A, 2007)
2. (Taylor L S, 1958a)
3. ibid
4. (Lindell B, Dunster H J and Valentin J, no date; Lindell B, 1996; Sowby F D and Valentin J, 2003)
5. (Pochin E E, 1983)
6. (Metivier H, 2007)
7. ibid
8. (Fischer D, 1997)
9. (*History of the BIR*, no date)
10. (X-ray and Radium Protection Committee, 1915)
11. (MRC, 1956)
12. (MRC, 1960)
13. (O'Riordan M, 2007)
14. (*History of the Radiological Society of North America*, no date)
15. (Taylor L S, 2002)
16. (NRC, no date; Buck A L, 1983; Walker J S and Wellock T R, 2010)
17. ibid
18. ibid
19. ibid
20. ibid
21. (Boerner A J and Kathren R L, 2005)
22. (Martin J H, 1988; Jackson J, 2003)

---

12 Changing Standards

The story of standards evolution is very much an international one revolving around the International Commission for Radiological Protection (ICRP) and its predecessor the International X-ray and Radium Protection Committee (IXRPC). Since their first publication the recommendations of the ICRP have, even though they have had no national legal force, had considerable influence as an international benchmark. Their recommendations have, in fact, been the basis for national regulations around the world, being incorporated into the national legal structures with varying enthusiasm and delays. This is an essential step in making the recommendations work but detailing national implementations is beyond the scope of this book: a sensible account would take us into details of national systems and relate more to the law than to radiological protection. So, here we avoid details of national implementation except where they seem relevant to the main international theme.

The first standards were basic rules of design and management of x-ray facilities. The German Röntgen Society published guidelines on shielding of x-ray tubes in 1915 and the recommendations of the British Roentgen Society were published as a one-page broadsheet in November 1915. They recommended that the tube should be shielded, that adjustments should be made from a protected space and that the hands should be protected and never used to test the hardness of the tube. These were followed,in 1921, by the first report from the UK X-ray and Radium Protection Committee (XRPC), termed a Preliminary Report. This was four pages long[1] and in the same vein but rather more detailed. The emphasis for x-rays was on shielding the tube, the use of protective screens and gloves and some minimum thicknesses of lead were suggested. A section on electrical protection recognised that there had been a number of accidents involving these high-tension supplies and there was some concern about ventilation to remove the fumes from coronal discharges. For radium, forceps were recommended for manipulations with storage in a shielded room with walls not less that the equivalent of 8 cm of lead. Personnel should work no more than seven hours a day and were to have Sundays and two half-days off duty each week "to be spent as much as possible out of doors". The walls and ceilings of dark rooms "are best painted some more cheerful hue than black". There was a practicality in these early days that was sometimes lost as protection principles became more complicated.

The first quantitative limit relating to exposure came from Arthur Mutscheller, a German who had gone to the USA as a physicist working for an x-ray equipment manufacturer, at a meeting of the American Rontgen Ray Society in September 1924. He based the recommendation on the experience of x-ray workers who had received—he estimated using calculations from tube currents and voltages, filtration and their patterns of working—about 1/10th of an erythema dose per month. None of them had shown any ill-effects so Mutscheller proposed that, applying a factor of safety of 10, a safe tolerance dose would be 0.01 erythema doses per month. He presented this in a paper[2] the following year and later[3]. In 1934 he proposed[4] the same value for more penetrating radiation equating to a tolerance dose of around 3 r/month[5]. The tolerance value was picked up by the German Committee on X-ray and Radiation Protection in 1925.

Rolf Sievert proposed a similar value to Mutscheller's in 1925 but expressed on a yearly basis—0.1 erythema doses per year. In 1928 Barclay and Cox[6] suggested that 0.00028 of an erythema dose could be tolerated as a daily exposure on the basis of a study of two people (a radiologist and a technician) who had been exposed to soft x-rays for several years. The technician, a woman, was judged to have received 0.007 of an erythema dose daily and the radiologist about one third of this. Neither had shown any effects of exposure. Barclay and Cox divided the technician's daily exposure by 25, as an arbitrary safety factor, to get the tolerance dose.

The First International Congress of Radiology (ICR) in 1925 supported the idea that there should be international agreement on standards of protection and at the 1928 ICR there was acceptance of the British XRPC Third Report with minor amendments[7] (although by now several nations had developed their own basic standards). This was essentially an evolution of their Preliminary Report of 1921 with more detail on shielding. While the recommendations had no explicit exposure-based criteria, Kaye reviewed them after the Congress[8] and looked at how they tied in with the tolerance levels that had been suggested. To those above he added a 1926 proposal from Solomon for a tolerance level of about 0.03 of an erythema dose per day and the value recommended by the Dutch Board of Health in May 1926 of an 0.001 erythema dose in 15 working days. He took a mean value of these as 0.001 erythema doses over 5 days and showed that the protective measures in the recommendations were reasonably consistent with such a tolerance dose and were, as he put it, "reasonable and practical".

The tolerance dose for radium gamma-rays was proposed by Failla[9] in 1928. He reviewed experience in a radium facility where technicians had been exposed to around 0.001 of an erythema dose per month without apparent ill effects. This was lower than the x-ray values but it was, as Failla pointed out, for more penetrating radiation. Since it was relatively easy to provide protection against the softer x-rays, Failla suggested his value be adopted for both.

The 1931 IXRPC[10] recommendations amended the previous ones slightly, extending the tables of lead shielding required for higher energy x-rays and increased amounts of radium (they also added the significant requirement that workers should be examined medically twice a year with blood tests). The tolerance dose thus remained implicit as it did in Handbook 15[11] produced by the US Advisory Committee on X-ray and Radium Protection just before the IXRPC meeting.

Of course, the "international" Rontgen (r) had been adopted at the 1928 ICR so that the tolerance dose could be expressed in rontgens (although there was some reluctance in the medical community who preferred the erythemas they could see to the rontgens they couldn't). The distinction between "exposure" and "dose" was yet to be fully appreciated.

The 1934 recommendations of IXRPC[12] contain the first explicit reference to a tolerance level:

> ... under satisfactory working conditions a person in normal health can tolerate exposures to X rays to an extent of about 0.2 international rontgens (r) per day. On the basis of continuous irradiation during a working day of seven hours, this corresponds to a dosage rate of 10^{-5} r per second. The protective values given in these recommendations are generally in harmony with this figure under average conditions. No similar tolerance dose is at present available in the case of radium gamma rays.

The protective values—the thickness of shielding required for x-ray tubes—were the same as those in 1928 and 1931.

The US Advisory Committee produced its second report in 1934 and also recommended a tolerance level. However this was lower at 0.1 r per day and applied to both hard x-rays and radium gammas. The rationale for the lower value is not clear but it seems to have been a result of more drastic rounding than that the IXRPC applied[13]. It was added as something of an afterthought and appears as a note under "Personnel". It appeared in the next report, published as NBS Handbook 20 in 1936[14], more prominently. Taylor[15] considered this the first appearance of a quantitative tolerance dose in US guidance.

The 1937 IXRPC recommendations[16] reiterated the 0.2 r per day tolerance level and extended it to radium work and after that there was no general international activity for the next 13 years. However, in the USA there was concern that the tolerance level of 0.1 r per day did not take account of genetic factors which were known to lack a threshold: mutation frequency seemed to be proportional to dose down to the lowest levels. In December 1940 there was discussion at the Committee[17] of a proposal to reduce the tolerance dose by a factor of five. Gioacchino Failla, who missed the meeting, subsequently persuaded them not to adopt the reduced level and the 0.1 r per day remained the recommendation and was the level adopted by the Manhattan Project.

Although the NBS Handbook 23 on handling radium[18] was issued in 1938 it had no guidance on limits on the analogue of tolerance dose, the permissible intakes of radioactive materials, or of body content[19]. This was at least partly due to the difficulties of measurement. As we have seen, Robley Evans was responsible for the first accurate measurement of radium body burden around 1934 at MIT using a combination of measurements of radon in breath, of gamma-emissions from the subject's body and of attenuation by the body of standard sources[20]. Using these techniques he measured the body burden of several patients who had been dial painters (and were being treated to see if the radium could be removed by chemical means) and he provided similar gamma-measuring equipment to Dr H Martland, so that from about 1938 he could make similar measurements on his dial painters.

By the end of 1940, as war approached for the US, radium dial instruments were being produced in large numbers and the US military (notably the Navy) pressed for safety standards for dial painting. Evans was a natural person to approach (he was threatened with forcible induction into the Navy as an alternative to volunteering)[21]. An NBS committee was set up at the end of 1940 with nine members (including Martland) and by February 1941, thanks to Evans and Martland, they had information on the body burden and state of health of 27 people who had been exposed to radium. Although there had been some work with rats at MIT between 1936 and 1940 to establish toxicity, the skeletal concentrations needed to produce sarcomas were so much higher than those known for humans that the Committee concluded they were irrelevant. Conclusions on permissible levels would have to be drawn from Evans's and Martland's data. The 20 patients with body burdens of between 1.2-23 µCi all showed some associated injuries but the 7 people with less than 0.5 µCi seemed quite healthy. Evans proposed that the body burden limit should be set "at such a level that we would feel perfectly comfortable if our own wife or daughter were the subject" So the Committee set the tolerance level at 0.1 µCi (or 0.1 µg)—with a safety factor of about 10—on 2 May 1941. In fact, since the figure was usually used as a standard immediately after an intake rather than for residual activity, there was another factor of safety of 10-100.

It was around the end of 1943, with the prospect of kilogram amounts of plutonium being handled, that concern began to be expressed about its safety and how to monitor it. The first proposal for a tolerance level appears to have come from Robert Stone at a Manhattan Project Council Meeting at Clinton on 19 January 1944. Based simply on the differences in half-life and alpha-particle energies, Stone suggested that plutonium should, mass-for-mass, be fifty times less hazardous than radium and the Council endorsed a tolerance level of 5 µg. At the same meeting Arthur Compton authorised the diversion of 10 mg of the material (just becoming available in gram quantities) to Joseph Hamilton at Berkeley for animal experiments that were to provide the initial biokinetic data for plutonium and, specifically the data that formed the basis for urine monitoring. They showed that the behaviour of the element differed from that of radium in several important ways. Notably, it deposited in the liver as well as the bone and excretion was much lower than for radium. For urine monitoring the data suggested a working principle that, after an initial period of high clearance, plutonium was found in urine at about 0.01% per day of the initial intake and this was widely used in the Project.

The sense of urgency had been increased in mid-1945 when animal experiments showed a different distribution of plutonium in bone from that of radium: it irradiated the growing parts rather than becoming incorporated into the older, mineralised, areas of the bone. It was therefore probably significantly more harmful than anticipated and Wright Langham insisted that the tolerance limit should be reduced by a factor of five to reflect this. So, in July 1945, it became 1 µg[22].

This remained the tolerance level until the the Tripartite Permissible Dose Conference at Chalk River, Canada, in September 1949. Here the results of Austin Brues with rats and mice were presented, showing that plutonium was considerably more toxic than previously thought. The conference recommended a reduction in the permissible body burden by a factor of 10, to 0.1 µg. This came at a bad time for the US programme: the new limit would make operations at Los Alamos extremely difficult at a time when the USSR had just exploded its first atom bomb and thoughts were turning to the hydrogen bomb. Reflection after the conference by Brues suggested that respectable arguments could be made for a relaxation of this new proposed limit by a factor of six and in 1950 the AEC adopted a limit of 0.5 µg for Pu-239. In 1951 ICRP recommended 0.6 µg(0.04 µCi) and this was adopted by the US in the same year[23].

This was based on the assumption that the bone was the organ most at risk from intakes: if this was within the necessary tolerance limits then all other organs would be too. This "critical organ" concept was a logical consequence of the tolerance approach. For plutonium in soluble form, that could readily get into the bloodstream, the bone was indeed the critical organ. However, it was found that when plutonium in insoluble form was inhaled (the most usual route of intake) a large proportion remained in the lung and irradiated this organ. The lung then became the critical organ and the maximum permissible body burden was determined by this.

During the Manhattan Project the metabolism of some of the more important fission products and a few other elements was studied in animal experiments. If anything, with the concern about fallout, the work on isotopes such as Sr-90 increased immediately after the war. Maximum permitted body burdens were agreed for around a dozen isotopes at the first Chalk River Tripartite conference in September 1949, and these were published in 1951 as part of the ICRP recommendations[24]. These values, for other than radium and plutonium, were based on tracer studies supplemented with the results of animal experiments to determine movement of material around the body and excretion. This had to be complemented by knowledge of radiation damage from external radiation sources, radium intakes and animals. The maximum permissible total body burden for around 100 isotopes was published in the USA in 1953[25] and by ICRP (originating from its 1953 meeting in Copenhagen) in 1955[26].

The external radiation tolerance dose of Handbook 20 lasted through the war but by the time of the Tripartite conference in 1949 the NCRP (formed from the Advisory Committee in 1946) had a new proposal. In a preliminary report[27] their Subcommittee on Permissible Dose from External Radiation proposed a permissible bone marrow tissue dose (the term "tolerance dose" was losing favour because it was recognised that any dose might cause genetic harm) expressed as 0.3 r per week.

When after the thirteen year gap, the ICRP (as it now was) met in London in 1950[28] they revised the maximum permissible dose for whole body irradiation down because 0.2 r per day was thought "very close to the probable threshold for adverse effects": the x-ray or gamma-ray dose from radiation of less than 3 MeV received by the surface of the body should be less than 0.5 r in any one week. This corresponds to 0.3 r per week in free air and was considered to be in line with the US value of 0.3 r per week to the bone marrow. The limit for fast neutron was provisionally set at one tenth the energy absorption of the x-rays and gamma-rays allowing for an RBE of ten (this and other RBEs are listed in an Appendix). There were also recommendations for maximum permissible exposure to betas. Most of this was taken directly from the US NCRP work presented and discussed at the Chalk River Tripartite meeting the previous year. Sievert apparently argued for an even lower value of the maximum permissible dose; it should be 0.1 rather than 0.3 r per week.

The concept of the critical tissue was introduced and the limit for each (except the skin) was set at 0.3 r per week. The Commission remarked that 0.3 r per week was "the fundamental figure for the irradiation of any critical tissue". As was noted above, for internal emitters the permissible amounts of eleven nuclides in the body for occupational exposure derived from the 1949 Tripartite Conference values were listed with maximum permissible concentrations in air and water. Basic biokinetic data for each nuclide were given and Standard Man made his international debut.

They added an overall caveat:

> Whilst the values proposed for maximum permissible exposures are such as to involve a risk which is small compared to the other hazards of life, nevertheless in view of the unsatisfactory nature of much of the evidence on which our judgements must be based, coupled with the knowledge that certain radiation effects are irreversible and cumulative, it is strongly recommended that every effort be made to reduce exposures to all types of ionizing radiations to the lowest possible level.

ICRP next met in Copenhagen and their recommendations were published as a 95 page Supplement to the *British Journal of Radiology* in 1955[29]. The critical tissue concept was made dominant. The 0.3 rem per week limit (the first appearance of the new unit) to critical tissues was now the primary one and the critical tissues for whole body irradiation were the bone marrow and gonads. The maximum permissible concentrations in air and water were calculated for around 90 radionuclides. The caveat that doses should be reduced to the lowest possible level was repeated.

By the time of the next formal ICRP meeting in Geneva in 1956 attention had focussed on exposure of the public from nuclear fallout. The Bikini tests had resulted in fallout in the Pacific (including the Lucky Dragon incident) and measurements had shown Cs-137 in the bodies of people living in the USA. The possibility of the serious genetic damage that was thought likely from widespread exposure was reason enough to limit public exposure but news was coming from Japan of the increased level of leukaemia among bomb survivors. In their recommendations ICRP therefore set a level for permissible public exposure at one-tenth the occupational level and, in view of the perceived genetic hazard, demanded a limitation of public exposures to close to natural background. The recommendations, which were not published until 1958[30], also reduced the permissible weekly dose to gonads to 0.1 rem.

While the 1956 recommendations were working their way through the system, the Commission had been working on a more comprehensive report and this was published in 1959 and is generally known as Publication 1[31]. The primary occupational limit was now expressed as a limit on the cumulative dose equivalent of 5(N-18) rem where N is the age of the worker with not more than 3 rem per quarter. This was determined by the gonads and the bone marrow; there were more relaxed limits for specific irradiation of other organs. The dose limit for the public was recommended as 0.5 rem/y with a genetic dose limit of 5 rem per generation. The internal dose issues were dealt with in a Subcommittee report the following year[32].

The 1964 report[33] was essentially a revised version of Publication 1 with perhaps the most significant theoretical change being that of substituting Quality Factor for RBE, but it did make important recommendations on the exposure of women of reproductive capacity.

Publication 9[34] in 1966 was the first one to fully accept that there probably was not a threshold for late effects. They made:

12 Changing Standards

the cautious assumption that any exposure to radiation may carry some risk for the development of somatic effects, including leukaemia and other malignancies, and of hereditary effects. The assumption is made that, down to the lowest levels of dose, the risk of inducing disease or disability increases with the dose accumulated by the individual. The assumption implies that there is no wholly 'safe' dose of radiation.

Although this might be seen as dramatic turning point in ICRP philosophy—the espousal at least cautiously of the linear no-threshold hypothesis—it did not result in any real change to the dose limits proposed; the 5(N-18) formula was replaced with a simple 5 rem/y limit but not much else changed. What did change was the caveat that doses should be kept to "the lowest possible level", which had been a feature since 1950. This became apparently more relaxed, requiring that : "... all doses be kept as low as is readily achievable, economic and social consequences being taken into account."

The Commission had taken an important initiative as it prepared Publication 9 by asking R Scott Russell to chair a working group to look at the extent to which the risks of radiation could be expressed quantitatively. The working party report appeared as Publication 8[35] also in 1966. It concluded that the risk of leukaemia following a dose of 1 rad of gamma radiation was about 20 per million people exposed, only a fraction of the value later accepted. It assumed that the risks of all other cancers together would be similar, an even more significant underestimate as it turned out.

In the following ten years there was plenty of work filling in the technical background to the recommendation and in minor revisions but by 1977 the Commission was ready to bring the recommendations into line with the assumption of a no-threshold linear dose response and to accept risk as the key parameter. Publication 26[36] was the turning point.

The threshold effects (now called deterministic or non-stochastic) still had to be protected against and limits were proposed to control them but the stochastic effects—malignancy and to a lesser extent genetic effects—were the central challenge.

A scheme for protection had to be developed that acknowledged that there was no longer a level of exposure that could be described as completely safe. The approaches they could use had been investigated earlier. Publication 22[37] had set out a way, based on cost-benefit analysis, to judge what would be acceptable in its impact on society. The Scott Russell report and subsequent work had provided a link between exposure and risk that could provide a way to judge what might be an acceptable level of risk for individuals

The recommendations were therefore built on three principles that came to be called: Justification, Optimisation and Dose Limitation. Justification required that any practice should have a positive net benefit: unless a practice could leap this hurdle it would be disqualified immediately. It excluded frivolous uses of radioactivity—such as incorporation in jewellery—that brought no real benefit. The Optimisation principle stated that all dose should be kept "as low as reasonably achievable, economic and social factors being taken into account" (ALARA) and was clearly rooted in earlier ideas, with some adjustment in wording. Dose Limitation restricted the dose that any individual should receive and could be based on individual risk considerations.

Optimisation was interpreted as a differential cost benefit analysis (CBA), building on the ideas that had been developed in Publication 22 a few years earlier in interpreting "readily achievable" (the authors thought "reasonably achievable" a better term). As in Publication 22, the Publication 26 formulation was restricted to an analysis of the level of protection—the question of the net benefits of a practice was covered by the Justification principle. The essential of optimisation was to compare the cost of increasing the level of protection with the benefits that gave in terms of collective dose reduction. When the two are the same, the total cost protection is optimised.

To undertake such an analysis the benefits of dose reduction have to be expressed in monetary terms: a monetary value had to be set for the manSievert. This was something that ICRP did not do but many national organisations and individuals did and, using several different approaches, a range of values appeared and were used in a range of optimisation calculations. The basic technique was refined to take some account of so-called risk aversion—the increasing unwillingness of someone to accept increments of risk as the risk rises—with a varying manSievert value. In the UK there were attempts to align it with the long-established legally-binding concept of reasonable practicability, where there was judicial guidance that implied some kind of cost-benefit analysis should be made. Reasonable practicability required spending more money on protection improvements, stopping only when this was grossly disproportionate to the dose reduction achieved—in principle a more onerous requirement than the ALARA one.

The application of the CBA technique was useful but most regulators were unwilling to accept it as binding, particularly since it often suggested that measures they had promoted or wished to see went far beyond being optimised in the cost-benefit sense. Operational health physicists found that it did not help in existing ALARA programmes which were designed to encourage a good safety culture and apply considered common-sense measures to dose reduction. It also became clear that it was difficult to take account of factors such as regulatory pressure, doses delivered far into the future and public perception (some important economic and social factors) and the pure cost-benefit approach began to be seen as just one way towards optimisation. Supplementary and alternative approaches such as multi-attribute utility analysis, based on helping decision-makers take account of all the relevant factors, were promoted. ICRP offered detailed and practical guidance on cost-benefit analysis in Publication 37^{38} in 1983 and on a range of other optimisation techniques in Publication 55^{39} in 1989.

The Dose Limitation recommendations of Publication 26 were now set firmly on a risk basis. There was reason to believe that the existing occupational dose limit of 50 mSv led to a mean dose to the workforce of around 5 mSv. With the fatal risk factor for uniform whole-body irradiation taken as 0.0165 Sv^{-1} this meant that the existing occupational dose limit of 50 mSv/y corresponded to a mean annual fatal risk of about 8x10^{-5} to the workforce, a value comparable with that in other safe industries. Similarly, it was argued the pre-existing dose limit for members of the public in critical groups at 5 mSv would result in lower average doses and lead to "acceptable" levels of risk in the range 10^{-6} to 10^{-5} per year. The formulation in this way (and Lindell, who wrote the text, later said[40] it was "almost incomprehensible") meant that the Sellafield discharges did not exceed the limit. Even when the public limit was reduced to 1 mSv in a year in a 1985 Statement[41] there was still some flexibility and doses up to 5 mSv were allowed for a few years provided the lifetime average was below 1 mSv/y.

The concept of critical organ was strictly relevant only when threshold effects were the problem. As soon as stochastic effects dominated, a different approach was required and sufficient data were available to base this on risks from irradiation of particular organs and tissues. This was done through a system proposed by Wolfgang Jacobi in 1974[42] of weighting of different organs according to their sensitivity and obtaining an effective dose equivalent—the dose equivalent which, if delivered uniformly to the whole body, would give the same fatal risk.

This gave a new basis for calculating limits on intake of radioactive materials, based on the limitation that the intakes in a year should not give rise to a dose over a lifetime greater than the annual dose limit, subject to not exceeding any non-stochastic limits. Using these principles, the Annual Limits on Intake of a wide range of isotopes were calculated and published over the next few years as Publication 30, based on the various biokinetic models that had been developed.

During the 1980s the data coming from Japan showed that the risk estimates used in Publication 26 were underestimates and pressure grew on ICRP to issue revised recommendations to respond to this. This was done in 1990 and appeared in the 1991 Publication 60[43] which also introduced some new ideas. For the first time potential exposures were included in the optimisation framework and a distinction was made between dose control in a practice and an intervention (when action was taken to reduce pre-existing doses).

The most immediately obvious change was in the occupational dose limit which was reduced from 50 mSv/y to 20 mSv/y averaged over five years with a maximum of 50 mSv in any year. The public limit remained at 1 mSv/y.

Optimisation remained central to the recommendations but an additional restriction was now placed on the outcome: individual doses were subject to constraints for individual practices that were more restrictive than the dose limits. This reflected not just the concern to protect individuals but an emerging principle that equity

12 Changing Standards

in consequences was important. Similarly, potential exposures were subject to the optimisation process with risk constraints.

Publication 60 was some kind of summit for cost-benefit analysis and collective dose. After 1990 attention—or at least its focus—turned to the individual and his dose and risk so, in Publication 77[44] on waste management in 1997, the link with cost-benefit and collective dose was weakened. Thoughts turned to a thoroughgoing revision of the Commission's approach. It moved from the ethics of utilitarianism—the basis of a societal approach to protection—to others with an individual focus. By the end of the century ICRP was preparing to consult on recommendations that emphasised "deontological and equity-based" ethics.

As well as the general dose limitation recommendations the ICRP has also been concerned with protecting special groups. Infants, young workers, miners, pregnant and potentially pregnant women and patients have all featured. There have been, as well, technical reports ranging from the description of metabolic and dosimetric models to discussion of the possibility of creating an index of harm that would allow a fairer comparison between the delayed effects of radiation on nuclear workers and the rather prompt effects of fatal accidents in other occupations. By the end of the century the Commission had been responsible for some 82 numbered publications.

Notes Chapter 12

1 (X-ray and Radium Protection Committee, 1915)
2 (Mutscheller A, 1925)
3 (Mutscheller A, 1928)
4 (Mutscheller A, 1934)
5 (Cantril S T and Parker H M, 1945)
6 (Barclay A S and Cox S, 1928)
7 (IXRPC, 1928)
8 (Kaye G W C, 1927)
9 (Failla G, 1932)
10 (IXRPC, 1931)
11 (NBS, 1931)
12 (IXRPC, 1934)
13 (Hacker B C, 1987)
14 (NBS, 1936)
15 (Taylor L S, 2002)
16 (IXRPC, 1937)
17 (Hacker B C, 1987)
18 (NBS, 1936)
19 (Rowland R E, 1994)
20 (Evans R D, 1981)
21 ibid
22 (Moss W and Eckhardt R, 1995)
23 ibid
24 (ICRP, 1951)
25 (NBS, 1953)
26 (ICRP, 1955)
27 (Taylor L S, 1984)
28 (ICRP, 1951)
29 (ICRP, 1955)
30 (ICRP, 1958) also Radiology 70, p261
31 (ICRP, 1959)
32 (ICRP, 1960)
33 (ICRP, 1964)
34 (ICRP, 1966b)
35 (ICRP, 1966a)
36 (ICRP, 1977)
37 (ICRP, 1973)
38 (ICRP, 1983)

39 (ICRP, 1989a)
40 (Lindell B, 1996; Sowby F D and Valentin J, 2003)
41 (ICRP, 1985)
42 (Jacobi W, 1975)
43 (ICRP, 1991a)
44 (ICRP, 1998)

13 Criticality Safety

The discovery of fission and the possibility of a chain reaction based on it were exploited more quickly than most scientific discoveries and certainly with more terrible results. The key to this was the estimation of the critical mass needed for a chain reaction—and particularly the critical mass for a bomb. Most of the physical principles were established quickly in the wartime environment through the development of the weapon and through the design of the reactors that were to produce plutonium. This chapter spends some time looking at the bomb development to see how the understanding of the critical mass developed through experimental and theoretical investigations. However, the main interest here is in safety when the key questions is: how close to a critical configuration are we? The evolution of the methods of criticality safety that have been developed to answer this question is our central topic.

Given the diversity of situations in experimental facilities, reactor management systems, industrial-scale reprocessing plants and fissile material transport, the techniques from the early days proved inaccurate and their results demanded great caution in application. The history since has been one of improving the assessment techniques to give greater confidence and remove excessive conservatism in design. Since the 1960s much of this effort has gone into developing and validating computer codes and the histories of the two principal Monte Carlo codes (KENO and MONK) with wide international use have been traced in some detail. Of course, Monte Carlo codes have emerged from other nations –notably France—and alternative deterministic computational techniques have been used by many. Some of these are outlined but, given the complex and dynamic nature of the field, it would have been impossible to do justice to all of them without creating a long and elaborate list.

Safety in criticality arises not just from improved assessment but from better understanding of what might cause an accident and how big that accident might be. These lead to principles for design and operation of facilities with fissile materials. While the principal aim is to trace the evolution of assessment techniques, accident consequences and management principles are briefly mentioned for completeness.

When a uranium-235 nucleus fissions, on average, 2.4 neutrons are emitted with an average energy of 2MeV[1]. At this energy each neutron has three possible fates: it may collide with a nucleus and lose energy by inelastic scattering, it may be absorbed and lost or it may cause another fission.

If we first consider natural uranium, with approximately 0.7% U-235, then the most likely fate, happening on about 80% of occasions, is inelastic scattering by a U-238 nucleus. Of the remaining 20% of neutrons almost all will cause fission in U-238 with just 0.3% will causing fission in U-235. The fissions will not produce enough neutrons to sustain a chain reaction so we should look at the scattered 80%. These will lose a substantial fraction of their energy and further inelastic collisions are still the most likely fate but, because the neutron energy is now below the threshold for fission of U-238, U-238 fission will not happen. However, because U-235 is more susceptible to fission at this energy and in spite of its low content, U-235 fission becomes more important with about 15% of neutrons causing its fission. This is still not a high enough proportion to cause a chain reaction and the dominant effect is that the neutrons continue to lose energy by collisions. This continues until the energy falls to about 1keV when a new process becomes important. This is resonance capture when the neutrons are most likely to be absorbed by the U-238 nucleus and therefore to be removed completely from the system. A few neutrons escape capture, and they do reach an energy region where U-235 fission is much more likely than any other process. However, in the U-238—dominated natural uranium, their number is so small that a chain reaction cannot be sustained.

If the uranium is enriched and the U-235 content increased to 50%, the picture changes dramatically because at energies between

0.3 and 2MeV the probability of U-235 fission becomes about equal to that of non-elastic scattering. Since an average of about 2.4 neutrons are released on each fission, there is now the possibility of a chain reaction. This is fast fission and is the basis for the fission bomb and the fast reactor.

An alternative approach route to a chain reaction is to get the energy of the neutrons below the U-238 resonance capture region before too many are lost and this depends on the presence of a moderator. A moderator is a material composed of lighter elements such as hydrogen or carbon with low neutron absorption characteristics. When neutrons collide with nuclei of these lighter elements they lose a high proportion of energy in each collision and their energy is quickly reduced through the resonance region. In fact the energy falls until it is comparable with the thermal energy of the nuclei of the moderator. In this region the probability of fission of U-235 is at its highest and, provided the moderator is well-chosen, a chain reaction becomes possible even with natural uranium. This is the basic process of the thermal reactor and makes possible a chain reaction with little or no enrichment.

These factors are summarised for a chain reaction in a thermal reactor by the Four Factor Formula to give k_∞, the infinite multiplication factor:

$$k_\infty = \varepsilon p f \eta$$

where

ε is the fraction of fission neutrons slowing down below 1MeV(the fast fission factor)

p is the resonance escape probability (the fraction of these that escape resonance capture)

f is the fraction of neutrons absorbed in fissile material rather than elsewhere (the thermal utilisation)

and η is the number of neutrons produced per neutron absorbed in the fissile material

If a chain reaction is to take place in an infinite medium then $k_\infty \geq 1$.

13 Criticality Safety

In practice media are not infinite and there is another way for neutrons to be lost from the potential chain reaction; they simply escape from the surface of material. If the material is spherical then the larger the radius, the less likely it is that neutrons will escape before they go on and cause a fission. The radius of the sphere just large enough to allow the chain reaction to progress in a particular material is its critical radius; its mass is the critical mass of the sphere.

Perhaps the most important parameter in criticality is the effective multiplication factor k_{eff} (called k effective):

k_{eff} = k_∞ x probability of neutrons not escaping from the system

When k_{eff}=1 the non-infinite material will just be critical and sustain a chain reaction. Because of the different probabilities of neutron escape different geometric shapes of the same material will have different critical masses. The critical mass will also depend on the surroundings because these may act as a reflector and direct escaping neutrons back into the fissile material. They may also introduce a moderating effect. These factors will reduce the amount of fissile material needed to make a critical mass. So, for example, a sphere of U-235 in water will have a lower critical mass than one in air.

So there are two distinct aspects to assessing whether criticality can occur in uranium or any other fissile material such as plutonium-239:

- Whether the material is intrinsically able to support a chain reaction
- Whether there is enough of it, given its environment, to limit neutron escape and actually support one

The first of these is not generally difficult to deal with: given the data some simple arithmetic will suffice to give an answer. The second presents a much greater challenge and this is the central theoretical problem of criticality safety.

TAMING THE RAYS

There is one further complication that is important for reactor control and criticality accident evolution: the time spectrum of fission neutron emission. While most fission neutrons are emitted very promptly, a small fraction are released up to a minute later. Although this delayed fraction is small it is important in reactor control because, with careful design, it gives time for mechanical systems to control the chain reaction and the reactor power. It is also significant in criticality accidents. If a system is critical with just the prompt neutrons (prompt critical) the excursion evolves extremely quickly and will be exceptionally violent. If the system is not made critical by the prompt neutrons but is once the delayed neutrons arrive (delayed critical) then, while the energy released may still be enormous, it will happen over a longer period.

In a series of five lectures between 5 and 14 April 1943, Robert Serber summarised[2] what was then known of all aspects of the bomb. He reviewed the available nuclear data and gave an analysis, based on diffusion theory, of the critical size of a naked sphere of fissile material. He calculated that the critical radius would be given by

$R_c^2 = \pi^2 D \tau /(v-1)$

Where D is the diffusion coefficient, τ the mean time between fissions and v the number of neutrons produced per fission. For U-235 metal the calculation gave a critical radius of 13.5 cm and a critical mass of 200 kg.

Serber recognised that the simple diffusion theory did not apply because the mean free path at 5 cm was comparable with the dimensions of the system. He quoted the result of a "more exact diffusion theory" as leading to a critical mass of 60kg. He later indicated that this difference was largely accounted for by better accounting for the effects at the sphere boundary.

In the practical weapon the fissile core was to be surrounded by a thick uranium metal shell (the tamper) which would reflect some of the escaping neutrons back into the core and also hold the core together for the briefest of times while the chain reaction built up. With a uranium tamper "best available calculations" gave R_c as 6 cm with a critical mass of 15 kg for uranium. Serber estimated the critical mass of a tampered sphere of plutonium as 5 kg.

Diffusion theory continued to form the basis for estimating the weapon critical mass throughout the project but it was extended in the most complex ways to account for the fact of a spectrum of neutron energies, the role of inelastic scattering and the slowing down process, and the anisotropic nature of the collisions. The need to handle a distribution of neutron energies led to the introduction in late 1943 of a multi-group approach in which neutrons were treated as having energies in three or four discrete bands. Also, since one model of the weapon involved a core of uranium hydride, the complexities introduced by hydrogen scattering had to be addressed. Of course, this was all complicated by the fact that the weapon was blowing itself apart while the chain reaction built up.

The calculations for carbon moderated reactors were somewhat more straightforward because Fermi's age theory, based on diffusion and continuous slowing-down, could be used. However the experimental Water Boiler facility demanded a three energy group approach because of the large discrete losses of energy that occurred in neutron collisions with hydrogen nuclei. Diffusion theory was also used in plant criticality calculations and an example can be found in a report written by Edward Teller with Oak Ridge and Chicago staff[3].

Diffusion theory was not just useful for the prediction of critical mass. It could also extend the value of experiments, a procedure best explained through the concept of "buckling". The "material buckling" (B_m^2) depends only upon the properties of the nuclear material under consideration and is defined as:

$$B_m^2 = (k_\infty - 1)/ M^2$$

Where k_∞ is the infinite multiplication factor in the material and M^2 is the mean square distance from the point where a neutron is released in fission to the point where it is absorbed in an infinite medium. M is known as the Migration Length; M^2 as the Migration Area. The material buckling is therefore a measure of the net production rate of neutrons combined with an indication of how far they spread from their origins.

From the diffusion calculation comes a factor that depends only on the geometry of the fissile assembly: the "geometric buckling" (B_g^2).

For a bare sphere of radius R, for example:

$B_g^2 = (\pi/(R+\delta))^2$

Where δ is a small correction factor, the extrapolation length, which is usually determined experimentally

Diffusion theory says that, when the two bucklings are equal, the sphere becomes critical i.e. when

$B_m^2 = B_g^2$

$(k_\infty-1)/M^2 = (\pi/(R+\delta))^2$

A key factor is the ratio between M and R. If M is large compared with R then the neutrons must wander a large distance before causing another fission and this may take them out of the sphere before they do so.

The term "buckling" is a strange one. It derives from the change in shape of the neutron flux profile in a material as it passes criticality. Buckling or Shape Conversion is a method still recommended for preliminary or check calculations and proved, as will be seen, a useful way to extend the applicability of experimental results.

For example, the geometric buckling of a cylinder is calculated as:

$(2.405/(r+\delta))^2 + (\pi/(h+2\delta))^2$

where h is the height and r is the radius. If these are the critical dimensions known from some experiment or calculation on the cylinder then the critical radius R of a sphere of the same material will be given by equating the two geometric bucklings.

$(\pi/(R+\delta))^2 = (2.405/(r+\delta))^2 + (\pi/(h+2\delta))^2$

The technique probably originates from Paxton[4].

These methods, developed before and during the Manhattan Project, when combined with criticality experimental data, provided the foundation for criticality safety in the early nuclear programme[5].

While calculations could be made, there was considerable uncertainty about their accuracy and they could deal with only relatively simple shapes. In the early years of criticality safety direct measurement was the basis for most decisions and a considerable body of data for many materials and configurations was built up.

The aim in most such experiments is to approach criticality closely enough to be sure where it is but to avoid it and the damage that might be caused by the associated energy release. The general procedure involves measuring, with a suitable neutron detector, the neutron multiplication (m) produced in the system from a fixed neutron source. At criticality m will become infinite so if 1/m is plotted as the system is brought incrementally towards criticality it will be found to tend towards zero. It is then possible, at least in principle, to extrapolate to zero and find the critical point[6].

Some early experiments on enriched U-235 have been described by Reider[7]. The Water Boiler was an assembly in which enriched (14.7%) uranyl nitrate solution was pumped into a sphere to investigate the approach to criticality. The Dragon Experiment, dating from early 1945, dropped a piece of fissile material through a hole in another piece to give a 1/100 second burst of prompt criticality. The "Drop Leaf" assembly was used just after the end of the war and allowed a reflector to be placed remotely around the enriched uranium. The Moveable Table, from 1946, could be used to move two pieces of material together to reach criticality. This is just a selection: there were many additional tests at Los Alamos for safety reasons including a number on the effects of flooding. Some of these experiments were directed solely at the bomb design but they began to give general information on criticality.

Arrays are important for storage and transport of fissile material and they present special problems because, while the individual units may be safely sub-critical, when they are put close together they may reach criticality. This is because when neutrons escape from one unit they may enter another and this interaction may just tip the balance of criticality.

The need to understand arrays meant that they were the subject of many experiments from the 1940s onwards at several US facilities[8]. However, such experiments are expensive and some (relatively) simple methods were needed for evaluating proposed arrays and eliminating clearly unsafe ones. Even when computer codes became available such methods could absorb their results and make useful guides. Several methods, each with various formulations, were developed and they fall into two broad classes: those that seek correlations from experimental data and those that start from a more fundamental basis. The four methods in the first category are the Surface Density, NBN2, Density Analogue and Solid Angle methods. The Interaction Parameter method is more fundamental.

The *Surface Density* method was originated by Paxton at Los Alamos in the 1950's and involves imagining the array of fissile material projected onto a surface, merging into a slab and then flooding with water. In one formulation the array is regarded as safe, based on experimental results, if the surface density of the slab σ

$$\sigma = 0.54\ \sigma_0\ (1 - 1.37f)$$

where σ_0 is the surface density of a critical water-reflected infinite slab of the same material and

f = mass of array unit/critical mass of un-reflected sphere of same material

The *Limiting Surface Density* (NB_N^2) method, developed by JT Thomas at Oak Ridge in the late 1960s (18) can be seen as an extension of the Surface Density method that takes explicit account of the finite number of elements in a real array. It is based on the observation that the neutron non-leakage fraction from the whole array, which must be the same for different critical arrays of the same units, can be represented by the expression:

$$1/(1 + NB_N^2)$$

where N is the number of units and B_N^2 is a geometric buckling independent of the fissile material involved. The material composition is brought in through a simple empirical relation based on experiments and Monte Carlo calculations. Combining the geometrical and material factors defines the critical conditions.

The *Density Analogue* method, proposed by Paxton in 1975, defines limits that are independent of the storage arrangement and is based on regularities in data from critical arrays. In one formulation it gives the sub-critical limit for storage arrays of any shape as N units as:

$$N = (2.1\ \sigma_0(1 - 1.37f)/m)^3 V^2$$

Where V is the volume occupied by a unit in the array, m is the mass of a unit and N is the total number of units.

The *Solid Angle* method was developed at Oak Ridge in the 1950's as a fairly quick method for evaluating interactions between individual fissile units. It depends on the observation that, with no absorption or reflection, the probability of a neutron emerging from one unit reaching another one depends on the solid angle subtended by the target unit at the originating one. The method, as described by Knief (19), is based on calculating the total solid angles subtended at the centre of one unit by all the other units. Then the solid angle (in steradians) for that unit must be less than or equal to:

$\Omega_{allowed} = 9-10\ k_{eff}$ where keff is the effective multiplication factor for that unit

This is then repeated for all the units and all of them must satisfy the condition. There are some restrictions on the method including that the keff of individual units must be less than 0.8 and that no solid angles may be greater than 6 steradians. With these restrictions the method is conservative—and some times very so.

The *Interaction Parameter* method embodies a more fundamental approach and[9] originated with A F Thomas at AWRE Aldermaston in the UK in the 1950s and was subsequently developed into a powerful tool. It starts by considering the neutron balance of the array and notes that the condition for criticality is:

$F_i = \Sigma q_{ij} F_j$ (i=1,2,3.....n) summed over j

Where F_i is the number of neutrons leaving unit i per unit time and qij (the interaction parameter) is the number of neutrons leaving unit i as a result of one neutron leaving unit j. It is assumed that the interaction between i and j is not affected by the presence of the other units.

Then a sufficient condition for sub-criticality is that :

Maximum value Σq_{ij} for all i should be less than 1

The key to the method is then efficiently and conservatively estimating the q's. This may come from direct experimental work on a single array element, measuring its response to a neutron source representing another unit. It may also be estimated from the geometry of the array and the surface multiplication factor of individual units. The method and some applications are described in great detail by Thomas and Abbey.

All these methods have slipped into obscurity as computer codes have become easier to use, more accurate and cheaper to run.

The Oak Ridge Criticality Measurements Laboratory performed many experiments on U-235 and U-233 until its closure in 1975. Some of the early ones addressed criticality limits for particular facilities but others (including the array studies mentioned below) made a vital input to code validation. The Battelle (PNL) facility, the Plutonium Critical Mass Laboratory, was designed specifically for experiments on plutonium and has conducted experiments on thousands of different system of many kinds[10].

The most comprehensive measurements have probably been made at the Los Alamos Critical Assemblies Laboratory. The laboratory has experimented with U-235, U-233 and Pu-239 in all physical forms. The facilities there (Godiva, Jezebel, Flattop, Comet, Sheba and others) provided basic data for design, codes and criticality alarm detectors and continued to do so under Lawrence Livermore management.

Measurements like these answered the immediate problem but many were not taken near enough to criticality to give data supporting the computer calculations that were beginning to be developed in the early 1960's. Higher quality experiments were made at ORNL in 1963-4 on U(93.2) metal cylinders with various reflection and interspersed materials and these gave the first precise data for metal arrays. They were very important for establishing standards for storage as well as providing an input to computer codes. Good data for plutonium metal arrays came from the Lawrence Livermore series mentioned above.

Rather smaller critical measurement programmes occurred in the United Kingdom and France. The UK programme included high-enriched uranium and plutonium solution measurements at Dounreay in the 1950's and 1960's, and the SCAMP programme at Aldermaston in the 1970s on mixed plutonium-uranium. There were also numerous low energy reactor experiments performed at Harwell and Winfrith (after the DIMPLE reactor was moved there from Harwell) as part of data gathering for the UK reactor development

13 Criticality Safety

programme and at the Atomic Weapons Research Establishment, Aldermaston for more basic research. The UK programme, which effectively ended in the 1980s, has been described by Simister and Clemson[11]

The French have had a much greater sustained commitment to criticality experimentation. Their first experiments (ALECTO) at Saclay were concerned with U-235, U-233 and Pu-239 nitrate solutions. They then opened the Station de Criticité at Valduc in 1963 and conducted experiments on high enrichment uranium and plutonium solutions including the effects of neutron poisons and interaction effects.

As criticality data accumulated and computers became more powerful the emphasis changed from collecting more data to developing better models for criticality calculation—although these models had to be supported by benchmarking experiments.

With the increasing availability of more powerful computers it became practicable to carry out more refined diffusion calculations using multigroup diffusion theory—as had been done at Los Alamos for the weapon. The first recorded application[12] was by Ehrlich and Hurwitz in 1949 at Knoll's Atomic Power Laboratory using the IBM 604 Electronic Calculating Punch. Similar one-dimensional multigroup calculations were undertaken on IBM Card Programmed Calculators at many AEC labs in the early 1950s. However, at about this time stored-program computers were becoming available, notably the ORACLE at Oak Ridge and the MANIAC I at Los Alamos and these allowed the calculations to be made more quickly and permitted features like automatic searches for the most critical configuration of a system. With the beginning of the nuclear power industry, such codes were a major development in reactor design and, as computing power and fast access memory increased, more detailed nuclear data could be made available to them. However, they were designed for reactor situations with high moderation, low leakage and a minimum of materials with high absorption losses and these were the conditions where the diffusion approximation worked best. Criticality safety often involves low moderation, high leakage and the use of materials with high absorption cross-sections—just where the diffusion approach was least reliable. Alternative techniques were required[13].

The Monte Carlo approach tries to simulate an actual process that depends on chance with a computer programme that uses a random number generator to select the options for the process. It is applicable to fission because the interaction processes of neutrons with nuclei are random ones. It has been claimed[14] that Enrico Fermi used a hand version of the technique in the mid-1930s to help him in work on neutron diffusion but its serious application to physical processes had to await the digital computer. ENIAC, one of the first computers, had been constructed towards the end of the war for ballistics calculations and had been used for thermonuclear problems associated with the Super (the hydrogen bomb). It appears to have been John Von Neumann who suggested, in 1947, that ENIAC could be used for a Monte Carlo solution of the neutron transport problems associated with a bomb and it was agreed that this would be done once ENIAC had been moved 200 miles to its permanent base in Maryland. While this move was being made (and there were those who thought its 18,000 valves would never work again) Fermi devised a simple mechanical analogue device[15] (a sort of trolley later christened the Fermiac) that could be used to trace out the path of neutrons on a scale drawing of the fissile assembly. Externally-generated random numbers determined whether the neutron was fast or slow and the direction and distance travelled after each collision. It could account for boundary crossings and was in use for about two years[16].

However, the future lay with computers and when ENIAC survived its move it was used successfully for the calculations on neutron transport, handling complex geometries and a realistic neutron velocity spectrum. Other workers ran other problems and by mid-1949 there was enough evidence and interest for a conference in Los Angeles[17].

The principle of the Monte Carlo technique is easy enough to understand but, in practice, there are many technical issues to address to get a practical code:

> *the representation of the cross-section data as a function of neutron energy*

13 Criticality Safety

the cross-sections of relevance—particularly those for resonance capture—vary with neutron energy and this must somehow be factored into the code. Both point and group representations have been used. A significant development was the provision of cross-sections in six and sixteen groups for fast and intermediate energy neutrons by Hansen and Roach in 1961[18]. The detail of cross-section data libraries is beyond our scope (although their increasingly international character is touched upon later) but they are now available with such fine structure that the distinction between point and group data sometime becomes blurred.

the representation of the geometry

The geometry of systems can be quite complex—far from the simple spheres, cylinders and slabs. To be useful for criticality assessments it must be possible to construct the required geometry of the system in a way that is straightforward and readily checked.

the neutron tracking system

The simplest Monte Carlo routine would try to simulate directly the behaviour of each neutron: it would be born from a fission, collide a number of times and then finally be absorbed or escape from the system. In practice this is wasteful because the history of neutrons that leave the system, which contains valuable information, is lost. Rather than losing it, a more efficient method is to continue to track the neutrons and adjust the weight given to them in the calculation.

variance reduction

Suitable biasing techniques allow the user to steer the calculation so that neutrons are guided towards regions of importance to the calculations. Such non-analogue strategies can reduce calculation times and/or statistical errors.

The first general Monte Carlo code[19] for neutron problems was the O5R code from Oak Ridge, designed to run on the ORACLE mainframe computer. Initially conceived as a reactor code, it used pseudo-point neutron cross-sections and could deal with a range of geometries. However, the complexity of the cross section data made it difficult to validate and the geometry was difficult to set for complicated arrangements. It was difficult to use with confidence for criticality calculations[20].

These shortcomings led to the development of the KENO code at Oak Ridge beginning with Elliott Whitesides' work in 1963. He started with a small Monte Carlo testing programme and steadily expanded this to make a complete criticality assessment code with a geometry package. This has continued to be enhanced by Whitesides, his co-workers and successors and remains one of the world's principal assessment codes today.

By the end of the 1950's the civil criticality safety code development effort in the UK was concentrated around Ed Woodcock at United Kingdom Atomic Energy Authority (UKAEA) Risley. Woodcock had been drawn from the Meteorological Office for his expertise in diffusion theory and was responsible with his team for the GEM code[21]. This ran on the IBM 7090, first appeared in 1962 and was designed specifically for criticality applications. It had an unusual neutron tracking scheme and, while easy to use, its results (the British rather disliked the keff concept at this stage and expressed the code predictions in a more complex way) were rather difficult to interpret. However, it had the capability to perform calculations on assemblies of several materials made up of spheres, cylinders and regular parallelepipeds. This geometric capability could be refined with the so-called HOLE routine which, adapting an approach developed by John von Neumann to handle the problem of voids in fissile material, introduced a notional collision that had no effect on the neutron. This reduced the problem of tracking the neutron across boundaries to the much simpler one of knowing the material in which the interaction, notional or otherwise, occurred. The cross-section data capability was high but, in practice, the data were simplified and approximate.

The successor to GEM in the UK was MONK[22] originally developed in the 1970's. This, like KENO, has been updated and enhanced regularly since and is the first-choice code today in the UK. MONK originally used point data and this continues to be the preferred option for most users. However, group data was introduced in MONK-G in the mid-1970's and this was later consolidated as the better-known MONK-5W. Both KENO and MONK have become much more flexible codes with options on the nuclear data available. MONK offered options on the tracking regime used and included "superhistory tracking" in the 1990's. It retained the HOLE option for geometry specification to the end of the century.

13 Criticality Safety

Other Monte Carlo codes have been developed and used for criticality safety assessment. The principal competitor for KENO in the USA is MCNP, developed at Los Alamos and a direct descendent of the original Monte Carlo work there (Los Alamos claim that over 300 person-years have been invested in it). It is a much more general code than either KENO or MONK since it includes both neutron and photon transport and it has been widely used for physics problems related to radiation including dose calculations and radiation damage studies[23]. For criticality work its comprehensive geometry package and variance reduction techniques are key features. It finds some use in the UK where its complete independence of MONK and KENO makes it of particular interest to the UK nuclear regulator. The French code MORET is also a well-validated and internationally-recognised multigroup Monte Carlo code developed by the Institut de Radioprotection et de Sûreté Nucléaire (IRSN) and is the approved Monte Carlo assessment code in France. As with TRIPOLI, a continuous energy Monte Carlo code from the French CEA, it has not, so far, established itself elsewhere.

The increase in computing power has changed the availability of Monte Carlo techniques immensely. Calculations that could be only be run on a mainframe could, by the 1990's, be performed to sufficient precision on stand-alone workstations and personal computers. The power and dominance of the Monte Carlo codes by the end of the century led to concerns among some regulators that no independent checking of their predictions was possible. Also, even though so powerful, the Monte Carlo method still requires considerable experience from the practitioner if reliable results are to be obtained efficiently. The main weakness derives from its statistical nature; Whitesides in 1971 pointed out a particular problem leading to underestimation of keff in loosely coupled systems[24] that remains a classic challenge.

An alternative to the Monte Carlo approach is to solve the transport equation that describe the generation, movement and capture of neutrons. This complex equation can be simplified by breaking the nuclear data down into energy groups—the multi-group approach used in both diffusion and Monte Carlo. The resulting equations can then be solved numerically on a spatial mesh in a finite difference approach. However, one of the most difficult things to manage in such calculations is the directional anisotropy—the colliding neutron has a preference to continue in its original direction of travel. The flux at one point therefore depends in a very complex way on the flux at many others. One way to proceed in numerical computations is to divide the possible directions into a discrete number of groups and then to weight these to reflect the directional behaviour.

The method (often known as the Sn method) emerged from the nuclear weapons design programme at Los Alamos in the mid-1950s[25] and was recognised quickly as a more accurate tool than diffusion theory for most criticality situations. Initially as a one-dimensional tool it was able to handle spheres, infinite cylinders and planes but it was extended to two dimensions in the late 1950's. By the 1960's it was a standard computational tool for criticality computations with computers with the two-dimensional version giving a capability for some simple arrays. The one-dimensional codes were DTF at Los Alamos and ANISN at Oak Ridge. The well-known two-dimensional codes DOT from Oak Ridge (later developed as DOT2B by General Electric) and TWOTRAN (Gulf General Atomic) appeared in the late 60's. Even then the methods were restricted to relatively simple geometries and even the two dimensional codes could not be fully exploited until computer power increased and became cheaper. However, they were a useful as a cross-check on other methods.

The Winfrith Improved Multigroup Scheme (WIMS) code began to be developed by the UKAEA at Winfrith in England in the 1960s as a deterministic thermal reactor physics code. Although primarily aimed at reactor calculations, some of its later versions—much enhanced by reactor research and development—found extensive use in the UK as a safety assessment tool when simplifying assumptions on geometry could be made and advantage taken of the lattice geometry. Such deterministic approaches have particular value in scoping calculations and sensitivity studies where they give results free of the uncertainty associated with Monte Carlo methods. Similarly, the development of the deterministic APOLLO code by the CEA in France began in the 1970s and continued to the end of the century. The entry of new deterministic codes such as Imperial College, London's EVENT in the last decade of the century seemed to mark a resurgence of interest in deterministic techniques, driven perhaps by cheaper and more powerful computing resources.

13 Criticality Safety

The US and UK codes (Monte Carlo and deterministic) mentioned above are now generally operated on a strictly commercial basis for the use of the current and latest editions. The not inconsiderable cost of this leads some smaller users to employ older versions of some of the codes which are available (free of charge to some non- commercial users) primarily through the OECD Nuclear Energy Agency (OECD/NEA) Databank in Paris (www.nea.fr).

The Monte Carlo and deterministic approaches such as diffusion and discrete ordinates developed quite independently and indeed have been to some degree in competition. In the 1990s there was an interest in combining the best features of each to make a more efficient overall tool and provide a complete package for criticality assessment. Typically, these use the deterministic approach, perhaps on a simplified geometry, to estimate the flux distribution and then use this to ensure that the Monte Carlo neutrons are concentrated in regions of greatest importance.

Most of the major codes were, by the end of the century, incorporated into code suites which performed criticality calculations and other related ones such as shielding. Provision was made for easier and more reliable data management and, in some cases, the use of a hybrid methods. For example, the UK WIMS code (managed by the ANSWERS Service of Serco Assurance www.sercoassurance.com) has developed into a code suite built around MONK. Similarly the Oak Ridge SCALE suite is built around KENO and the French nuclear industry began, in the mid-1990s, to construct an integrated suite of programs called CRISTAL based on the MORET/APOLLO/TRIPOLI combination.

The computer codes for criticality are complex and demand some considerable skill in use if they are to be accurate and efficient. They have allowed designers to decrease the conservatism demanded by the cruder methods of earlier years so their accuracy is crucial to safety and this depends on the nuclear data that is used. This is a complex topic rather beyond the scope of this paper but at least one trend is worth noting: the increasingly global nature of the databases used. In the early work it was necessary for individual national groups to develop their own data sources—although there was always much sharing of information [26]—and structure it for their particular requirements. Progressively this cooperative approach has become more international with general availability through the OECD/NEA Databank of US (ENDF) and other national nuclear data libraries. This cooperative approach culminated in an international project, sponsored by the OECD/NEA, to produce the Joint Evaluated File (JEF) data. This began in 1995 and the first evaluated data became available in 2002.

Although there has never been an accident as a result of a failing in a code, the need for validation against benchmark experiments has long been recognised. For example in the USA, although there had been comparisons by individual workers for many years, the requirement for validation was formalised in a Nuclear Standard in 1975[27]. This was re-iterated in ANS 8.1 in 1983 and subsequent revisions.

An essential part of the validation process has been international benchmarking and the most significant of these is the International Criticality Safety Benchmark Evaluation Project (ICSBEP) that was initiated by OECD/NEA in 1992. The International Handbook of Evaluated Criticality Safety Benchmark Experiments, prepared under the project, contains over 400 evaluations of nearly 4000 critical or near critical configurations and is widely used around the world in code validation.

The data from experiments and code calculations have been made available to practitioners through handbooks, where they were generally combined with guidance on criticality safety principles. The earliest guidance came with the AEC's Nuclear Safety Guide TID 7016 in 1957[28], the declassified LA-2063 from the previous year. This was not intended as a handbook but, in its 24 pages, it sets down the basic nuclear safety problem, some rules and data and then goes on to consider applications to processing plants.

It begins by pointing out the challenges. First, the nuclear data were not sufficient and the theoretical methods not well enough understood to calculate critical masses to better than 15 to 20 %. Reliance had to be placed on experimental measurements of critical mass and extension of these by theory. Second, the disposition of fissile material may not be well-known in a process, particularly in

13 Criticality Safety

off-standard conditions. Third, administrative controls, based on operating rules, must be rigorously applied.

After reviewing the factors that contribute to criticality, the need for conservatism in design criteria is stated. The inability of instrumentation to warn, in practical situations, of the approach to criticality was noted. However, criticality accident detectors that warn that an excursion has occurred (already installed in many operations) were important. A short section considers the consequences of an accident.

The preference was for protection through safe geometry by limiting pipe and container diameters and keeping them far enough apart to avoid significant interaction. Where safe geometry was impracticable, protection could be through control of quantities or concentrations but this must require two or more simultaneous and independent contingencies before a chain reaction could occur. These key principles have survived to the present day.

Basic limits are given for individual units and then for arrays. These are handled with the two-group diffusion theory and a limit for the total solid angle subtended at any unit by all the others. This was set at 1 steradian. The revision in 1961[29] corrected a few mistakes and made comments on mishaps. It dealt with neutron poisons more completely and expanded on the solid angle method.

TID 7016 continued to develop and in the 1978 revision[30] (by now the document had reached nearly 130 pages and came from the Nuclear Regulatory Commission) there was mention of computer codes and reference to the 1975 ANSI standard N16.1(36)[31], first adopted in 1964. The methods for control were reiterated as the double contingency principle, control through safe geometry and the use of neutron poisons and the sub-critical limit was introduced from the Standard. The surface density and density analogue methods were offered for arrays in addition to the solid angle approach.

The document thus evolved (the document says "matured") over the 20 years as a much broader data base became available. The dramatic development was the increased reliance on computer codes: "The advance of calculational capability has permitted validated calculations to extend and substitute for experimental data." However, the principles of protection remained much the same[32].

Two other US documents have found wide use. The important "Criticality Handbook" ARH-600 was published in 1968[33]. While it too expressed the principles of criticality safety, its significance was probably in the wide range of critical data drawn from experiment and code calculations (mainly GEM and KENO). Although intended for use at the Hanford site it found wide use around the world as a source of authoritative data that was, on the whole, less conservative than TID 7016. Similarly the Los Alamos report LA-10860[34], published about the same time, presented a range of data that made it a useful reference document, after a 1986 update, to the end of the century.

In the UK the earliest principles and criteria were prepared by UKAEA and are associated with staff at the Authority Health and Safety Branch Safeguards Division. The Branch was formed, with Safeguards and Radiological Protection Divisions after the Windscale reactor accident in 1957 as the focus for safety within the organisation. The Safeguards Division was, as has already been noted, responsible for the development of criticality methods but it was also responsible for criticality inspection and it was from this part, led initially by Jack Chalmers, that the guidance came[35]. The most significant early collection of data was probably the Manual of Criticality Data[36].

The Handbooks and Manuals were originally intended primarily for designers and as guides for safe operation and their roles in this and in safety assessment were overtaken by codes. However, they are still used by assessors for scoping calculations and cross checking.

A particularly complex and important aspect of guidance and criteria has been the role of regulating authorities in criticality safety. Detailed description of this is beyond the scope of this paper. Suffice it to note that regulators around the world have recognised, within their different frameworks, the importance of criticality control and have themselves maintained a high degree of technical expertise while taking an active interest in the competence of criticality assessors. There is a general agreement on the broad structure of protection that flows from the early handbooks. For example, controls which are passive—such a safe-by-geometry—are preferred to those which depend upon active engineering and these are better than those that rely on human management control.

13 Criticality Safety

We should distinguish between assuring that defined assemblies and systems are subcritical, using handbooks and computer codes, and assuring that they remain subcritical under abnormal or accident conditions. The latter is the bigger challenge because it requires the identification and assessment for criticality of the abnormal states—of which there could be very many.

The approach that has been adopted was formalised in the US standard ANSI/ANS 8.1, first published in 1964, which requires that:

> "Before a new operation with fissionable material is begun or before an existing operation is changed, it shall be determined that the entire process will be subcritical under both normal and credible abnormal conditions."

This became known as "Process Analysis"(PA) in the USA. A similar approach (if not the actual name) was adopted by operators elsewhere —and imposed as a requirement by many regulators. The standard has been reissued several times with no substantial change of wording up to the time of writing— perhaps making it the most venerable of all nuclear standards. That is not to say that what counts as Process Analysis has remained the same: the techniques for identifying and assessing abnormal states have evolved to include standard methods such as HAZOPS and, towards the end of the century, PRA.

Mentioned within ANS-8.1, and widely used in the criticality community is the Double Contingency Principle(DCP):

> "Process designs should incorporate sufficient factors of safety to require that at least two unlikely, independent and concurrent changes in process conditions before a criticality accident is possible."

In the original ANSI/ANS-8.1-1964 standard of 1964 PA was a requirement and DCP a recommendation, with DCP therefore clearly subservient to PA. Nonetheless the DCP seemed a much more definite standard and became, for some, dominant or at least an alternative to Process Analysis. This interpretation was encouraged when USDOE, in drafting a new safety order, turned all the ANS-8 standards into requirements (all the "shoulds" became "shalls"). All the original recommendations were turned back into recommendations before the Order was published—with the exception of DCP. The debate that went on in the 1990s and extended well into the 21st century did re-establish the primacy of PA but emphasised the continued importance of the DCP.[37]

Criticality safety has also not been divorced from the risk-based approach that dominated safety thinking in the final decades of the century. This extended from the application of structured techniques such as HAZOP to identify criticality hazards to efforts to apply Probabilistic Risk Assessment methodologies in situations where this was appropriate. In the UK such an approach has been promoted by the Nuclear Installations Inspectorate with, for example, the setting of frequency targets for criticality events which may trigger the installation of Criticality Incident Detector (CID) systems. As a result of adopting such approaches, criticality safety has become more integrated into the overall radiological risk assessments.

Although we have focussed on the techniques used to establish how close a fissile system is to criticality, it would not be appropriate to ignore the work that has been done on establishing the consequences of criticality accidents. This falls into two parts: understanding the accidents that have happened and devising methods to calculate the consequences of possible future accidental criticalities.

The first fatal criticality accidents occurred at Los Alamos in 1945 and 1946 and both involved the same 6.2 kg sphere of plutonium—just below the critical mass in air. In the 1945 accident, on 21 August, the lone experimenter, Harry Daghlian, was adding tungsten carbide bricks as a reflector around the sphere. As he was about to place the last brick he realised from the response of neutron counter that it would make the assembly critical but, as he withdrew his hand, the brick slipped onto the assembly and which became critical. He managed to push off the final brick to bring the excursion to an end but received a dose of 510 rem and died 28 days later.

In the second accident a senior scientist, the group leader Louis Slotin, was using the sphere to demonstrate to a group of six people how criticality experiments were performed. This time the sphere was surrounded by two beryllium hemispheres as reflectors. The scientist had removed two safety spacers and was holding the top

13 Criticality Safety

hemisphere away from the sphere with his thumb and a screwdriver. The screwdriver slipped and the hemisphere fell onto the sphere, its neutron-reflecting capability making the assembly critical. The demonstrator received some 2100 rem and died nine days later. The others received non-fatal doses between 360 and 37 rem[38].

Both these accidents resulted in a chain reaction without the involvement of the delayed neutrons. This accounts for the speed with which they developed. This kind of excursion is referred to as prompt criticality.

A third, much later, accident in a Los Alamos plant on 30 December 1958 illustrates how things can go wrong with solutions. More generally it shows the need to understand the physics and chemistry of systems if accidents are to be avoided. The plant was designed to recover plutonium from a variety of laboratory operations and, for reasons that are not clear, the residues from several process tanks were transferred into a single large tank. Here two layers formed. At the bottom was acid/water and on top of it a layer of organic solvent. Dissolved in the organic solvent was just over 3kg of plutonium. The arrangement was sub-critical until an operator turned on a stirrer when there was an excursion that gave him a dose of over 10,000 rem that killed him 36 hours later. It seems that the stirring had made the organic layer thicker at the centre of the vessel and this, with the change in shape of the reflecting water layer, created a criticality excursion.

The systematic study of criticality accidents began with WR Stratton in 1960 [39] and the revisions and updates he made to produce report LA-3611 in 1967[40]. That report was revised by David R Smith in 1989 as NCT-04 and then by joint US and Russian Federation authors in 2000 as LA-13638[41]. This most recent report lists 22 criticality accidents in process operations (with over half in the Russian Federation and 21 of them involving solutions) and another 38 in experiments. In the process accidents nine people died and a further three had limbs amputated. Most of the accidents occurring in experiments actually happened in facilities designed to study criticality. In the process accidents there was very little (if any) damage to equipment but there was of course disruption of activities. Altogether 12 people are known to have died in these experiments but members of the public were exposed in only one accident and this at a relatively low level.

The evolution of a criticality accident is complex and depends on the specific circumstances. However, the general pattern is that criticality excursion creates changes in the system that quench the excursion (at least temporarily) to bring the system back to sub-criticality. Experiments have been conducted with specialised reactor systems in the USA and France and these have shown several quenching mechanisms. In LA-3611 in 1967 Stratton was able to refer to results in the mid to late 1950s from a number of reactor systems: Dragon, Borax, Godiva-I and Godiva-II, the Spert and Kewb reactors,Triga, Treat and Snaptran. He was then able to list the significant quenching mechanisms (thermal expansion, rise in neutron temperature, boiling, U-238 Doppler effect and radiolytic bubble formation) and indicate how these came into play in experiments with the reactors.

He used the energy model suggested by Klaus Fuchs (the atom spy) in 1946[42] and developed by Hansen[43] to make some estimates of likely energy release profiles and his calculations showed some patterns useful in interpreting the accidents. The first key parameter was the size of reactivity step injected in the accident. Where this took the reactivity from delayed critical to significantly above prompt criticality there was a characteristic energy spike. The spike would be more energetic and shorter in unmoderated systems and be followed by a plateau in power generation lasting up to a second. The moderated systems showed a smaller spike and no plateau. When the excess reactivity injected is lower and just takes it to prompt criticality, neither unmoderated nor moderated systems showed the short spike and a pulse lasting a few seconds is predicted.

An indication of Stratton's strength, and perhaps of the maturity of the topic, is that, 33 years later, in the second revision (LA-13638) his calculations and wording about power excursions and quenching mechanisms could be repeated with little change. The additional data that could be produced related to the French CRAC studies undertaken at Valduc between 1967 and 1972 (and, incidentally, continued to the end of the century) that were specifically aimed at understanding criticality accidents in process solutions and these did not cause any changes.

One of the first calculations of criticality consequences, for an aqueous solution, may well have been that of Christy and Wheeler

13 Criticality Safety

(this was the Wheeler who coined the term "buckling") in January 1943[44]. They imagined a solution of Pu-239 that evaporated until it just reached criticality in a closed vessel. The temperature rise that followed, they estimated, would reduce the density of the solution and increase the leakage of neutrons enough to drop the effective multiplication factor below 1. Fission energy generation would stop and the solution would cool down enough for the process to repeat. This would go on until a steady state was reached. The system would be a potential self-regulating power plant. If the vessel were open, they thought that the solution would boil down until the solution was so concentrated that it was not critical. They estimated that the boiling away of 1 litre of solution 10m from an unshielded individual would give them an "integrated dosage of 44 roentgen units". A figure, in itself, "not dangerous".

The development of codes designed to predict the development and consequences of accidental criticality excursions was a natural extension of the early work. In the UK, Don Mather of UKAEA was responsible from the 1970s to the 1990s for codes that predicted transients in both solutions (CRITEX) and powders (POWDER). By the end of the century the FETCH code from Imperial College, London was a world leader.

Since the earliest days of experimentation, devices that warn of criticality have been in place. The earliest were probably devices that measured neutron generation as part of experiments (many experiments tested the approach to criticality after all). Harry Daghlian had some warning from such a detector—but too late.

Enrico Fermi had anticipated the problem of accidental criticality with his first pile, CP-1. A control rod that was guaranteed to shut down the pile dangled from a rope. If neutron monitors indicated that the reaction was heading out of control, at Fermi's command, a man was to sever the rope with an axe, plunging the control rod into the pile and stopping the reaction. The man, Norman Hilberry, was known as the Safety Control Rod Axe Man and this gave rise to the term "SCRAM", still used today for a rapid reactor shutdown. Or so the story goes.[45]

The detectors—Criticality Incident Detectors (CIDs) and other names—that evolved specifically for warning of accidental criticality were (and are) predominantly gamma-ray sensitive. Their characteristics and the ways in which they should be deployed and used were codified through the 1950s and early 1960s, not least in a comprehensive and elegant report by Aspinall and Daniels[46] of the UKAEA in 1965. Difficult to obtain but easy to read, in its 51 closely-argued pages it expands and justifies some key principles[47].

> alarms should be designed taking account of the time profiles of possible excursions
>
> they should respond rapidly so that the time between the dose passing the trigger threshold and the sounding of an alarm is short
>
> rapid evacuation of the area covered by the alarm is essential
>
> they should continue to function after high doses
>
> false alarms should be avoided—perhaps by combining detectors in "X out of N" voting arrays—since evacuation is not without risk
>
> increased complexity may lead to lower reliability; the increased maintenance required for more complex devices may well lead to more false alarms
>
> the alarms should be fail-safe, self-monitoring and subject to routine examination
>
> alarms should be installed unless a strong case can be made for omitting them

Importantly, they remind us that the criticality safety case or plant analysis should cover all foreseeable circumstances. That leaves the exercise of "assessing the unforeseeable, which must at least be a difficult one". It is then that the CID system comes into its own as the "last-ditch defence designed to operate when all else has failed." Aspinall and Daniels has remained a prime reference and guide for 50 years.

The criticality community is an international one and similar principles have been developed by other nations. Aspinall and Daniels compared their recommendations with those current in 1965 in the USA and France and found them quite similar.

One topic that became particularly active towards the end of the century was that of when a criticality alarm could properly be omitted[48]. Aspinall and Daniels had proposed that this could be in situations where an excursion could not reasonably be expected to occur or when the maximum foreseeable event could not result in a dose to man of more then 250mSv[49]. Some operators considered that alarms were often specified because they had historically been specified and that what constituted a case for omission was unclear. Given the costs of such systems and the fact that false alarms are not without risks, considerable effort was put into clarification of when they were needed—even employing PRA and cost benefit analysis— but it was not obvious that matters had been much clarified[50].

The systems themselves have usually been based on detecting gamma-rays. In the UK the MkIV system developed and made by UKAEA (later BNFL) Springfields was widely used. Based on Geiger-Muller counters, its hallmark was simplicity. The reputation for reliability it acquired led to its survival as the obvious choice for most users from the 1960s to nearly the end of the century. When BNFL came to update their criticality alarm systems and install new ones, they based their Mark X CIDAS on the Mk IV detectors with upgraded electronics.

Ionisation chambers were used in the first French systems but they switched to scintillation detectors for the CEA-designed systems of the early 1960s and 1970s. These systems were taken up and developed by both the Japanese (who redesigned the electronics in the late 1970s) and by commercial suppliers. Some of these systems had both gamma and neutron detectors[51].

The characteristics of the development of all systems has been a conservatism. Given the extensive testing they must undergo before acceptance and the value of long experience of actual use, the core systems would not be expected to have changed much. The major developments have been in enhancing the peripheral electronics without compromising the essential alarm function. Thus the detectors, which should respond to and survive extremely large doses, may be based on quite old technology with a reliance on discrete components, the peripheral electronics may be more modern but rather remote and shielded from possible excursions.

While the aim of early systems was to avoid fatal doses and deterministic threshold effects, as stochastic effects came to the fore the aim of criticality emergency planning changed. One question that then grew in importance was whether evacuation zones should be extended to include areas of potential lower doses.

A general technical procedure developed to approach the question which involved specifying, on the basis of review of past accidents, a reference incident fission yield (for example, 10^{18} fissions, the likely maximum size of a criticality event in a solid system). The radiation doses (neutron+gamma) from this were then calculated using the calculational methods available at the time, usually taking account of shielding and perhaps of neutron moderation and reflection. A dose contour would then be specified and an evacuation zone and evacuation plan would be prescribed based on this.

The dose contour would generally represent the dose that would be experienced if the person being exposed did not evacuate and might be chosen at, say, 250 mGy. Evacuation zones would then be based on this taking account of practicalities such as building construction and potential evacuation routes.

As the aim moved from just preventing fatalities and serious injuries to limiting later stochastic effects the attention moved from absorbed dose to dose equivalent and the role of neutrons became greater. When this was combined with the pressures to extend evacuation to even lower dose equivalent levels, operators began to find that criticality emergency management might well extend across wide areas of their sites. They looked again at the arguments using, for example, ALARP, cost-benefit techniques and probabilistic risk analysis

Workers in areas where there is deemed to be a risk of accidental criticality usually wear dosimeters which will allow estimates to be made of their exposures should the worst happen. The exposures are likely to be, in fairly equal measures, due to gamma-rays and neutrons. While the gamma doses can be estimated using film or TLD technology– provided it is designed to cope with the much higher doses (several Gray) than are routine—the neutrons present particular problems.

The main challenge is the range of energies of the neutrons because no single detector can cover the whole range of energies that may be present.

13 Criticality Safety

The most common detector type relies on activation of foils by the neutrons and these are combined in the dosimeter to achieve the required energy response. In a typical dosimeter there may be sulphur disc which is activated by neutrons of energy greater than 2.7 MeV. Indium may be used for intermediate energy and gold for thermal neutrons. With combinations of these detectors is possible to construct a rough energy spectrum and estimate the dose.

While activation detectors predominate, others have been used. Nuclear track detectors record the fission products produced when neutrons interact with fissile material (often Np-237). Changes in the properties of silicon diodes caused by intermediate and fast neutron irradiation are relatively easy to measure and provide a quick estimate of doses.

The various detectors adopted—and there has been a surprising variety—are usually combined in a single small package, charmingly called a "criticality locket" in the UK. An exploded view of a locket is shown in the figure.

Since the neutrons interact with the human body, the neutron response of a detector is strongly direction-dependent and it is usual to wear several detectors, perhaps spaced on a belt, for a better estimate of dose.

The human body itself can act as an accident dosimeter. Activation of the sodium in the body can give an indication of neutron dose (without the problem of directional-dependency) and studies of the damage to the chromosomes of exposed individuals can also allow estimates of neutron and gamma dose to be made. Both methods are particularly useful if, for any reason, the exposed individuals were not wearing suitable dosimetry.[52]

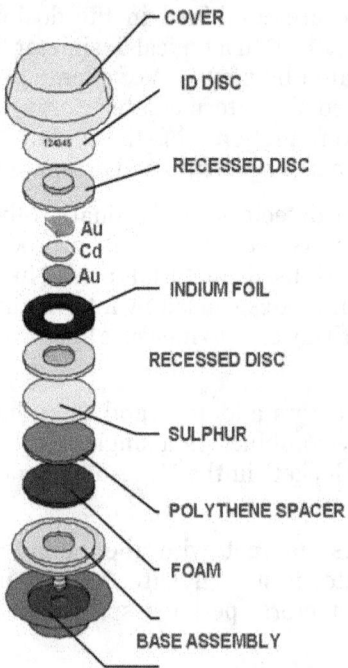

Figure 18: Harwell Criticality Locket (adapted from Majborn 1980)

Notes Chapter 13

1 Much of what follows about criticality assessment methods is derived from(Meggitt G C, 2006)
2 (Serber R, 1992)
3 (Konopolis E, Metropolis N, Teller E and Woods L, 1943)
4 (Callihan A D, Ozeroff W J, Paxton H C and Schuske C L, 1957)
5 (Whitesides G E, 2003)
6 (Knief R A, 1985)
7 (Reider R, 1971)
8 (Schuske C L and Paxton H C, 1976)
9 (Thomas A F and Abbey F, 1973)
10 (Knief R A, 1985)
11 (Simister D N and Clemson P D, 2003)
12 (Greenspan H G, Kelber C N and Okrent D(eds), 1968)
13 (Whitesides G E, 2003)
14 (Metropolis N, 1987)
15 ('Fermi invention rediscovered at LASL', 1966)
16 (Metropolis N, 1987)
17 ibid
18 (Hansen G E and Roach W H, 1961)
19 (Whitesides G E, 2004)
20 (Whitesides G E, 2003)
21 (Longworth T C, 1965, 1968; Woodcock E R and Murphy T, 1965; Hemmings P J, 1967)

22 (Moore J G, 1974)
23 (Greenspan H G, Kelber C N and Okrent D(eds), 1968)
24 (Whitesides G E, 1971)
25 (Whitesides G E, 2003)
26 (Hansen G E and Roach W H, 1961)
27 (ANS, 1975a)
28 (Callihan A D, Ozeroff W J, Paxton H C and Schuske C L, 1957)
29 AEC, 1961)
30 (Thomas J T (ed), 1978)
31 (ANS, 1975b)
32 (ANS, 1983)
33 (Carter R D, 1968)
34 (Paxton H C and Pruvost, 1986)
35 (Jordan G M, 1963)
36 (Abbey F, 1967)

37 Interestingly, the 2015 DOE Order made presentation of an argument based on DCP a requirement in criticality safety evaluations(USDOE, 2015)
38 (McLaughlin P M, 2000)
39 (Stratton W R, 1960)
40 (Stratton W R, 1967)
41 (McLaughlin P M, 2000)
42 (Fuchs K, 1946)
43 (Hansen G E, 1952)
44 (Christy R F and Wheeler J A, 1943)
45 An alternative version has SCRAM arising from the need to get out of the place quickly
46 (Aspinall K J and Daniels J T, 1965) An update by Delafield and Clifton (Delafield H J and Clifton J J, 1984) was produced in 1984 and included consideration of neutron detecting systems but Aspinall and Daniels seems preferred as the authoritative document
47 These are my key principles not Aspinall and Daniels's
48 (Reverdy L et al, 2003)
49 This is a paraphrase of the wording on p13 and 14 of the report. They go on to propose that such excursions should not go undetected but that prompt evacuation should not be required.
50 (Thorne P R, Bowden R L and Venner, J., 2003)
51 (Barbry F, Grivot P et al, 2003)
52 Useful summaries of accident dosimetry can be found in (Majborn B, 1980). (Heinrichs, D et al, 2014) contains details of a range of dosimeters (admittedly modern ones).

---

14 Reactor Safety

The early government reactors in the USA were located in sparsely populated areas[1] so reducing greatly the risks to the public. In 1947 the AEC set up the Reactor Safeguard Committee[2], an independent group who were to review the safety aspects of proposed test reactors and particularly their siting. However the Hanford weapons production reactors had developed problems that jeopardised the existing emergency plans and were seen as the major safety issue of the time. A concerned RSC, under the chairmanship of Edward Teller, proposed that there should be an exclusion zone around un-contained reactors with a radius $R= 0.01 \sqrt{P}$, where R is in miles and P is the thermal power in kW.

This was based on the likely distance at which fatalities might be expected from a large release. For the Hanford situation, with several reactors strung out along the Columbia River, this area extended out to beyond the northern boundaries of the enormous Hanford Reservation. This might have been manageable but there would be little allowance for uncertainties in calculations. If this was the case for the smallish AEC reactors in a very remote location it was not encouraging for the future. When utility companies considered entering the nuclear business one of their major concerns would be to locate their power plants close to their customers, so minimising distribution costs. For a 400 MWt reactor the exclusion radius came out at about six miles and, even then, more powerful reactors were being considered. The criterion just had to be too conservative.

In 1947 the AEC contracted with General Electric for a study at Knolls Atomic Power Laboratory, Schenectady, NY of the possibility of a so-called Intermediate Power Breeder Reactor ("Intermediate" referred to the energy of the neutrons involved) which would produce significant amounts of power (which might have been fed into the grid) but also breed plutonium. The engineering design work was well-advanced and site work had begun when the AEC terminated the project in the spring of 1950 in favour of work directed towards an improved submarine reactor.[3] Two reasons that emerged. More uranium ore was being discovered in the USA than expected— making breeding less of a priority for the future of power generation —and experimental data was showing that the IPBR was unlikely to be as efficient in producing plutonium and power as expected. It could be added that Hyman G Rickover (soon to be appointed Rear-admiral) applied immense skill and pressure to have the effort switched to Navy work. The AEC downplayed the effect of this action on the development of nuclear power reactors but it meant that at this point the Commission had no work in hand leading to development, other than as a possible spin-off, of nuclear power.

The Knolls IPBR may not have been built but it did prompt one important development. The reactor, it had been decided, had to be located fairly close to Schenectady, NY to retain scientists at the Knolls Atomic Power Laboratory just east of the city. The proposed West Milton site was hardly a remote location: Schenectady is just 10 miles away. When the proposal was put to the the Reactor Safeguards Advisory Committee in 1947 they concluded "unenthusiastically" that it might be acceptable. At the first formal meeting of the RSC (in November 1947) the proposal had been modified considerably and now included a gas-tight steel containment sphere and this was accepted. The 225 ft sphere was subsequently constructed in 1953 but it housed the prototype of the *Seawolf* submarine reactor rather than the IPBR[4] The first containment actually to be built was thus the gas-tight concrete structure that surrounded the CP-5 research reactor at Argonne. That started construction in late1951.

As the civil programme got started the there was no lack of appreciation of the risks. McCullough listed some of them[5]:

the lack of experience of nuclear systems

the control of exposure of workers

risks of administrative control

reactor run-away

decay heat and melt-down

large amount of radioactivity involved

potential widespread public hazard

escape of radioactivity

To sharpen the point he quoted (on p137) Edward Teller from 1953[6]:

The hazard is crudely analogous to conducting both explosive and virulent poison production under the same roof.

In the very early days decisions on siting and safety were taken with worst case accident scenario but there were always questions about what a worst case scenario actually was. Someone could usually think of something worse. By the late 1950s(the time of WASH-740, see below) the benchmark had become called the worst conceivable accident but when the consequences of this were analysed—and found to be probably unacceptable—voices called for something more realistic. Thus was born, in 1960, the maximum credible accident(MCA).

The general rule was that a credible accident was one caused by a single equipment or operational failure that resulted in a release of fission products to the environment. The maximum credible accident would be the credible accident with the worst consequences. In LWRs it was reckoned that the MCA would be a rupture of a major coolant pipe with a complete loss of coolant leading to fuel meltdown and partial release of fission product inventory to the secondary containment—which would remain intact.

Perhaps in response to questions about the meaning of "credible" the MCA slowly morphed into the Design Basis Accident. It rather shifted the whole idea from being a tricky descriptive one to being a normative one. The DBA was, to take a modern USNRC definition, "a postulated accident that a nuclear facility must be designed to withstand without loss to the systems, structure and components necessary to ensure public health and safety." The accident actually chosen as the DBA for most reactors turned out to be what was previously called the MCA; for LWRs a major coolant pipe break.

A persistent principle of reactor safety has been "defence in depth". Sorensen[7] found the first use of the term in a statement submitted by Clifford Beck, the Deputy Director of Regulation to the Joint Committee on Atomic Energy in April 1967. Beck takes a seemingly rather restricted view of defence in depth, illustrating it with the successive barriers that prevent the escape of fission products from fuel. These barriers include the nature of the fuel matrix itself through to the reactor containment building.

Beck goes on to describe three "lines of defense" built into the physical systems of power reactors for safety reasons.

The first of these includes superior quality of design, construction and operation. The "principles of fail-safe design, redundancy and backup, defense-in-depth, and extra margins of safety at key points" should be employed. The importance of regular maintenance and inspection is emphasised as well as investigation and correction of faults. It also includes good management, with well-trained staff, good procedures and periodic audit.

The second line of defence is the accident prevention system designed into the facility to stop mishaps and upsets turning into more serious accidents. It includes redundancy in control and shutdown systems and emergency power.

The third line consists of consequence-limiting safety systems that limit the escape of fission products in an accident. These include the building spray and washdown system, filtration and, of course, the containment building itself.

Over the following decades the defence in depth concept was broadened and reflected more and more what Beck called lines of defense. It marked a holistic approach based on sound design of plant and safety systems based on thorough hazard assessment, careful construction and well-managed operations. It encompassed the idea of multiple independent barriers—including, in operations, audit, performance review and accident investigation.

By the end of the century INSAG-10[8] identified five levels of defence in depth (p6) that were apparently rather different to Beck's three lines but which expressed very much the same ideas.

Defence in depth remained a fundamental principle of nuclear safety—emphasising, for example, the importance of safety-critical systems, both physical and organisational, that are robust, fail-safe, independent and redundant.

The Design Basis Accident in a PWR is a break in the primary coolant circuit leading to a loss of coolant; a Loss of Coolant Accident, commonly a LOCA. If no action is taken this could lead to a serious loss of water from the primary circuit which might allow the core of the reactor to overheat. Fuel elements would then leak and release fission products which would accumulate in the secondary containment and be liable to escape from this and expose members of the public.

If the accident progressed, the fuel might reach its melting point and the core form a very hot and very radioactive molten mass that could slump to the bottom of the primary containment, breach it and then continue burrowing on. While it was in a critical configuration it would be self-heating, so when would this molten mass stop? Well, no-one can be quite sure but it is not nicknamed the "China Syndrome" for nothing.

How can such an accident be avoided?

The preferred option is not to have a LOCA in the first place by, for example, designing in systems that monitor the primary containment pressure and provide a safe relief pathway. Selection of high-grade materials, careful manufacture and integrity monitoring during service also play a part. The last of these might well spot anomalies and even leaks that precede something catastrophic.

The designer then has to ask whether, once the LOCA has happened, the consequences can be mitigated by finding some other means of cooling the core. In our PWR it has become (since the rather early days) standard to provide an Emergency Core Cooling System—ECCS. This forces water into the reactor to cool the core until some more permanent arrangement can be made to do this. The water is borated to reduce the reactivity of the core and is stored in tanks ready to be injected (at high pressure if necessary) into the pressure vessel. There is also usually a provision to provide cooling longer-term by recirculating the used water in the building sump through a cooling system.

Steam will be released into the reactor containment building and this will be very hot and radioactive. It will pressurise the building. To reduce the pressure and temperature two systems are used. The first is a fan-driven one that recirculates the steam through heat exchangers to cool. The second system will automatically spray water through spray rings into the building at a high level reducing the temperature, condensing the steam and so lowering the pressure. Reactor containments are designed to withstand the temperature and pressure expected but they may crack if the raised levels persist and fission products would then be released. A stainless steel liner forms a gas tight membrane to reduce this.

When the non-hardware elements associated with design control, construction audits and operational and maintenance procedures and training are added in there is seen to be an implementation of the defence-in-depth approach. The emergency response plans provide another level.

An Emergency Core Cooling System (ECCS) was first seriously considered in the mid-1960s as proposed reactors grew in size to the 1000 MW range. It was realised that there was in fact little information available about the possible issues and an experimental programme (LOFT) was set up (or rather modified) to look at the issues. This was in 1969 but before this got going (it was delayed for various reasons until the mid-1970s) some disturbing results came from semi-scale experiment at the Idaho Testing Station in 1970/71. All of the five trials the ECCS system were suggesting that that pumping water into a reactor full of steam and hot water might not be as easy as everyone seemed to think. This led the AEC to set up a series of hearings in 1972 that lasted 18 months and generated

22,000 pages of transcripts. It also became apparent in the course of the hearings that AEC staff and other experts were divided on the demonstrated effectiveness of ECCS. Nuclear sceptics had something of a field day and the AEC lost some of its credibility

However, the real outcome was that a research programme on LOCAs and ECCSs was finally set up and the whole safety assessment process was reviewed and strengthened through the enactment of Federal Code 10CFR50.

One of the early principles of reactor safety was that it should be possible to shut the reactor down quickly in the event of a problem[9]. In Fermi's first reactor experiments this was achieved by dropping a rod of a neutron absorbing material into the core—a worker was designated to chop the rope suspending the rod on command—and so ending the reaction. Such an emergency shutdown became called a Scram. The emergency shutdown arrangements quickly (and thankfully) became automatic but there were reliability concerns and questions about the consequences of a failure to operate. With high reliability a necessary requirement, it became a focus as the ATWS issue

The acronym stands for "Anticipated Transient without Scram".

> An ATWS is one of the "worst case" accidents, consideration of which frequently motivates the NRC to take regulatory action. Such an accident could happen if the scram system (which provides a highly reliable means of shutting down the reactor) fails to work during a reactor event (anticipated transient). The types of events considered are those used for designing the plant.[10]

The concern about ATWS started formally (there must have been misgivings about reliance on the scram system particularly after a failure under test of a system at the Kahl reactor in Germany in 1963) in 1969 with a letter from a consultant to the ACRS. This predicted that failure on demand of the system in BWRs could hardly be better than 10^{-4} per demand, largely because of the possibility of common cause failure. The problem at Kahl had arisen because of a change to the protective coating used on the relays which prevented them operating.

The letter was enough to prompt the ACRS to ask a number of LWR reactor manufacturers to consider how the frequency of common cause failures might be reduced and how the results of failures might be mitigated. From the replies it was clear that a major concern was that the response to a failure would have to be quick: just a few seconds were available for successful mitigation. But one responder, General Electric, had the germ of an idea that might work. This was the summer of 1970.

There were discussion which centred around the predicted reliability of scram system. Most vendors claimed much better values than 10^{-4} per demand. Somewhere between 10^{-6} and 10^{-7} was popular but General Electric claimed it was around an astonishing 10^{-15} per demand.

Soon after this GE developed their mitigation idea. This was that, in the event of a failure to scram, the main coolant water circulation pumps should trip. While this might seem to add to the problems rather than solve them, it did prove to be a quick and effective response. As a result of the pump trip the coolant flow would be reduced and steam would be created in the core. Steam being an appalling moderator compared to water, the reactivity of the core would be reduced and the reactor would shut down. Or at least it would buy some time for operator actions. This was agreed as a way forward by the AEC regulators.

Some alternative back-up systems were developed as the matter progressed.

By now the ATWS was firmly on the agenda. In 1972 the regulators demanded an assessment of consequences of ATWS to show they would be acceptable for their designs. If not they would be required to improve the designs or improve the reliability of scram systems. In 1973 WASH-1270 was issued with requirements that new plants would need to have two independent scram systems but need not have to withstand the consequences of one failing. The goal was set that the probability of a serious accident (a core melt or similar) following an ATWS should be less than 10^{-7} per reactor-year.

Many thought this had resolved the ATWS issue but it had not. While new BWRs incorporated a recirculation pump trip system as a back-up, in spite of demands from the regulators, nothing had really been done about back-fitting to older reactors.

14 Reactor Safety

The new regulator, the NRC, reviewed the area in 1977 and concluded that the provisions in place were adequate. However, in 1978, influenced by the conclusions of WASH-1400, they weakened the safety goal for the probability of a major accident by a factor of ten. Some provisions were introduced that would guide back-fitting: improving scram system reliability through better circuitry was a path more more likely to bring results than hardware changes. Hardware changes to reduce consequences was still the aim for new plant.

A near-ATWS occurred at Browns Ferry Unit 3 BWR when 76 of 185 control rods failed to fully insert in a scram demand. Two ATWS events occurred at Salem 1, a PWR, in February 1983. These increased the urgency of getting a final resolution of the ATWS issue.

The complex issue was addressed in a rule entered in the Federal Register in 1984—fifteen tears after it was first raised.

Plainly if a reactor pressure vessel failed it would be a very serious matter. If it failed suddenly with the reactor at power it would very likely be catastrophic as it would most likely destroy the secondary containment and all the safety systems. Fission products would be released directly to the environment and large numbers of people would die. From a design point of view there is nothing that can really be done to mitigate the consequences. The only defence is to make it very unlikely that the pressure vessel will fail catastrophically in the first place.

This is done by careful and conservative design using a material which will stand up well to the conditions of high pressure, temperature and radiation flux, minute attention to quality during manufacture, careful inspection of the finished product by non-destructive techniques and close control of installation.

It is very likely that there would be some indication through regular inspection in service of a potential problem. A crack would be found long before it became critical; the vessel would leak before it broke open.

So far this has worked. Minor cracks have been found in several pressure vessels, usually associated with penetrations, and repaired. In 2002 a cavity the size of an American football was found in the head of a reactor at Davis-Besse plant. This was caused by corrosion associated with a nearby penetration which had leaked for years before discovery. The pressure vessels may be 250 mm thick but a cavity is a serious matter.

There was an interesting interval in the mid-1960s when British experts concluded that, even at operating temperatures above the nominal brittle-ductile transition range where sudden failure was not supposed to happen, it (the "incredible") could occur. It took a 10-year research programme at Oak Ridge to reassure the regulators that it couldn't.[11]

One of the first events leading to nuclear Probabilistic Risk Assessment (PRA) was a four-page Memo written by the statistics director of Hanford in 1953 titled "The Evaluation of Probability of Disaster". It said that there was a "chain of events" originating in some minor malfunction and compounded by faults and errors that might lead to a disaster. The individual events in the chain could be evaluated and combined to arrive at the probability of disaster.

After several months work along these lines GE admitted defeat: it was difficult to imagine all the things that could go wrong, there was no extant methodology for combining them into a chain and the data available on failure rates were inadequate. Although it continued to work on the general probabilistic approach to reliability and possible failure models for another decade, GE did not manage to create a practical methodology. However in the late 1960s (when there were emerging problems with purely deterministic approaches to commercial reactor safety) they began to promote probabilistic methods. With other contractors they became interested in the FTA developed by other industries.[12]

The Price-Anderson Act of 1957 addressed the problem that commercial insurers were unwilling to provide cover for the consequences of large accidents at premiums any utility could afford. This was seen as a significant barrier to the development of commercial nuclear energy. Price-Anderson required that constructors and operators take out insurance for the first $60 million but that above that the government would provide cover for free (above $500 million Congress would have to make special arrangements). Other industries, particularly coal who saw nuclear as a future competitor, thought this gave nuclear an unfair commercial advantage and persistently opposed the provisions.

The original act was supposed to be temporary and was scheduled to last only until 1967 but as that date approached it became clear that utilities were very unlikely to obtain suitable cover and that it would be necessary to extend the act.

The Price-Anderson Act became a prime driver in the development of nuclear safety assessment as people tried to understand the scale of the consequences of a serious accident and then the likelihood of it happening.

The 1957 AEC report WASH-740[13]—also known as the Brookhaven Report—was the first significant one to look at the impact of a major accident at a civilian reactor. It was ground-breaking in its estimation of the consequences.

It took a large 500MW(t) reactor and postulated that 50% of the fission products contained in the core were released—after a reactor runaway or LOCA—to the atmosphere as a cloud of gas and inhalable particles. This dispersed in the atmosphere and was inhaled by and irradiated the surrounding population—a representative distribution of people for a site in the USA was assumed—causing death and injury. Different possible meteorological conditions were considered using the best atmospheric models of the time; the effects were calculated for both hot and cold clouds. In short the study went about its business in a way that many later studies were to follow. The headline results were that there would be as many as 3400 deaths, 43,000 injuries and put the affected area at somewhere between 18 and 150,000 square miles. The actual numbers would depend on the meteorological conditions and these worst case figures were possible in only about 10% of accidents.

The report then set about the task of putting things in proportion by showing that such an accident was highly unlikely. All the experts agreed that this was so but not all of them would express it as a number or even range of numbers. However, in a section called "The Best Judgment of the Most Knowledgeable Experts" the report did its best to extract some numbers while admitting that they had "no demonstrable basis in fact and have no validity of application beyond a reflection of the degree of their confidence in the low likelihood of occurrence of such reactor accidents."(p6) The risk of a major release was concluded, on the basis of expert judgment, to be between 10^{-9} and 10^{-5} per reactor per year.

In July 1964 the AEC agreed to update WASH-740 on the basis that reactors were already significantly larger than the one considered there and analytical techniques had developed significantly. It was driven also by the fact that the Price-Anderson negotiations were drawing to a close. Clifford Beck was put in charge and Brookhaven were approached once more. They agreed to conduct the study but refused to consider the effects of engineered safeguards or to have anything to do with probabilities. It was to be consequences only on a similar basis to the original work but at least some of the numbers would be refined.

By December 1964 it was clear that the consequence numbers would be refined—but upwards. They were, in some cases, somewhere between 40 and 100 times worse. Beck naturally turned to probabilities to put these numbers in perspective and turned to a report he had commissioned from Planning Research Corporation, another AEC contractor who had been working on methods for judging probabilities. Using available data from the 1500 years of reactor experience that existed PRC concluded that they were 95% certain that the probability of an accident was no more than 1 in 500 per reactor year. This was hardly good news.

Beck understood that little had changed since WASH-740. Expert judgement was needed again. He thought that far from being 1 in 500, when all was taken into account, it could be as low as 10^{-9} —the figure of WASH-740. Drafts of the Brookhaven consequence report and Beck's summary of the position on probabilities were circulated around the AEC. Almost everyone thought neither should be published. The exceptions were Beck himself, who thought there should be a short bland and vague report to the JCAE. In contrast, John G Palfrey was adamant that the full report of the Brookhaven work should be released. Over the following months the AEC became progressively less forthright about the reports conclusions. In the end the AEC revealed just that a review had taken place but said that the results were not much different from WASH-740's and that no report was planned.[14]

The frailty of the situation was this. Although considerable advances had been made in the consequence side of the analysis (and

14 Reactor Safety

further improvements could be envisaged in the calculation of source terms, dispersion and health effects) there had really been no substantial progress in the estimation of likelihood of failure. To some degree the situation had not changed since people asked what "Credible" actually meant. The estimation of probabilities was done by asking experts their views. The only data available on actual accidents was too sparse to be reassuring. The Task Force on Nuclear Safety, Licensing and Risk set up by the AEC under Malcolm Ernst, when it reported in 1973 (The Ernst Report)[15], concluded that the complexity of reactors made quantitative risk assessment impossible for them. Not everyone agreed and some looked at the successes of a relatively new field, reliability engineering, and the possibilities of techniques for analysing complex systems that had been developed there.

The development of reliability engineering[16] started in the late 1930s with the recognition that many systems were in fact particularly unreliable. But it was found also that it required virtually continuous maintenance to keep complex systems based on vacuum tubes (radio valves) operational. It has been claimed that for every valve in equipment there had to be one on the shelf and seven in transit; there had to be a technician for every 250 valves. More than 50% of airborne electronics was not fully working when taken out of store. Just after the War, in the early 1950s, the Royal Air Force was spending half its budget on maintenance.

By the early 50s, as the ballistic missile programme developed in the USA, it was recognised that one key to improved reliability was the collection of data on system failure frequency across high tech defence industries and this activity grew and spread. The statistical tools for data interpretation were developing at the same time. It was clear that design and manufacturing quality were important issues but that there needed to be an analysis tool that linked individual component failures with whole system failures. The first of these to be formalised was Failure Mode and Effects Criticality Analysis (FMECA) with the publication of the US DoD Standard MIL-P-1629 in 1949. This combined techniques which had, no doubt, been used to improve design for a long time and promoted a systematic examination of the design to evaluate the consequences of individual component failure on the performance of the total system. The approach, also called FMEA, found use in military aviation and space systems but was soon taken up by other industries, like automotive. Since it started with individual components it was described as "bottom up". It was the technique chosen to evaluate systems in the Apollo programme in the 1960s and in the automotive industry in the wake of the Ford Pinto scandal in the 1970s

Another analysis method, a top-down one, was developed by H A Watson of the Bell Telephone Labs around 1961. Called Fault Tree Analysis (FTA) it was used initially for the assessment of reliability and safety of Minuteman missile launch systems. It takes possible system failures ("top events") and looks for the faults in sub-systems and components that might cause themeres. If there is data on the individual component failure rates then it can determine a frequency for the occurrence of the top event. It was, like FMEA, systematic but it proved to give greater insight into the vulnerabilities of complex systems and to be more amenable to quantitative analysis; on the other hand it was much more time consuming. As a result, the two techniques—and there were several others—generally found different places in reliability engineering.

Event Tree Analysis answers a different question from FTA: if a particular component failure occurs in a system, what will the consequences be? It takes the failure (the initiating event) and constructs a binary logic tree to see how it propagates and results in mishap or disaster, taking account of the performance of any protective systems[17]. It derived from business decision theory.[18]

It was Fault Tree Analysis and Event Tree Analysis that, in combination, were to find an important place in nuclear reactor safety analysis.

The safety of the first reactors was down to good and (generally) conservative design, careful operation and remote location. These principles still play an important part but there has always been a nagging thought that they were not enough. Among the nuclear community it was quite widely believed that they would naturally lead to over-conservatism and extra expense. On the other hand people could see that this approach (which was, they thought, largely based on trial and error) was hardly appropriate with this fast growing and phenomenally hazardous industry.

14 Reactor Safety

The search for a more rational and balanced approach to the whole problem seems to have started with Ernie Siddall[19], an engineer at AECL's Chalk River establishment, in 1956. Siddall set down some of the essential principles of the use of redundancy and coincidence in reactor safety systems. The first of these recognises that having two instruments monitoring a critical reactor safety parameter is likely to be safer that having just one: if either instrument detects an unsafe condition the reactor is shut down. Of course the other side of the coin is that it makes the system more unreliable in the sense that spurious trips are more likely.

Spurious trips are unwelcome so a more elaborate system making use of coincidence, is probably justified. In this three (or more) instruments may be used and the reactor tripped if any two show an unsafe condition. Such voting systems make spurious trips more unlikely, increasing reliability, without compromising safety.

1973 saw the publication, on a fairly limited distribution, of the 500+ page report WASH-1250[20]. This dealt with many aspects of the safety of nuclear power from the effects of routine emissions and the problem of nuclear waste to accidents, including severe accidents. On the accident issue it could report and expand on the various analyses that had already taken place and describe the experiments and initiatives that had been taken on consequence mitigation. However it had little or nothing to say about the likelihood of any serious accidents—largely because nothing of significance had been done on the subject.

In 1972 Senator John O Pastore, Chairman of the JCAE wrote a letter to James Schlesinger, Chairman of AEC, prompted by the prospect of an another extension of Price Anderson and the failure of WASH-1250 (presumably he had seen a draft) to give convincing answers to questions of accident likelihood. This combined with the LOFT results, which showed that ECCS might not be as reliable and effective as previously thought, prompted the AEC to start a study, the Reactor Safety Study, led by Prof Norman Rasmussen. The study, which involved some 40 engineers and scientists started in 1972[21] and reported as WASH-1400 in 1975.

Peach Bottom Unit2 (BWR) and Surrey Unit1 (PWR) were chosen as reference cases and fault trees were initially developed for most of the major safety-related features but it was soon realised that a comprehensive analysis based on creating and integrating the fault trees for a complete reactor system was not feasible. The system was so complicated and time and resources were anyway limited. There was a short cut (well, a shorter one) if Event Trees, which could be developed reasonably quickly, were used to model the development of accident sequences[22]. The time-consuming Fault Trees were then needed only for the protection sub-systems that were involved in the accident sequences that led to significant health effect consequences for members of the public.[23]

The number of individual accident sequences generated was about 1000 and these were then allocated to one of 15 Release Categories (10 for PWRs and five for BWRs) chosen to be representative of releases to the atmosphere, in terms of isotopic composition, elevation, energy and time profile. From a study of the event trees of the dominant accident sequences in each Category the probability per reactor-year of a release in each Release Category occurring was estimated. The approach drastically reduced the scale of the subsequent calculations—in fact, making them tractable.

Thus far the work had generated estimates of the risks of releases occurring—and this, we have to remember, was the first time that had been done through proper analysis rather than through asking experts to guess. However what was really wanted was estimates of the risk to the public posed by reactors and this required estimates of the consequences of the releases. To some extent this had been done already by the earlier studies for WASH-740 and WASH-1250 but the Reactor Safety Study deepened and extended the analyses. The three requirements were: to estimate the physical and chemical nature of the releases (the source term), to model the dispersion in the atmosphere and to account for the various pathways by which populations could be exposed.[24]

Two source terms were used in WASH-740[25] to represent the releases from reactors that had undergone thermal runaway or LOCAs and suffered what was called a meltdown. Experiments had indicated that it was likely that a substantial fraction of noble gas fission products (xenon and krypton) and halogens (iodine and bromine) would escape from molten fuel along with a smaller fraction of the important element strontium if it simply oxidised. A significant fire could lead to the release of a large proportion of all the fission products. To represent these two cases conservatively the calculations assumed that, in the first one, all the noble gases and halogens escaped from the containment along with 1% of the strontium. For the second, with the fire, it was assumed that 50% of all the fission products were released to the atmosphere.[26]

There was also some assessment of the sizes of the particulate that might be generated and the temperature of the released cloud.

By the time of WASH-1400 there was a considerable amount of data so, while the principles did not change, the amount of detail and the complexity increased greatly. For the assessment the releases were divided into eight groups of isotopes and the proportion of the likely to be released in each Release Category was estimated. The eight groups were noble gases (Kr, Xe), noble metals(Ru, Rh etc), halogens (both total and organic iodine were evaluated), alkali metals(Cs, Rb), alkali earths(Sr,Ba) rare earths (this included Np and Pu) and refractory oxides (Zr, Nb). As well as the increase in numbers of groups the striking feature was the detailed character of the assessment.

The general WASH-1400 methodology was to persist in future similar assessment.

The next problem facing the analysis was to determine how the material released in the accident spread out in the atmosphere, so that the exposure of people could be estimated.

The atmosphere, under the influence of sun and wind, is always in some degree in turmoil and any small parcel of it is constantly being teased apart by eddies that range in size from the microscopic to the vast. The intensity of this turbulence depends on several factors but its effect is always to cause any stream of buoyant pollutant released into it to spread out into a plume as it is swept downwind. Although the process is essentially random and it is impossible to predict the instantaneous concentration of the pollutant, averaged over a sensible period, the concentration across the wind is fitted quite well (with a few restrictions) by a Gaussian distribution, the bell-shaped curve that crops up so often when random processes are present. Atmospheric turbulence being different in the horizontal and vertical directions, the crosswind and vertical standard deviations are different but once you know them it is possible to estimate the (averaged) concentration of the pollutant anywhere in the plume. The Gaussian plume model, as it is generally known, made the whole dispersion calculation tractable—although of course there had to be a method for determining the standard deviations.

This came from some work that had been done not too many years before by Frank Pasquill of the UK Meteorological Office. In the late 1950s Pasquill developed a classification of atmospheric stability (and so turbulence) based on wind speed, solar radiation, cloud cover, and time of day. He related the the six categories A(very unstable) to F(very stable) to what were effectively the standard deviations. Simple formulae to relate them were devised, notably by Briggs in the early 1970s. This meant that calculations could be made of airborne concentrations of pollutant from any release under a realistic set of weather conditions. Just what was needed for a risk assessment.

There were some complications that had to be taken into account. First, at the bottom of the atmosphere, because of the vertical temperature profile, there is a mixing layer and the pollutant tends to be restricted in its vertical dispersion to within this so, eventually, the vertical concentration profile becomes uniform there. Then there is the tendency of hot plumes to rise, reducing concentrations at ground level close to the release. The situation close to the source is also complicated by pollutant being entrained in the wakes of buildings. A further important factor is the effects of any rain which can both deplete the plume and increase the amount of pollutant deposited on the ground where the rain occurs. The composition of the plume also changes because of radioactive decay.

In all, calculations were made of ground level concentrations and ground contamination for 90 different representative weather sequences at six composite sites around the USA for 54 radiologically important isotopes[27] With titles like "Atlantic coastal site", "Southeast river valley influenced by Bermuda high" and "Great Lakes shore" these "composite" sites could be expected to have rather similar weather patterns.

The pathways of exposure assessed were many. They included internal doses from inhalation directly from the plume and from resuspension of deposited material. Doses from ingestion through foodstuffs of material deposited on the ground, taking account of the range of isotopes released, were considered. External exposure from the plume as it passed and from ground deposits were also estimated. The dosimetric and risk models were, of course, those of the time but they were subject to extensive review in the report.

One hundred reactors on 68 sites were considered. Rather than trying to make separate calculations for each population each of the 100 reactors was assigned to one of the six representative "composite" sites. Representative population data for each site was then obtained from enumeration district returns in the 1970 census and assigned to 22.5 degree annular sectors centred on the site.

As well as assessing the health effects the study also estimated the economic consequences of the accidents—one driver for the study was, after all, Price-Anderson. The RSS model[28] attempted to include all the significant direct financial costs of accidents. These included those of evacuation of residents and of the temporary accommodation they would require, the value of any condemned goods, the cost of property decontamination and restoration and the loss of value of any property subject to restriction after the accident. His was done at some level of detail. We learn, for example, it would cost $0.11-0.14 per square foot to replace a lawn but only $0.05 per square foot to fire-hose paving. Deep ploughing would be the best way to deal with agricultural land in general but would not be suitable for orchards where it would be better to scrape off the top layer off the soil and bury it.

The rather nice point is made that the assessment is confined to costs and that any compensating benefits from the accident are not considered. The only example given[29] of a benefit is the possible employment of previously unemployed teachers in the areas to which families might be relocated. One suspects that there would be rather few other examples that were not politically inflammatory.

The report aimed to estimate two types of risk: individual risks to people living quite close to reactors and societal risks arising from a programme of 100 LWRs.

The headline value[30] was that the early fatality risk to individuals living within 25 miles of a reactor (some 15 million people) was 2×10^{-10} per person year.

The societal risk was expressed using f-N curves which showed, for example, the early fatalities from accidents at the 100 reactors versus the frequency per year of this number being exceeded. This example is shown below and shows that 100 or more early fatalities were predicted to occur, on average, every 100,000 years. More than 3000 early deaths were expected every 10,000,000 years.

This was a massive undertaking and the report, issued in final form in 1975[31], runs to over 2500 pages with a main section of 210 pages supported by eleven Appendices. Appendix XI is 150 pages long and contains an analysis of the comments received on the draft issued the previous year. There is an 18 page Executive Summary—which proved to be the most contentious part of the Report. Representative Morris Udell conducted hearings in March 1977 and concluded that the Executive Summary which presented the conclusions of the Main Report gave a "misleading impression of the certainty and comprehensiveness of it conclusions"[32]

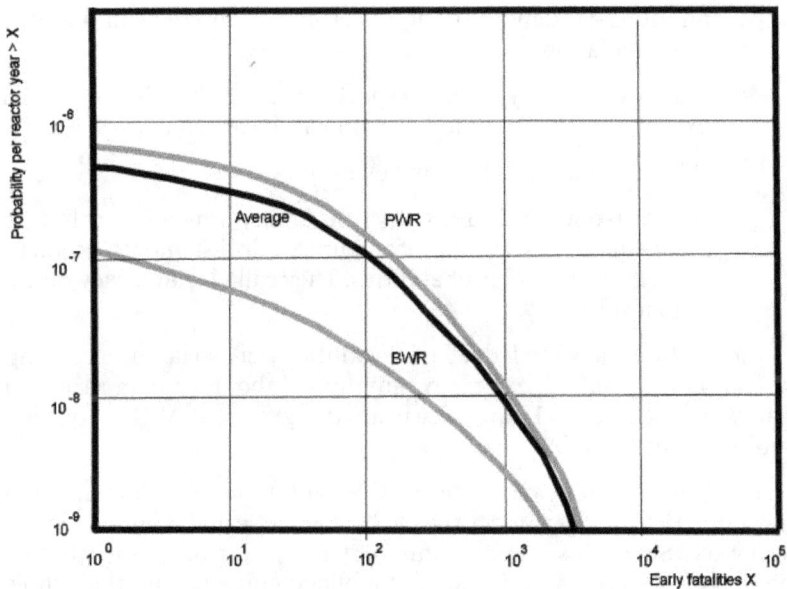

Figure 19: f-N curve for expected early fatalities (WASH-1400 Fig 5-3)

Udall asked the NRC to prepare a new Summary but, instead, in late 1977 they set up a technical peer review group for the whole report: the Ad Hoc Risk Assessment Review Group, chaired by physicist Hal Lewis and known ever after as the Lewis Committee. They reported in 1978[33].

In some respects they praised WASH-1400. For:

    its integrity and honesty

    the potential it revealed for fault tree analysis

    the methodology that pointed to a possible new and more rational direction in regulation.

On the negative side they criticised:

    the adequacy of the dispersion and health effects models

    the presumption that external events and human factors made a negligible contribution to the overall risk

> an under-statement of the uncertainties inherent in many of the calculation
>
> the inscrutability of the report—even at this distance it is extremely difficult to follow some of the arguments
>
> the inadequacy of the peer review process
>
> the over-optimistic Executive Summary which failed to give an adequate description of the uncertainties in both accident consequences and probabilities. They called it advocacy rather than a summary.

Indeed the unacknowledged uncertainties were so large, according to Hal Lewis that "I can't say anything [about civilian reactors] that would be useful. I can't learn anything from WASH- 1400 that would help me on it."[34]

The Lewis Report caused the NRC to withdraw its endorsement of the numerical risk values the RSS and rejected the Executive Summary. Since this was the only part that most people (including the press) had read, this shook public confidence in the whole enterprise.

As everyone paused and wondered what to do about this immense piece of work, so poorly and unwisely presented, there was an accident at the Three Mile Island reactor which changed almost everything.

Things developed rather differently in the UK. One of the early publications about siting considerations was that of Marley and Fry of Harwell at the first Geneva Conference in 1955[35]. They set down some principles which had been developed earlier by Fry:

> Very few people should be exposed to extreme risk (plans should be prepared for the urgent evacuation of nearby people in the downwind direction).
>
> Protracted evacuation or severe restriction on normal living should not be imposed on any but small population centres.
>
> Temporary evacuation or restriction should not be necessary for more than 10,000 people in any but exceptional weather conditions.

If an accident were to coincide with exceptional weather conditions not more than 100,000 people should ultimately be affected.

These implied, on the basis of consequence calculations, that in any 10 degree sector around the plant the population would have to be less than 500 within 1.5 miles, less than 10,000 within 5 miles and less than 100,000 within 10 miles. Population limits all around the site would have to be less than six times the 10 degree limits[36].

F R Farmer and J R Beattie[37] of the UKAEA considered that this paper was the first significant step forward because it dealt in some detail with the radiological consequences to derive this implication. Marley and Fry were the first to delineate publicly the various pathways of exposure of members of the public and to discuss the way these might be managed. A radioactive cloud can expose people by direct exposure to beta and gamma rays from the cloud, by inhalation and by the effects of activity deposited on the ground and vegetation which remains there after the cloud has passed[38].

While it would be difficult to avoid the exposures as the cloud passed—unless there was plenty of warning or the release was extended over a long period—it would be possible to limit the exposure derived from deposited material. This might be temporary evacuation and restriction on the consumption of certain food crops and water. Notably, given the quantity of I-131 expected in releases, there might have to be widespread restrictions on milk. The analysis was limited in several ways—for example the release considered (a "full" or "mixed" one) was composed of radionuclides in the proportions that existed in the reactor. Nonetheless in many ways the treatment was quite comprehensive.

There was no consideration of the quantitative risks to the population. After all there was no sensible agreed methodology for assessing the risks of an accident and the risks associated with radiation were hardly understood outside the early death regime. It was nonetheless a marked move forward in general understanding.

Two papers to the 1967 IAEA conference on containment and siting are of particular relevance:

Adams and Stone[39] from the CEGB first noted that the differences in term of total casualties from reactor accident was not greatly different if they were located on any of the Class I to Class IV sites being considered; casualties differed by no more than a factor four between category II and IV sites. It was, in their view individual risks that mattered. The calculated the risk of thyroid cancer per year at different distances downwind from a release of C curies of iodine, which had a probability per year of p, and estimated that this was $C \times P \times 1.05 \times 10^{-5}$ at 100 metres and $C \times P \times 1.9 \times 10^{-7}$ at 1000 m. This could be reduced quite substantially if countermeasures were considered too.

Speculating that a risk of about 10^{-6} per year might be acceptable, given the risk to life encountered every day, they concluded that they had found a fairly simple procedure for judging the acceptability of a reactor that offered, for an well-designed plant, more flexibility for siting than was currently available.

Charlesworth and Gronow[40] from the Nuclear Inspectorate referred to the inadequacies of relying on remoteness as a safety factor and the ideal that reactors would be so safe that they could be put virtually anywhere. The early criteria had been based on an assessment of sites based on their surrounding population distributions weighted with a factor that represented the dispersion of iodine in an accident. Possible sites that met other criteria were then classified from I to IV on this factor. The most desirable sites were class I and least class IV and only I and II were considered suitable.

They calculated thyroid doses from a notional release (ground level of 1000 Ci of iodine) at each existing or proposed reactor site in the UK and went on to estimate: the collective thyroid dose, the collective thyroid dose when the release was into the most densely populated sector and the individual dose to the most vulnerable people. The details are somewhat more complex than I have suggested here but the essential point was that the three different parameters each gave rise to a different order of merit of the sites. Rather few sites scored high on all three, certainly not enough to accommodate the proposed programme. So, they said:

> "The time has not yet come when it can be pretended that decisions on the safety of a nuclear plant can be based more firmly on objective assessment than on judgement."(p160)

That future probably depended, they suggested, on developing and embracing the probabilistic approach to safety. It might be a more rational basis for safety assessment.

14 Reactor Safety

The final paragraph of their conclusions is worth quoting in full:

> "An examination of the practical sites demonstrates that siting alone is an inadequate means of providing proper safeguards for the public. Therefore greater emphasis must be placed on the safety aspects of the design, construction and operation of a nuclear plant. Criteria proposed for doing this have been mentioned in this paper and are further discussed elsewhere in the Symposium. At present no conclusions have been reached although work is in hand to define standards which will meet the immediately foreseeable siting requirements."

In 1967 F R (Reg) Farmer of the UKAEA presented a paper to the 1967 IAEA Vienna conference which changed the direction of reactor safety assessment quite dramatically. He pointed out that the "obsessive preoccupation" with major accident consequence studies had diverted attention from the likelihood of accidents happening and, what actually mattered, the risks to people. With a side-swipe at the notion of "credible" accidents as a basis for decision making, he set about developing a risk-based approach. It depended, of course, on the existence of a reliable and tractable method for estimating the probabilities of accidents in a complex system based on the probability of individual component failures. This was already being used in the form of fault tree analysis, (a term he did not use).

It was accepted at the time that the major hazard from a reactor accident came from I-131 which would be inhaled. The risks of thyroid cancer developing after an intake had been published by ICRP so the quantity of I-131 could be used as a surrogate for the health consequences and the link with health effects could be made as required.

So accident sequences could be analysed using FTA to obtain the frequencies of various outcomes and then estimates could be made of the quantity of I-131 likely to be released. The memorable thing Farmer did was to plot these two quantities on a graph. Frequency (well actually the average time between accidents) as the ordinate and and the Curies of I-131 released as the abscissa.

TAMING THE RAYS

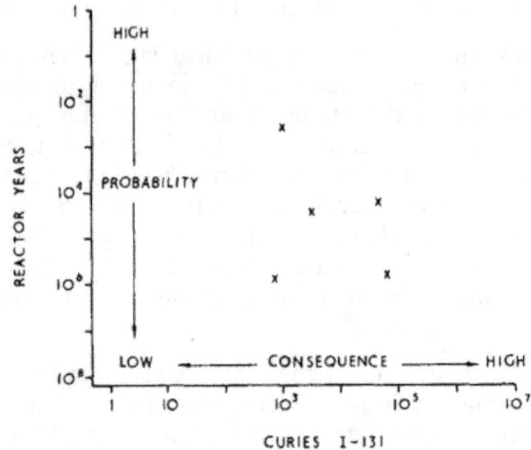

FIG. 1. Diagram of probabilities and consequences from a full safety evaluation.

Figure 20: Probability consequence diagram Farmer 1967

It was immediately clear that accident sequences that plotted into the bottom-left of the graph were less serious in risk terms than ones that finished up at the top-right. This much must surely have been done before. What was new was that Farmer had the audacity to suggest where the boundary between an accident being acceptable and unacceptable lay.

14 Reactor Safety

The first step was to note that a line with a slope of -1 on this log-log

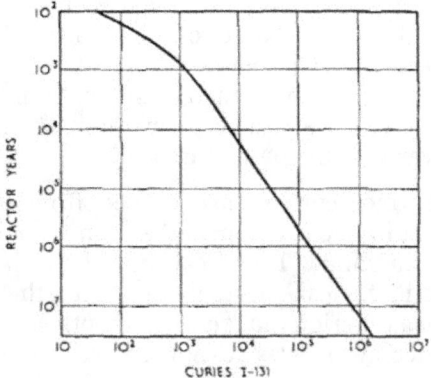

FIG. 12.   Proposed release criterion

Figure 21: Farmer release criterion

plot would represent points of equal risk of Curies released per year. It may not represent an equal risk of thyroid cancers. Farmer also considered that most people would wish to apply a relatively greater penalty against larger releases than the -1 slope represented. He therefore based his criterion on a slope of -1.5 meaning that a release 100 times bigger than another one had to be 1000 times less likely.

Having set this as the form of the criterion it was then necessary to define just one point of the boundary of acceptability and this was done recognising that it should reflect: public reaction to an accident, the number of casualties (thyroid cancers) caused by the release and the increased risk to any individual.

Farmer took as his defining point a release of a few thousand Curies. It would not cause many casualties but he thought that it would hardly be acceptable for this to happen within the lifetime of the reactor[41]. Indeed he concluded that it should not happen often in an entire reactor programme of 1000 reactor years and he decided to set the maximum acceptable frequency of a release of a few thousand Curies at 10^{-3} per reactor year. This, with the -1.5 slope, set the acceptability criterion with one minor adjustment: Farmer thought that there should be an extra restriction on smaller release (10 to 1000 Ci) because of their nuisance value. The final Farmer curve was therefore curved around as shown in the figure.

The implications of the criterion in terms of risk were worked out in an Appendix to the paper by Jack Beattie. He concluded that, for a population distribution of 4 million people within 10 miles and a representative weather mix (based on the Pasquill scheme), the reactor would pose an extra collective risk of 0.01 extra thyroid cancers per year. This was to be seen against a natural thyroid cancer risk of 72 cases per year in the same population. The individual risk to a child at 1000 yards would be 1×10^{-7} per year.

Reactions to Farmer curves have ranged from thinking them almost breathtaking in their vision and elegance to deeming them arbitrary and oversimplified. Their real impact on what followed in reactor studies is still unclear—beyond their being the first coherent public statement of a criterion that addressed both consequences and probabilities. Farmer's work was acknowledged in WASH-1400 and the spirit of his curves was used in representing its results as f-N plots.

The Reactor Safety Study was followed up quite quickly with two PSAs on the Zion and Indian Point reactors. These were required by NRC in 1976 as a response to a petition to close the Indian Point reactors. The three-year studies followed the RSS methodologies in most respects but greatly extended the consideration of external events (including fire, wind, flooding and earthquakes). These were now seen as significant contributors to risk and even dominant for the risk of core melt.

Hayns's useful Table 5[42] lists the 17 PSAs undertaken from the RSS in 1975 to the mid-1980s. There were significant differences in approaches. For example some included external events; others did not. BWRs and PWRs were included and there was some variation in reactor size. Comparisons may make little sense but the following risks for four eventualities were estimated. Sizewell gave uniformly the best numbers; Calvert Cliffs the worst

14 Reactor Safety

Event	Risk per year		
	Surry (RSS) 1975	Sizewell B 1982(best)	Calvert Cliffs (worst)
Core melt	6 x 10^{-5}	2 x 10^{-8}	2 x 10^{-3}
Major release	1 x 10^{-5}	2 x 10^{-10}	1 x 10^{-4}
Early fatality <1 mile	2 x 10^{-7}	2 x 10^{-10}	9 x 10^{-6}
Cancer fatality < 1 mile	1 x 10^{-7}	6 x10^{-11}	2 x 10^{-5}
The Sizewell study excluded external events			

Table 9: PSA study results (from Hayns 1999)

The Sizewell B study was undertaken specifically to investigate whether there might be a cliff-edge just beyond the DBA region. One conclusions that can be drawn from these studies as a whole is that, probably because reactor design has improved from a safety perspective so much, the severe accidents that are beyond design basis are the principal contributors to public risk.

PSA had developed a good deal in a decade. Hayns thought the areas that needed to be improved were the handling of human actions and the treatment of dependent failures. But perhaps most important of all was a much better understanding of the uncertainties in PSA. He suggested that "risk analysts are only at the threshold of performing comprehensive uncertainty analyses."

He could say at the end of the century that nowhere had PSA become the all-embracing and flexible panacea that had once seemed possible. In the UK and many other places it was acknowledged by regulators as an important technique in producing safe and balanced designs. In the USA the NRC adopted the term "risk-informed regulation." So, in some ways, PRA became just another layer of defence-in-depth.

One of the reactions to WASH-1400 was to raise the question of just how safe nuclear reactors should be anyway. The report had compared f-N curves for a reactor programme with those for other industrial activities and natural disasters. However, the results from the RSS were (allegedly) so far below the rest that it was hardly necessary to discuss in detail what was acceptable and what not. In the wake of the scathing criticisms of NRC in the Kemeny Commission Report[43] on TMI in 1979 the Commission decided that this was a matter that should be addressed and it published its "Plan for Developing a Safety Goal" in 1980[44]. After six years of thought and consultation the NRC published its proposals in 1986[45]. The two safety goals adopted related to individual risk and societal risk (paraphrasing):

> individual members of the public should bear no significant additional risk to life and health from nuclear power plant operations

> societal risks to life and health from nuclear should comparable to or less that those from   competing energy
> technologies and should not be a significant addition to other societal risks.

These were translated into "quantitative objectives", later to become "quantitative health objectives, (QHOs)".[46]

> The risk to an average individual in the vicinity of a nuclear power plant of prompt fatalities that might result from reactor accidents should not exceed one-tenth of one percent (0.1 percent) of the sum of prompt fatality risks resulting from other accidents to which the members of the U.S population are generally exposed.

> The risk to the population in the area near a nuclear power plant of cancer fatalities that might result from nuclear power plant operation should not exceed one-tenth of one percent (0.1 percent) of the sum of the cancer fatality risks resulting from all other causes.

The negotiations that led to the goals and objectives involved compromises: the industry generally thought 0.1% overly restrictive and suggested a level ten times higher; objectors thought 0.01% would be better. Discussions on cost-benefit analysis seemed to lead nowhere and the idea that there should be an objective for frequency of core damage of 10^{-6} per reactor year was referred to NRC staff. This latter objective was taken up in another form later.

There was a good deal of discussion in the document of how to define the location of the individuals (within one mile of the plant) and the population (within 10 miles) but there was no discussion of what the the background fatalities might be. This had taken place earlier in the consultation document NUREG-0880[47]. Here the average risk of accidental fatalities across the USA was taken as 5×10^{-4} and the fatal cancer risk as 1.9×10^{-3} /y. These values did not make it into the final proposals and the NRC preferred to stay with the percentage expression above. However researchers and others took to expressing the objectives as 5×10^{-7} and 2×10^{-6} respectively.

The implementation of PRA and the safety goals and objectives[48] brought forth a mass of explanatory and other documentation. One very important step was the definition of so-called surrogate (later Subsidiary) objectives. These were:

> Core damage frequency (CDF) of $<10^{-4}$ per year as surrogate for the latent cancer objective
>
> Large early release frequency (LERF) of <10-5 per year as surrogate for the early fatality objective

The LERF was intended to take account of the fact that releases which took place quickly—before evacuation could be organised—were much more likely to result in early deaths than releases which were delayed from the inception of the accident. Defining these two objectives meant that there should often be no need to extend the consequence into atmospheric dispersion and health consequence calculations. It perhaps reflected an understanding that details of population distributions had, in practice, a rather small effect on the consequences compared with the uncertainties in the analysis.[49] It certainly simplified things.

The plan to build a PWR at Sizewell was announced in 1980 and the expected objections led the Government to set up a public inquiry under Sir Frank Layfield. The inquiry reported in 1987 with a recommendation that the plant should go ahead subject to a satisfactory safety case. During the Inquiry it had become clear that, while the CEGB had been required to undertake a severe accident

PRA and calculate the risk posed, the NII had failed to provide any guidance on what an acceptable level of risk might be. Layfield therefore called on the Inspectorate to organise an extensive consultation on this and to then to come up with some authoritative advice. This the NII did and published "The Tolerability of Risks from Nuclear Power Stations" in 1988.[50]

This outlined the legal and structural arrangement for health and safety in the UK including the key concept of ALARP[51]—that risks should be as low as is reasonably practicable. The risk acceptability scale of fatality risk for members of the public had three regions. The highest was the unacceptable region and this ran up from 10^{-5} per year; practices that fell in this region were very unlikely to be allowed. Any risk that fell in the region below 10^{-6} per year was deemed broadly acceptable and so would probably be permitted. Risks between these two values were in the "tolerable" region. Practices that fell in this region would be permitted if a suitable ALARP argument could be made that they could not be reduced. The document implied that this would be more stringent at the top of the region than at the bottom.

The numerical values came from comparisons with the risk of other activities. Some weight was given to the risks associated with worker limits on radiation dose originating from ICRP and it was asserted that risks to the public should be ten times lower.

Societal risks were limited by expectations about the frequency of releases. Design assessments were to be based on the frequency of large releases from a reactor being no more than 10^{-7} per year (subject to ALARP considerations) and the frequency of releases delivering doses greater than 1 Sv being below 10^{-6} per year (again, subject to ALARP). The 1988 version had indicated that the risk of "one considerable accident" of 10^{-4} per reactor year was just tolerable.

The conclusion was the comforting one that the the proper application of the NII's Safety Assessment Principles would ensure that all the goals would be met.

14 Reactor Safety

Notes Chapter 14

1 (Morone J G and Woodhouse E J, 1986)p26
2 The RSC later evolved into the Advisory Committee on Reactor Safeguards and became a Federal body
3 (Joint Committee on Atomic Energy, 1952) JCAE, 1952) p5, 39, 114
4 The IPBR story is taken from (Morone J G and Woodhouse E J, 1986) p38, (Hewlett R G and Duncan F 1972) p213 and (Russell C R, 1962)
5 (McCullough C R, 1957)Ch7 "Nuclear Disaster Effects"

6 (Teller E, 1953)
7 (Sorensen J N, 1997)
8 (INSAG, 1996) p6
9 See (Giachetti R T, 1989) for a summary of the history of the issue
10 Definition from Glossary on USNRC website Accessed 29.11.17
11 (Tanguy P, 1988)
12 (Bartel R, 2016)

13 (Theoretical Possibilities and Consequences of Major Accidents in Large Nuclear Power Plants, 1957)-

14 Most of this section from (Walker J S, 1992) but see also (Apostolakis G and Mosleh A, 2017)
15 (Ernst M, 1973) and see e.g. (Garrick B J, 2014) p48 and (Keller W and Modarres M, 2005) p274
16 (Ericson C A, 2005), (Bartel R, 2016), (NRC, 2014), (Jones C G, 2005) (McLinn, J, 2010)
17 See (Ericson C A, 2005) Ch 12
18 See *Reactor Safety Study An Assessment of Accident Risks in U.S.* from (Verma A K, Ajit S and Kranki D R, 2010).
19 (Siddall E, 1957a, 1957b). An article by Laurence(Laurence G C, 1960) gives a broader view of the Canadian position as it developed. Also see (Farmer F R and Beattie J R, 1976)p55
20 WASH-1250(USAEC, 1973)
21 Most of this from (Keller W and Modarres M, 2005) p275.
22 Keller etc p275 has argued that this use of event trees was key to making PRA a practical reality. See (Ericson C A, 2005) for examples.
23 See *Reactor Safety Study An Assessment of Accident Risks in U.S. Commercial Nuclear Power Plants,* 1975)
24 (Bartel R, 2016)

25 See Appendix IV of (*Theoretical Possibilities and Consequences of Major Accidents in Large Nuclear Power Plants*, 1957)
26 There was a third case in which nothing escaped from the containment and the hazard was therefore just from direct radiation. To represent this the calculations assumed a complete release into the containment building.
27 The basis for this section and those that follow is Appendix VI of the Reactor Safety Study (RSS)
28 The information here is from RSS App VI Section 12
29 RSS App VI Section 12.1.1
30 RSS Main report p104

31 (*Reactor Safety Study An Assessment of Accident Risks in U.S. Commercial Nuclear Power Plants*, 1975)
32 (Bartel R, 2016) p27
33 (Lewis H, 1978)
34 Quoted by (Bartel R,2016) p30
35 (Marley W G and Fry T M, 1956)

36 The wording is based on that of (Grimston M C and Nuttall W J, 2013).The original paper is quite difficult to find
37 (Farmer F R and Beattie J R, 1976)
38 The part of the paper concerning radiological consequences is presented in (McCullough C R, 1957).
39 Adams C A and Stone C N, 1967)
40 (Charlesworth F R and Gronow W S, 1967)
41 It was calculated, in an Appendix by Jack Beattie, that a release of 10^4 Ci would give rise to 33 thyroid cancers within the following 10 years
42 (Hayns M R, 1999)
43 (Kemeny J G, 1979)
44 (*Plan for developing a safety goal*, 1980)
45 (NRC, 1986)
46 (*Use of probabilistic safety assessment methods in nuclear activities: Final policy statement*, 1995)
47 (NRC, 1983)
48 The objectives came to be known as Quantified Health Objectives
49 Some of the background to this and other aspects of PRA as they were seen at the beginning of the next century can be found in NUREG-1860
50 (HSE, 1988). A revised version with fairly minor changes was issued in 1992. It is this that is referred to here.

51 ALARP is the requirement to reduce risk until the cost of doing so is grossly disproportionate to the risk averted.

15 Post Script

The striking feature of the early years from a protection perspective is perhaps rashness. Now it seems extraordinarily irresponsible to irradiate yourself, patients and anyone else with obviously potent rays without fulling acknowledging the risks. Of course, on the plus side, the fact that x-rays could cure conditions, especially those of the skin, was recognised within a few years of their discovery but the negatives were known quickly too: serious effects appeared promptly in workers and devastating effects were not far behind. Yet it took more than a decade to convince workers of the need for protection and three decades before there were effective standards in place. There was, of course, the confident belief that all effects, even the serious ones, could be avoided if high doses were avoided—a reasonable assumption at the time—and the medical benefits were so great (even given the gung ho approach of some physicians) that this belief was not scrutinised too closely anyway. Cancer was recognised as one of the serious effects but it was always seen to be associated with traumatic effects. There was even an idea that trauma and wounds were general causes of cancer and there was no possibility of epidemiological evidence to link it to lower levels of radiation. That came many years later. The insidious stochastic effects were not fully recognised until the 1970s. We cannot blame the pioneers for missing them. Particularly since many of them paid the price of their ignorance.

Their patients paid a price too and one more difficult to compute.

15 Post Script

The spectacular cures that undoubtedly happened probably concealed much pain, suffering and death among less fortunate patients. The pressure to reduce physician doses was not matched for a long time by concern for the longer term effects on patients. Of course, while the belief prevailed that there was some threshold below which no effects would be seen, the matter was rather simple: if there seemed to be a benefit with no obvious ill-effects then doctors could continue to use x-rays and radioactivity in diagnosis and therapy. As appreciation of the real risks grew and exposures of patients were measured and compared, concern began to grow too. However, general pressure to reduce doses really only started towards the end of the century. Even at the end there was still controversy over, for example, the use of CT scans, which delivered rather high doses and were sometimes used for routine, non-essential purposes.

It was not just in medicine that emphasis changed. While a whole galaxy of biological effects came to be linked, for better or worse, to radiation, the first to be given a firm experimental basis were probably the genetic effects, first seen by Muller. Although initially these were welcomed as a new tool for the study of mutations, the potential consequences for us all began to be considered. The step from fruit fly to man was a large one but the extrapolation was made and the genetic consequences of population doses in a possible nuclear conflict and later the consequences of weapons testing, made genetics the central concern in the 1940s, 1950s and 1960s. Population gonad doses and genetically significant doses were critical.

But when the Japanese bomb data was analysed, no true genetic effects were seen—and weren't for the rest of the century. What were found were first leukaemias—there had been earlier expectations that they might occur—and then solid cancers. Although the genetic risks were still assumed to be there for humans—because they had been seen in other species—the existence of the somatic risks was confirmed in increasing numbers of studies of the bombs and occupational exposures. The focus in the 1970s thus turned away from genetic to somatic effects and specifically to cancer. And this was not simply because the cancers had been found but because they seemed to increase steadily with dose. Any notion of a tolerance dose rapidly disappeared and the official line was that all doses carried risks.

There was too a growing understanding of the role of DNA in genetics and life. Pictures of the impact radiation had on this essential molecule developed and initially simple models became more complex. It was difficult to be sure that it was much better understood from a radiological viewpoint at the end. However these models became important beyond mere science: they were the tools that might allow the interpolation of the epidemiological data to low doses. Knowledge of effects comes from doses greater than about 0.1 Sv from the Japanese bombs and there is now some prospect of extending that down to 10 mSv by combining large epidemiological studies of workers. As direct evidence that is probably the best we will ever get, yet many people are exposed at much lower levels to man-made sources, so the effects of radiation on them will still have to be deduced using theoretical models of some kind. From these it is possible still to reach a variety of conclusions on the shape of the dose response curve: linear, supralinear, sublinear. Even curves that dip below the axis to give beneficial results can be deduced. Although there was always the suspicion that the supporting data were chosen with some care by proponents of particular views (it would be surprising if they weren't given the complexity of the subject and its highly-charged nature), there remained a range of opinions that could be plausibly-enough supported to keep controversy boiling.

The linear no-threshold response, popular since the 1970s, was admirable in may ways. Its slope might increase as more evidence became available but it could be defended as essentially conservative (although supralinearist elements did not accept that). However, it posed at least one problem. With the increasing availability of computers from the 1960s it became possible to construct on them increasingly complex models of the spread of radioactivity from, say, nuclear reactor discharges and its uptake by people. The consequence was the prediction of large numbers of people being exposed to minute doses: for a few nuclides the entire global population could be predicted to be exposed to unthinkably tiny doses. Adding these up for the exposed population could lead to significant numbers for the collective dose: multiply this by a risk factor and it was possible to predict a significant number of cancers. Add to this the fact that in some cases the doses are delivered over millennia—or more—and you have a moral dilemma of some magnitude; should such minuscule, diffused individual risks influence decisions on energy supply and security? All that from a simplifying, conservative assumption of no-threshold linearity. It makes ideas like effective dose possible—life would be almost impossible without such unifying concepts—but there is a cost.

At a fundamental level one concept has remained essentially unchanged since almost the beginning. The idea of dose—energy deposition—has been at the centre of radiation protection since the 1920s in various forms. It is surprising that something so blunt—with just the crude modification of LET—could survive given the subtlety and complexity of biological systems. The ideas of microdosimetry have much to tell us about what may actually happen when radiation crashes through tissue but dose concepts have become more complex only really to accommodate different macroscopic spatial energy distributions. With the slightest stretch of the imagination this could be taken to cover internal dosimetry where one of the great achievements of radiation protection has been the integration of internal and external pathways to give an overall risk estimate. It was only when this could be done that the true significance of radon could be seen for example. Now we believe that, given the metabolic data that exists, we can calculate the risks from an intake of almost any nuclide you care to mention. Something necessarily gets lost in such calculations: there are assumptions about (at least) the applicability of particular metabolic data and the distribution of energy is very different from that in cases where we have any firm human data. There are also areas where there are probably some special problems: hot particles and low-energy emitters may well be two. The persistence of the dose idea has had one remarkable consequence (and a consequence that may have helped give the idea its legs): the development of measurement techniques has been a steady one for more or less a century. The physical principles of many instruments at the end of the century would have been quite comprehensible to the pioneers, although they would have been astonished by the electronics associated with them and, no doubt, their robustness and reliability.

As well as all the technical problems like these there has been an intriguing set of ethical ones. The ones associated with medical uses have generally been approached through medical ethics, where the problems of balancing potential benefits and risks of procedures is an everyday one for practitioners and an inescapable professional imperative. It has remained that although, towards the end of the century, there was pressure to put this on a more transparent basis as the risks of radiation exposures became better understood quantitatively. Interesting as that development must prove to be, the bigger challenges came from the knowledge that radiation risk increased with dose without, according to the protection scheme current for the last several decades, any threshold. There were inequities even without that: people who suffered most from nuclear discharge did not receive the greatest benefits. But with it it was possible to predict doses—and consequent health effects—to people distant in both space and time. How to balance benefits now with doses now and in the distant future? It was a particularly tricky example of a more profound dilemma: how much protection to provide? Since it was possible to reduce radiation doses from almost all practices to any low level you cared to name, the question became when to stop. Reduction would of course cost time, trouble and money so there is a balance to be struck. Cost-benefit analysis seemed for a while to offer a solution (particularly since it seemed to be enshrined in common sense and English law) but it soon became apparent that it was only a technical one; it was not robust enough to accommodate all the social, political and commercial demands that pressed upon important decisions. Analysis fell back on decision analysis—which meant laying out the consequences of decisions as clearly as possible and then the right people making up their minds. It seemed, in some ways, to bring the medics and the rest of us together.

The second edition has two new chapters on two rather technical areas: criticality and reactor safety. They are in some senses peripheral and are not always seen as part of the mainstream of radiation protection. However they are quite definitely part of the framework of protection and illustrate the intellectual challenges of risk assessment, defining what is acceptable and demonstrating compliance. They can be taken as representative of a number of other specialised fields such as modelling discharges to seas and rivers and

15 Post Script

routinely to the atmosphere,a well as assessing the safety of waste repositories and chemical processing plants.

It would be improper to forget that radiation protection has had relatively modest aim of providing a framework for protecting people while extracting what benefits we can from nuclear power and medical applications. Even this modest aim has led to an extraordinary and complex system. There is still work in progress, still other voices at the gates but maybe it should be marked as a small but important achievement of the twentieth century—not least because nuclear power looks like playing an important role in this one.

And finally, the title. The rays have not been tamed; they are like wild and rather dangerous creatures in a zoo rather than domesticated animals. If we have managed to understand and use them it is because, after sometimes bitter experience, we have constrained and disciplined ourselves. In fact, we have been tamed, not them.

16 References

Abbatt J (1979) 'History of the use and toxicity of Thorotrast', *Environ Res*, pp. 6–12.

Abbey F (1967) *Manual of Criticality Data*. UKAEA / No: AHSB(S) Handbook 5.

Adams C A and Stone C N (1967) 'Safety and siting of nuclear power stations in the United Kingdom', in. *Containment and siting of nuclear power plants, 3-7 April 1967,* Vienna: IAEA, p. 129-.

AEC (1961) *Nuclear Safety Guide*. AEC / No: TID-7016 Rev1.

Alberts B, Johnson A et al (2002) *Molecular Biology of the Cell - 4th ed.* NY: Garland Science.

Allisy A (1996) 'Henri Becquerel: The discovery of radioactivity', *Rad Prot Dosimetry*, 68(1/2), pp. 3–10.

Alvarez R (2006) 'The risks of making nuclear weapons', *Institute for Policy Studies*.

Andersson I O and Braun J (1963) *A neutron rem counter with uniform sensitivity from 0.025 MeV to 10 Mev*. Vienna:IAEA.

Andersson I O and Braun J (1964) 'A neutron rem counter', *Nucleonik*, 6, pp. 237–241.

16 References

Anon (1896a) 'Notes', *Nature*, 53, p. 421.

Anon (1896b) 'Recent work with Rontgen Rays', *Nature*, 53, p. 613.

ANS (1975a) *Criticality safety in operations with fissionable material*. ANS / No: ANS-8.1/N16.1 -1975.

ANS (1975b) *Validation of calculational methods for nuclear criticality safety*. ANS / No: ANS-8.1/N16.9-1975.

ANS (1983) *Criticality safety in operations outside nuclear reactors*. ANS / No: ANS-8.1-1083.

Apostolakis G and Mosleh A (2017) *Risk-Informed Decision Making: A Survey of United States Experience*. Los Angeles: John B Garrick Inst.

Armitage P and Doll R (1954) 'The age distribution of cancer and a multi-stage theory of carcinogenesis', *Br J Cancer*, 8, pp. 1–12.

Armitage P and Doll R (1957) 'A two-stage theory of carcinogenesis in relation to the age distribution of human cancer', *Br J Cancer*, 11, pp. 161–169.

Arnold L (1995) *Windscale 1957: Anatomy of a nuclear accident*. Basingstoke: Macmillan(2ed).

Arnold L and Smith M (2006) *Britain, Australia and the Bomb*. Basingstoke: Palgrave MacMillan.

Aspinall K J and Daniels J T (1965) *Review of U.K.A.E.A. Criticality Detection and Alarm Systems 1963/64. PartT I. Provision and Design Principles*. AHSB(S)R-92. UKAEA.

Bacq Z M and Alexander P (1955) *Fundamentals of radiobiology*. NY;London: Pergamon.

Badash L (1978) 'Radium, radioactivity, and the popularity of scientific discovery', *Proc Am Phil Soc.* (122), 122, pp. 145–154.

Badash L (1979) *Radioactivity in America: growth and decay of a science*. Baltimore: Johns Hopkins Univ Press.

Baltimore D (1970) 'Viral RNA-dependent DNA polymerase', *Nature*, 226, pp. 1209–1211.

Barbry F, Grivot P et al (2003) 'Criticality Experiments Performed in Saclay and Valduc Centers, France (1958-2002)', *Nucl Sci Eng*, 145, pp. 29–63.

Barclay A S and Cox S (1928) 'The radiation risks of the roentgenologist', *Am J Roentgenol*, 19, p. 551.

Bardeen C R (1907) 'Abnormal development of toad ova fertilized by spermatozoa exposed to the Roentgen rays', *J Exptl Zool*, 4(1), pp. 1–44.

Barendsen G W, (first) (1962) 'Dose-survival cells of human cells in tissue culture irradiated with alpha, beta and x-radiation', *Nature*, 193, pp. 1153–1156.

Barker H H and Schlundt H (1930) 'Detection, estimation, and elimination of radium in living persons given radium chloride internally', *Am J Roentgenology Rad Ther*, 26, pp. 418–423.

Bartel R (2016) *WASH1400 The Reactor Safety Study: The Introduction of Risk Assessment to the Regulation of Nuclear Reactors*. NUREG/KM-0010. ONR.

Bauer K H (1928) *Mutationtheorie der Geschwultstentstehung*. Berlin, Germany: Springer.

Bay Z (1938) 'Electron-multiplier as an electron-counting device', *Nature*, 141, p. 284.

BEAR Committee (1956) *The Biological Effects of Atomic Radiation*. Washington DC: National Academy of Sciences/NRC. Available at: http://hdl.handle.net/2027/mdp.39015049805065.

16 References

Becker K (1973) *Solid State Dosimetry*. Cleveland: CRC Press.

Becquerel H and Curie P (1901) 'Action physiologique des rayons du radium', *Comptes Rendus des Seances... Paris*, 132, pp. 1289–1291.

Beebe G W (1979) 'Reflections on the work of the ABCC in Japan', *Epidemiol Revs*, 1, pp. 184–210.

BEIR Committee (1972) *The effects on populations of exposure to low levels of ionizing radiation*. National Research Council.

BEIR Committee (1990) *Health effects of exposure to low levels of ionizing radiation: BEIR V*. Washington DC: National Academy Press.

BEIR Committee (2006) *Health Risks from Exposure to Low Levels of Ionizing Radiation BEIR VII Ph2*. National Research Council.

Bell G E (1936a) 'The photographic action of radium gamma-rays', *Br J Radiol*, 9, p. 743.

Bell G E (1936b) 'The photographic action of X-rays', *Br J Radiol*, 9, p. 578.

Bell P R (1948) 'The use of anthracene as a scintillation counter', *Phys Rev*, 73, pp. 1405–1406.

Bender M A and Gooch P C (1962) 'The kinetics of x-ray survival of mammalian cells in vitro', *Int J Radiat Biol*, 5, pp. 133–145.

Bennett B (1995) 'Worldwide releases of radionuclides', in. *Symposium on Radioactive Releases*, Vienna: IAEA-SM-339/185.

Benton E V and Oswald R (1980) *Proton recording neutron dosementer for personnel monitoring in Proc 10th Int Conf on Solid State Nuclear Track Detectors*. Oxford: Pergamon Press.

Beral V, Fraser P et al (1988) 'Mortality of employees of the Atomic Weapons Establishment, 1951-1982', *Br Med J*, 297, pp. 757–770.

Beral V, Inskip H et al (1985) 'Mortality of employees of the United Kingdom Atomic Energy Authority, 1946-1979', *Br Med J*, 291, pp. 440–447.

Bergonie J and Tribondeau L (1906) 'De quelques resultats de radiotherapie et essai de fixation d'une technique rationnelle', *C R Acad Sci(Paris)*, 143, pp. 983–985.

Berrington, Weiss H A and Doll R Darby S C (2001) '100 years of observation on British radiologists: mortality from cancer and other causes 1897-1997', *BJ Radiology*, 74, pp. 507–519.

Birks J B (1964) *The Theory and Practice of Scintillation Counting*. Oxford: Pergamon Press.

Bishop M B (1996) 'The discovery of proto-oncogenes', *FASEB Journal*, pp. 362–364.

Black D (1984) *Investigation of the possible increased incidence of cancer in Western Cumbria*. London: HMSO.

Blau M and Altenburger K (1922) 'Über einige Wirkungen von Strahlen II', *Zeit Phys*, 12, pp. 315–329.

Blau M and Dreyfus B (1945) 'The multiplier photo-tube in radioactive measurements', *Rev Sci Instrum*, 16, pp. 245–248.

Boerner A J and Kathren R L (2005) 'The Health Physics Society: A 50-year Chronology', *Health Physics*, 88, pp. 193–213213, 193.

Bouwers A and van der Tuuk J H (1930) 'Fortschritte auf dem Gebiete der Rontgenstrahlung', *Strahlenschutz*, 41, p. 767.

Boveri T (1914) *Zur Frage der Entstehung maligner Tumoren*. Jena, Germany: Gustav Fischer Verlag.

Bowers B (1970) *X-rays their discovery and applications*. London: HMSO.

16 References

Bragg W H (1912) *Studies in Radioactivity*. New York: MacMillan Publishing Co.

Bragg W H and Kleeman R D (1904) 'On the ionization curves of radium', *Phil Mag S6*, 8, pp. 726–738.

Bragg W H and Kleeman R D (1905) 'On the alpha particles of radium and their loss of range in passing through various atoms and molecules', *Phil Mag S6*, 10, pp. 318–340.

Brecher R and Brecher E (1969) *The Rays*. Baltimore, MA: Williams and Wilkins.

Bronson F (2006) *The evolution of laboratory instrumentation for the operational health physicist*. AAHP Conference, Providence RI 27 June 2006.

Broser I and Kallmann H (1947) 'Uber die anregung von leuchtstoffen durch schnelle korpuskularteilchen 1', *Z f Naturforschung*, 2(8), pp. 439–440.

Brown G I (2002) *Invisible Rays: the history of radioactivity*. Stroud: Sutton Publishing.

Buck A L (1983) *A History of the Atomic Energy Commission*. DOE/ES-0003/1. USDOE.

Burrows E H (1986) *Pioneers and Early Years: A History of British Radiology*. Alderney: Colophon Ltd.

Butler B (1995) 'X-rays, not radium, may have killed Curie', *Nature*, 377, p. 96.

Calder J F (2001) *The History of Radiology in Scotland*. Edinburgh: Dunedin Acad Press.

Caldwell G G, Kelley D B and Heath C W Jr (1980) 'Leukemia among participants in military manoeuvers at a nuclear bomb test. A preliminary report', *J Am Med Assoc*, 244, pp. 1575–1578.

Caldwell G G, Kelley D B, Zack M et al (1983) 'Mortality and cancer frequency among military nuclear test (Smoky) participants 1957 through 1959', *J Am Med Assoc*, 250, pp. 620–624.

Calkins J (1971) 'A method of analysis of radiation response based on enzyme kinetics', *Radiat Res*, 45, pp. 50–62.

Callihan A D, Ozeroff W J, Paxton H C and Schuske C L (1957) *Nuclear Safety Guide*. AEC / No: TID-7016.

Cantril S T and Parker H M (1945) *The tolerance dose*. US AEC / No: MDDC-100.

Cardis E, Gilbert E S, Capenter L et al (1995) 'Effects of low doses and low dose rates of external ionizing radiation: cancer mortality among nuclear industry workers in three countries', *Radiat Res*, 142, pp. 117–132.

Cardis E, Vrijheid M, Blettner M et a (2005) 'Risk of cancer after low doses of ionising radiation: retrospective cohort study in 15 countries', *Br Med J*, doi:10.1136/bmj.38499.599861.E0. doi: doi:10.1136/bmj.38499.599861.E0.

Cardwell D S L (1972) *The Organisation of Science in England*. London: Heinemann.

Carlson J G (1938) 'Some effects of X-rays on the neuroblast chromosomes of the grasshopper (Chortophaga viridifasciata)', *Genetics*, 23, p. 596.

Carlsson G A (2008) *Bragg-Gray dosimetry: theory of Burch*. Available at: huweb.hu.liu.se/inst/imv/radiofysik/pdfs/Burch.pdf.

Carpenter L M, Higgins C D, Douglas A J et al (1998) 'Cancer mortality in relation to monitoring for radionuclide exposure in three UK nuclear industry workforces', *Br J Cancer*, 78, pp. 1224–1232.

Carter R D and Kiel G R (1968) *Criticality Handbook Vol1*. Atlantic Richfield Hanford Co / No: ARH600.

16 References

Cartwright B G, Shirk E K and Price P B, (1978) 'A nuclear-track recording polymer of unique sensisitivy and resolution', *Nucl Instrum Methods*, 153, pp. 457–460.

CEC (2002) *Assessment of the radiological impact on the population of the European Union from European Union nuclear sites between 1987 and 1996*. Luxembourg:CEC / No: Radiation Protection 128.

Chadwick K H and Leenhouts H P (1978) 'The rejoining of DNA double-strand breaks and a model for the formation of chromosome rearrangements', *Int J Radiat Biol*, 33, pp. 517–529.

Chadwick K H and Leenhouts H P (1981) *The Molecular Theory of Radiation Biology*. Berlin: Springer.

Charlesworth F R and Gronow W S (1967) 'A summary of experience in the practical application of siting policy in the United Kingdom', in. *Containment and siting of nuclear power plants, 3-7 April 1967*, Vienna: IAEA, pp. 143–170.

Christy R F and Wheeler J A (1943) *Chain reaction of pure fissonable materials in solution*. USAEC / No: CP-400 (1 Jan 1943).

Clarke R H (1990) 'The 1957 Windscale accident revisited', in Ricks R C and Fry S A (ed.) *The Medical Basis for Radiation Accident Preparedness*. New York: Elsevier, pp. 281–289.

Cochran T B, Norris R S and Suokko K L (1993) 'Radioactive contamination at Chelyabinsk-65, Russia', *Ann Rev Energy Environ*, 8, pp. 507–528.

Cohen B (1995) 'Test of the linear-no threshold theory of radiation carcinogenesis for inhaled radon decay products', *Health Phys*, 68(2), pp. 157–174.

Coltman J W and Marshall F H (1947) 'The photo-multiplier radiation detector', *Phys Rev*, 72, p. 528.

Colwell H A and Russ S (1934) *X-ray and radium injuries. Prevention and treatment*. London: OUP.

COMARE (2005) *Tenth report:The incidence of childhood cancer around nuclear installations in Great Britain*. Didcot: Health Protection Agency.

Condon E U and Terrill H M (1927) 'Quantum phenomena in the biological action of X-rays', *J Cancer Res*, 11, pp. 324–333.

Coppes-Zantinga A R and Coppes M J (1998) 'The early years of radiation protection: a tribute to Madame Curie', *Can Med Assoc J*, 159, pp. 1389–1391.

Court Brown W M and Doll R (1957). MRC Special Report 295. London: HMSO.

Court Brown W M and Doll R (1958) 'Expectation of life and mortality from cancer among British radiologists', *BMJ*, 2(5090), pp. 181–187.

Crow J C and Abrahamson S (1997) 'Seventy years ago: mutation becomes experimental', *Genetics*, 147, pp. 1491–1496.

Crow J F and Denniston C (1981) 'The mutation component of genetic damage', *Science*, 212, pp. 888–893.

Crowther J A (1924) 'Some considerations relative to the action of x-rays on tissue cells', *Proc Roy Soc B*, 96, pp. 207–211.

Crowther J A (1926) 'Actions of x-rays on Colpidium Colpoda', *Proc Roy Soc B*, 100, pp. 390–404.

Crowther J A (1938) 'The biological action of x rays - a theoretical review', *Br J Radiol*, 11, pp. 132–145.

Curie E (1938) *Madame Curie*. London: Heinemann.

16 References

Curran S C and Baker W (1948) 'Photoelectric alpha-particle detector', *Rev Sci Instrum*, 19(2), p. 116.

Curtis S B (1986) 'Lethal and potentially lethal lesions induced by radiation - a unified repair model', *Radiat Res*, 106, pp. 252–270.

Daniel J (2005) *Investigation of releases from Santa Susana sodium reactor experiment in July 1959*. Crescent City, FL:Daniel & Assocs Inc / No: TDR-DA/0502.

Darby S C, Kendall G M et al (1993) 'Further follow-up of mortality and incidence of cancer inmen from the United Kingdom who participated in the UK's atmospheric nuclear weapon tests and experimental prgrammes', *Br Med J*, 307, pp. 1530–1535.

Darby S C,Kendall G etal (1988) 'A summary of mortality and incidence of cancer in men from the United Kingdom who participated in the UK's atmospheric nuclear weapon tests and experimental programmes', *Br Med J*, 296, pp. 332–338.

Darby S, Hill D et al (2005) 'Radon in homes and risk of lung cancer: collaborative analysis of individual data from 13 European case-control studies', *BMJ*, 330, p. 223.

Delafield H J and Clifton J J (1984) *Design Criteria and Principles for Criticality and Alarm Systems*. SRD 309. UKAEA Safety and Reliability Directorate.

Dessauer F (1922) 'Uber einige Wirkungen von Strahlen', *Zeit Phys*, 12, pp. 38–44.

Dessauer F (1923) 'Point-heat theory', *Z Phys*, 30, p. 288.

Deutsch M (1948) 'High efficiency, high speed scintillation counters for beta- and gamma- rays', *Phys Rev*, 73, p. 1240.

Doll R (1994) 'Paternal exposure not to blame', *Nature*, 367, pp. 678–680.

Doll R (1995) 'Hazards of ionizing radiation - 100 years of observation on man', *Brit J Cancer*, 72, pp. 1339–1349.

Doll R (2004) 'Commentary: The age distribution of cancer and a multistage theory of carcinogenesis', *Int J Epidemiol*, 33, pp. 1183–1184.

Dolphin G W, Megaw W J and Rundo J (1962) 'Health physics', *Rept Prog Phys*, 25, pp. 337–394.

Douglas A J, Omar R Z and Smith P G (1994) 'Cancer mortality and morbidity among workers at the Sellafield plant of British Nuclear Fuels', *Br J Cancer*, 70, pp. 1232–1243.

Duane W (1922) 'Roentgen rays of short-wavelengths and their measurement', *Am J Roentgenology*, 9, p. 165.

Duane W (1923) 'Measurement of dosages by means of ionization chambers', *Am J Roentgenology Rad Therapy*, 10, p. 399.

Dublin L I and Spiegelman M (1948) 'Mortality of medical specialists 1938-1942', *J Am Med Assoc*, 137, pp. 1519–1524.

DuBridge L A (1931) 'The amplification of small direct currents', *Phys Rev*, 27, pp. 392–400.

Duncan K P and Howell R W (1970) 'Health of workers in the United Kingdom Atomic Energy Authority', *Health Physics*, 19, pp. 285–291.

Dunning J R (1934) 'Amplifier systems for the measurement of single particles', *Rev Sci Instrum*, 5, pp. 387–394.

Dunster H J, Howells H and Templeton W L (1958) 'District surveys following the Windscale incident', in *Proc 2nd UN Int Conference Peaceful Uses of Atomic Energy Vol 18. 2nd UN Int Conference Peaceful Uses of Atomic Energy*, Geneva: UN.

Eccles W H and Jordan F W (1919) 'A Trigger Relay Utilising Three Element Thermionic Vacuum Tubes', *Radio Rev*, 1, pp. 143–146.

16 References

Edler L and Kopp-Schneider A (2005) 'Origins of the mutational orig inof cancer', *Int J Epidemiol*, 34, pp. 1168–1170.

Elkind M M and Sutton H (1960) 'Radiation response of mammalian cells grown in culture I Repair of X-ray damage in surviving Chinese hamster cells', *Radiat Res*, 13, pp. 556–593.

Ericson C A (2005) *Hazard Analysis Techniques*. Wiley. Available at: https://www.fer.unizg.hr/_download/repository/Event_Tree_Analy sis_from_Hazard_Analysis_Techniques_for_System_Safety,_Wiley _2005%5B1%5D.pdf.

Ernst M (1973) *Study of the Reactor Licensing Process*. AEC.

Evans R D (1933) 'Radium poisoning. A review of present knowledge', *Am J Pub Health*, pp. 1023, 1017.

Evans R D (1981) 'Inception of standards for internal emitters, radon, and radium', *Health Physics*, 41(3), pp. 437–448.

Evans R D and Alder R L (1939) 'Improved Counting Rate Meter', *Rev Sci Instrum*, 10, pp. 332–336.

Failla G (1932) 'Radium protection', *Radiology*, 19, pp. 12–21.

Falconer D S (1965) 'The inheritance of liability to certain diseases estimated from the incidence among relatives', *Ann Hum Genet*, 29, pp. 51–76.

Farabee W C (1903) *Hereditary and Sexual Influence in Meristic Variation: A Study of Digital Malformations in Man*. Thesis. Harvard, USA: Harvard University.

Farmer F R and Beattie J R (1976) 'Nuclear power reactors and the evaluation of population hazards', *Adv Nucl Sci Tech*, 9, pp. 2–73.

Farmer F T (1942) 'An electrometer for measurement of voltage on small ionization chambers', *Proc Phys Soc*, 54, pp. 435–438.

'Fermi invention rediscovered at LASL' (1966) *Atom*, (October), pp. 7–11.

Findlay A and Williams T (1965) *A Hundred Years of Chemistry*. London: Methuen.

Fischer D (1997) *History of the IAEA The first forty years*. Vienna: IAEA.

Fitzgerald J J (1969) *Applied Radiation Protection and Control*. New York: Gordon and Breach.

Flinn F B (1929) 'Some precautions to be taken when making tests for radioactivity in the living body', *Am J Roentgenology Rad Ther*, 22, pp. 554–556.

Folley J H, Borges W and Yamawaki T (1952) 'Incidence of leukemia in survivors of the atomic bomb in Hiroshima and Nagasaki,Japan', *Am J Med*, 13(3), pp. 311–321.

Frame P W (2004) 'A history of radiation detection instrumentation', *Health Physics*, 87, pp. 111–135.

Franke H (1928) 'Uber die Bestimmung der Toleranzdosis auf photographischen Wege', *Fortschritte a d Geb Roentgenstrahlen*, 38, p. 22.

Fricke H and Glasser O (1925) 'A theoretical and experimental study of the small ionization chamber', *Am J Roentgenol*, 13, p. 462.

Fricke R and Glasser O (1925) 'Eine theoretische und experimentelle Untersuchung der kleinen Ionisationskammer', *Fortschritte a. d. Geb. Roentgenstrahlen*, 33, p. 329.

Fuchs K (1946) *Efficiency for very slow assembly*. Los Alamos / No: LA-596.

Fullwood R R (1999) *Probabilistic Safety Analysis in the Chemical and Nuclear Industries*. Butterworth-Heinemann. Available at:

https://books.google.co.uk/books?
id=ozVTbvz6YRMC&dq=mulvihill+prc+r-657+nuclear.

G Jaffe (1932) 'Effects of alpha rays on the passage of electricity through crystal', *Phys Z*, 33, pp. 393–399.

Gardner M J, Hall A J, Downes S et al (1987a) 'Follow-up study of children born elsewhere but attending schools in Seascale, West Cumbria(schools cohort)', *Br Med J*, pp. 819–822.

Gardner M J, Hall A J, Downes S et al (1987b) 'Follow-up study of children born to mothers resident in Seascale West Cumbria (birth cohort)', *Br Med J*, 295, pp. 822–827.

Gardner M J, Snee M P, Hall A J et al (1990) 'Results of case-control study of leukaemia and lymphoma among young people near Sellafield nuclea plant in West Cumbria', *Br Med J*, 300, pp. 423–429.

Garrick B J (2014) 'PRA-based risk management: History and perspectives', *Nuclear News*, (July), pp. 48–53.

Garrod A E (1902) 'The incidence of Alkaptonuria: a study in chemical individuality', *The Lancet*, II, pp. 1616–1620.

Geiger H and Klemperer O (1928) 'Beitrag zur Wirkungweise des Spitzenzahlers', *Phys Zeits*, 49, pp. 753–760.

Geiger H and Marsden E (1913) 'The laws of deflexion of alpha particles through large angles', *Phil Mag*, S6, p. 25.

Geiger H and Muller W (1928) 'Elektronenzählrohr zur Messung schwächster Aktivitäten" (Electron counting tube for the measurement of the weakest radioactivities)', *Die Natwiss*, 16(31), pp. 617–618.

Geiger H and Muller W (1929) 'Technische Bemerkungen zum Elektronenzählrohr" (Technical notes on the electron counting tube)', *Phys Zeits*, 30, pp. 489–493.

Gerward L and Rassat A V (2000) 'La decouverte des rayons gamma', *Pour la Science*, 275, p. 8.

Giachetti R T (1989) *Contribution of Anticipated Transients Without Scram(ATWS) to Core Melt at United States Nuclear Power Plants*. DOE/OR/00033--T442. Available at: https://www.osti.gov/scitech/servlets/purl/6892419.

Giles N H Jr (1940) 'The effect of fast neutrons on the chromosomes of Tradescantia', *Proc Natl Acad Sci*, 26, p. 567.

Glasser O (1933) *Wilhelm Conrad Rontgen and the Early History of the Rontgen Rays*. London: John Bale Sons&Danielson.

Glasser O(ed) (1933) *The Science of Radiology*. London: Bailliere, Tindall and Cox.

Glasstone S (1950) *The Effects of Atomic Weapons*. US Dept of Defense.

Glocker R (1932) 'Quantum physics of the biological action of x-rays', *Zeit Phys*, 77, p. 653.

Goodhead D (Chairman) (2004) *Report of the Committee Examining Radiation Risks of Internal Emitters(CERRIE)*.

Goodhead D T (1982) 'An assessment of the role of microdosimetry in radiobiology', *Radiat Res*, 91, pp. 45–76.

Goodhead D T (1985) 'Saturable repair models of radiation action in mammalian cells', *Radiat Res*, 104, pp. S58–S67.

Gowing M (1964) *Britain and Atomic Energy 1939-1945*. London: St Martins Press.

Gray L H (1929) 'The absorption of penetrating radiation', *Proc Roy Soc A*, 122, pp. 647–668.

Gray L H (1936) 'An ionization method for the absolute measurement of gamma-ray energy', *Proc Roy Soc*, A156, pp. 578–596.

Gray L H (1937) 'Radiation dosimetry', *Br J Radiol*, 10, pp. 600–612 & 721–742.

Greenspan H G, Kelber C N and Okrent D(eds) (1968) *Computing Methods in Reactor Physics*. NY: Gordon and Breach.

Greinacher H (1924) 'Über die akustische Beobachtung und galvanometrische Registierung von Elementarstrahlen und Einzelionen', *Z Phys*, 23, pp. 361–378.

Grimston M C and Nuttall W J (2013) *The Siting of UK Nuclear Power Installations*. CWPE 1344 and EPRG 1321.

Gupton E D, Davies D M and Hart J C (1961) 'Criticality accident applications of the Oak Ridge National Laboratory badge dosimeter', *Health Physics*, 5, pp. 57–62.

Gusev B, Abylkassimova Z N and Apsalikov K N (1997) 'The Semipalatinsk nuclear test site: a first assessment of the radiological situation and the test-related radiation doses in the surrounding territories.', *Radiat Environ Biophys*, 36(3), pp. 201–204.

Guy J (1995) 'The X factor in X-rays', *History Today*, 45(11).

Hacker B C (1987) *The Dragon's Tail: radiation safety in the Manhattan Project 1942-1946*.

Hankins D E (1962) *Neutron monitoring instrument having a response approximately proportional to dose rate from thermal to 7.0 MeV*. Los Alamos Nuclear Labs / No: LA-2717.

Hansen G E (1952) *Burst characteristics associated with the slow assembly of fissionable material*. Los Alamos / No: LA-1441.

Hansen G E and Roach W H (1961) *Six and sixteen group cross sections for fast and intermediate critical assemblies*. LASL / No: LAMS-2543.

Hanson A O and McKibben J L (1947) 'A neutron detector having uniform sensitivity from 10kEV to 3MeV', *Phys Rev*, 72, pp. 673–677.

Hanson F B and Heys F (1929) 'An analysis of the effects of the different rays of radium in producing lethal mutations in Drosophila', *Amer Nat*, 63, pp. 201–213.

Haque A K M M, Collinson A J L and Blyth-Brook C O S (1965) 'Radon concentration in different environments and the factors affecting it', *Phys Med Biol*, 10, p. 514.

Hart D and Le Heron J C (1992) 'The distribution of medical x-ray doses amongst individuals in the British population', *Br J Radiol*, 65, pp. 996–1002.

Harvey E B, Boice Jr J D, Honeyman M and Flannery J T (1985) 'Prenatal x-ray exposure and childhood cancer in twins', *N Engl J Med*, 312, pp. 541–545.

Hayns M R (1999) 'The Evolution of Probabilistic Risk Assessment in the Nuclear Industry', *Trans I Chem E*, 77B, pp. 117–142.

Heckelsberg L F (1980) 'Thermoluminescent dosimetry(LiF) 1950-51', *Health Physics*, 39, pp. 391–393.

Hedge A R (1959) 'Can a single injury cause cancer?', *Calif Med*, 90, pp. 55–57.

Hemmings P J (1967) *The GEM code*. UKAEA / No: AHSB(S)R105.

Hems G (1966) 'Detection of effects of ionizing radiation by population study', *Br Med J*, 1(5484), pp. 393–396.

16 References

Hess V F (1965) *1936 Nobel Lecture - Unsolved Problems in Physics: Tasks for the Immediate Future in Cosmic Ray Studies*. Amsterdam: Elsevier.

Hess V F and McNiff W T (1947) 'Quantitative determination of the radium content of the human body and of the radon content of breath samples for the prevention and control of radium poisoning in persons employed in the radium industry', *Am J Roentgenology Rad Ther*, pp. 102, 91.

Hewlett R G and Duncan F (1972) *A History of the United States Atomic Energy Commission: Atomic Shield 1947/1952*. USAEC. Available at: http://energy.gov/sites/prod/files/Hewlett%20and%20Duncan%20-%20Atomic%20Shield%20(1%20of%204)_3.pdf.

Hiebert R D and Watts R J (1953) 'Fast-coincidence circuit for H-3 and C-14 measurements', *Nucleonics*, 11(12), pp. 38–41.

History of the BIR (no date). BIR / No:

History of the Radiological Society of North America (no date). / No:

Hofstadter R (1948) 'Alkali halide scintillation counters', *Phys Rev*, 74, pp. 100–101.

Hofstadter R (1949) 'The detection of gamma-rays with thallium-activated sodium iodide crystals', *Phys Rev*, 75, pp. 796–810.

Holweck F and Lacassagne A (1930) 'Action on yeasts of soft x-rays', *CR Soc Biol*, 103, pp. 60–62.

Hope-Stone H F (1999) *A History of Radiotherapy at the London Hospital*. London: Royal London Hospital Archives and Museum.

HSE (1988) *The tolerability of risks from nuclear power stations*. London: HSE.

Huebner R J and Todaro G J (1969) 'Oncogenes of RNA viruses as determinants of cancer', *Proc Natl Acad Sci USA*, 64, pp. 1087–1094.

Hultqvist B (1956) 'Studies on naturally occurring ionizing radiations', *Kgl Svenska Vetenskaps Handl*, 6(3).

Hunt G J (1997) 'Radiation doses to critical groups since the early 1950s due to discharges of liquid radioactive waste from Sellafield', *Health Physics*, 72, pp. 558–567.

IAEA (1988) *The radiological accident at Goiania*. Vienna: IAEA.

IAEA (1990) *The radiological accident in San Salvador*. Vienna: IAEA.

IAEA (1999a) *Inventory of radioactive waste disposals at sea*. IAEA-TECDOC-1105. Vienna: IAEA.

IAEA (1999b) *Status and trends in spent fuel reprocessing*.

ICRP (1951) 'International recommendations on radiological protection', *Br J Radiol*, 23, pp. 46–53.

ICRP (1955) 'Recommendations of the International Commission on Radiological Protection', *Br J Radiol*, Supp 6.

ICRP (1958) 'Report on Amendments during 1956 to the Recommendations of the International Commission on Radiological Protection', *Radiology*, 70, p. 261.

ICRP (1959) *Recommendations of the International Commission on Radiological Protection(adopted 9 September1958)*. London: Pergamon.

ICRP (1960) *Publication 2: Recommendations of the ICRP, Report of Committee II on Permissible Dose for Internal Radiation*. London: Pergamon.

16 References

ICRP (1964) *Recommendations of the International Commission on Radiological Protection Publication 6.* London: Pergamon.

ICRP (1966a) *Publication 8: The evaluation of risks from radiation: A report of Committee 1 of ICRP.* London: Pergamon.

ICRP (1966b) *Recommendations of the International Commission on Radiological Protection Publication 9.* London: Pergamon.

ICRP (1968) *Publication 10: Evaluation of radiation doses to body tissues from internal contamination due to occupational exposure.* Oxford: Pergamon.

ICRP (1971) *Publication 10A: The assessment of internal contamination resulting from recurrent or prolonged intakes.* Oxford: Pergamon.

ICRP (1973) *Publication 22: Implications of the Commission recommendations that doses be kept as low as readily achievable.* Oxford: Pergamon.

ICRP (1977) 'Recommendations of the International Commission on Radiological Protection Publication 26', *Ann ICRP*, 1(3).

ICRP (1979) *Publication 30: Limits for intakes of radionuclides by workers (with later supplements).* Oxford: Pergamon, p. 0.

ICRP (1983) 'ICRP Publication 37: Cost-Benefit Analysis in the Optimization of Radiation Protection', *Ann ICRP*, 10(2/3).

ICRP (1985) 'Statement from the 1985 Paris meeting of the ICRP', *Ann ICRP*, 15(3).

ICRP (1989a) 'Optimization and decision-making in radiological protection', *Ann ICRP*, 21(1).

ICRP (1989b) 'Publication 56: Age-dependent doses ot members of the public from intake of radionuclides: Part 1', *Ann ICRP*, 20(2).

ICRP (1991a) 'Publication 60: Recommendations of the International Commission on Radiological Protection', *Ann ICRP*, 21(1–3).

ICRP (1991b) 'Publication 61: Annual limits on intake of radionuclides by workers based on the 1990 recommendations', *Ann ICRP*, 21(4).

ICRP (1993) 'Publication 67: Age-dependent doses to members of the public from intake of radionuclides: Part 2 ingestion dose coefficients', *Ann ICRP*, 23(3/4).

ICRP (1994) 'Publication 66: Human respiratory tract model for radiological protection', *Ann ICRP*, 24(1–3).

ICRP (1995) 'Publication 69: Age-dependent doses to members of the public from intake of radionuclides: Part 3 ingestion dose coefficients', *Ann ICRP*, 26(1).

ICRP (1996a) 'Publication 71: Age-dependent doses to members of the public from intake of radionuclides: Part 4 inhalation dose coefficients', *Ann ICRP*, 25(3).

ICRP (1996b) 'Publication 72: Age-dependent doses to members of the public from intake of radionuclides: Part 5 compilation of ingestion and inhalation coefficients', *Ann ICRP*, 26(1).

ICRP (1998) 'Publication 77: Radiological protection policy for the disposal of radioactive waste', *Ann ICRP*, 27 Supp.

ICRP (1999) 'Publication 83: Risk estimation for multifactorial diseases', *Ann ICRP*, 29(3–4).

ICRP (2001) 'Prevention of accidents to patients undergoing radiation therapy', *Ann ICRP*, 30(3).

ICRP-ICRU (1963) 'Report of the RBE Committee to the International Commissions on Radiological Protection and Radiological(sic) Units and Measurements', *Health Physics*, 9, pp. 357–386.

Ing H and Birnboim H C (1984) 'A new bubble-damage polymer detector for neutrons', *Nucl Tracks Radiat Meas*, 8, pp. 285–288.

INSAG (1996) *Defence in Depth in Nuclear Safety*. INSAG-10. Vienna: IAEA.

Inskip H, Beral V, Fraser P, Booth M, Coleman D and Brown A (1987) 'Further assessment of the effects of occupational radiation exposure in the United Kingdom Atomic Energy Authority mortality study.', *Br J Ind Med*, 44, pp. 149–160.

Ives J E, Knowles F L and Britten R H (1933) 'Health aspects of radium dial painting III measurement of radioactivity in workers', *J Indust Hyg*, 15, pp. 433–446.

IXRPC (1928) 'International recommendations for X-ray and radium protection', *Br J Radiol*, 1, pp. 358–363.

IXRPC (1931) 'Recommendations of the 2nd International Congress of Radiolology', *Br J Radiol*, 4, pp. 485–487.

IXRPC (1934) 'International recommendations for X-ray and radium protection', *Br J Radiol*, 7, pp. 695–699.

IXRPC (1937) 'International recommendations for X-ray and radium protection', *Radiology*, 30, pp. 511–515.

Jackson J (2003) 'A meander through forty years of the Society for Radiological Protection', *J Radiol Prot*, 23, pp. 125–128.

Jacobi W (1975) *How shall we combine the doses to different body organs -problems and ideas*. Selected papers of the International Symposium at Aviemore ed P Recht P and Lakey J R A, CEC Luxembourg, 1975.

Jennings W A (2007) 'Evolution over the past century of quantities and units in radiation dosimetry', *J Radiol Prot*, 27(2), pp. 5–16.

Johnson J C, Thaul S, Page W F, Crawford H (1996) *Mortality of veteran participants in the CROSSROADS nuclear test*. Washington, DC: National Academy Press.

Joint Committee on Atomic Energy (1952) *Atomic Power and Private Enterprise*. JCAE.

Jones S (1993) *The Language of the Genes*. London: Flamingo.

Jordan G M (1963) *The identification of criteria for criticality assessment*. UKAEA / No: AHSB(S) R 56.

Kallmann H (1949) 'Quantitative measurements with scintillation counters', *Phys Rev*, 75, pp. 625–626.

Kallmann H and Furst M (1950) 'Fluorescence of solutions bombarded with high energy radiation (Energy transport in liquids) PartI', *Phys Rev*, 79, pp. 857–870.

Kallmann H and Furst M (1951) 'Fluorescence of solutions bombarded with high energy radiation (Energy transport in liquids) Part II', *Phys Rev*, 81, pp. 853–864.

Kallmann H and Furst M (1952) 'Fluorescence of solutions bombarded with high energy radiation (Energy transport in liquids) PartIII', *Phys Rev*, 85, pp. 816–825.

Kappos A and Pohlit W (1972) 'A cybernetic model for radiation reactions in living cells I Sparsely ionizing radiations;stationary cells', *Int J Radiat Biol*, 22, pp. 51–65.

Kathren R (1986) *Herbert M Parker, Publications and other contributions to radiological and health physics*. Battelle Press.

Kathren R L (1980) 'Before transistors, IC's and all those other good things: the first fifty years of radiation monitoring instromentation', in *Health Physics, a Backward Glance: 13 Original Papers on the History of Radiation Protection*. NY: Pergamon.

16 References

Kathren R L (1984) *Radioactivity in the environment: sources, distribution and suveillance*. Char, Switz: Harwood Academic Publishers.

Kato H and Schull W J (1982) 'Studies of the mortality of A-bomb survivors Report 7 Mortality 1950-1978 Part1 Cancer mortality', *Radiat Res*, 90(2), pp. 395–432.

Kaye G W C (1926) *X-rays*. London: Longmans Green and Co.

Kaye G W C (1927) 'Protection and working conditions in x-ray departments', *Br J Radiol*, 1(9), pp. 295–312.

Keller W and Modarres M (2005) 'A Historical Overview of Probabilistic Risk Assessment Development and Its Use in the Nuclear Power Industry; A Tribute to the Late Professor Norman Carl Rasmussen', *Rel Eng Syst Safety*, 89, pp. 271–285.

Kellerer A M and Rossi H H (1972) 'The theory of dual radiation action', *Curr Top Radiat Res Quart*, 8, pp. 85–158.

Kellerer A M and Rossi H H (1978) 'A generalized formulation of dual radiation action', *Radiat Res*, 75, pp. 471–488.

Kemeny J G (1979) *The President's Commission on the Accident at TMI*.

Kemerinck M et al (2011) 'Characteristics of a first-generation x-ray system', *Radiology*, 259, pp. 534–539.

Kendall G M, Muirhead C R, MacGibbon B H et al (1992) 'Mortality and occupational exposure to radiation: first analysis of the National Registry for Radiation Workers.', *Br Med J*, pp. 220–225.

Kevles B H (1997) *Naked to the Bone: Medical Images in the Twentieth Century*. New Brunswick: Rutgers UP.

Kinlen L J (1993) 'Can paternal preconceptional radiation account for the increase in leukaemia and non-Hodgkin's lymphoma at Seascale?', *Br Med J*, 306, pp. 1718–1721.

Knief R A (1985) *Nuclear Criticality Safety: Theory and Practice*. La Grange Park,IL: ANS.

Knudson A G (1971) 'Mutation and cancer: statistical studies of retinoblastoma', *Proc Nat Acad Sci USA*, 68, pp. 820–823.

Konopolis E, Metropolis N, Teller E and Woods L (1943). CF-548. USAEC.

Korff S A (1946) *Electron and Nuclear Counters*. NY: D Van Nostrand.

Korff S A and Danforth W E (1939) 'Neutron Measurements with Boron-Trifluoride Counters', *Phys Rev*, 55, p. 980.

Kusnetz H L (1956) 'Radon daughters in mine atmospheres. A field method for determining concentrations', *Am Ind Hyg Assoc J*, 17, pp. 85–88.

Ladu M, Pelliccioni M and Rotondi E, (1963) 'On the response to fast neutrons of a BF3 counterin a paraffin spherical-hollow moderator', *Nucl Instrum Methods*, 23, pp. 173–174.

Ladu M, Pelliccioni M and Rotondi E (1965) 'Flat response to neutrons between 20 keV and 14 MeV of a BF3 counter in a spherical hollow moderator', *Nucl Instrum Methods*, 32, pp. 175–176.

Laurence G C (1960) 'Reactor safety in Canada', *Nucleonics*, 18(10), pp. 73–77.

Laurie J, Orr J S and Foster C J (1972) 'Repair processes and cell survival', *Br J Radiol*, 45, pp. 362–368.

Lauritsen C C and Lauritsen T (1937) 'A simple quartz fiber electrometer', *Rev Sci Instrum*, 8, pp. 438–439.

Lea D E (1940), p. 0.

Lea D E (1955) *Actions of Radiations on Living Cells.* Cambridge: Cambridge Uni Press.

Lea D E, Haines R B and Coulson C A (1936) 'The mechanism of the bactericidal action of radioactive radiations I Theoretical', *Proc Roy Soc B*, 120, pp. 47–75.

Lea D E, Haines R B and Coulson C A (1937) 'The actions of radiations on bacteria', *Proc Roy Soc B*, 123, pp. 1–21.

Leake J W (1967) *Portable instruments for the measurement of neutron dose-equivalent rate in steady state and pulsed neutron fields.* Neutron Monitoring STI/PUB/136. Vienna: IAEA.

Leake J W (1968) 'An improved spherical dose equivalent neutron detector', *Nucl Instrum Methods*, 63, pp. 329–332.

Lehman R L (1961) *Energy response and physical properties of NTA personnel neutron dosimeter track film.* UCRL-9513 Rept LA-12625-M. Livermore: LLNL:

Lentle B and Aldrich J (1997) 'Radiological Sciences Past and Present', *Lancet*, 350(7073), pp. 280–285.

Lewis E B (1957) 'Leukemia and ionizing radiation', *Science*, 125, pp. 965–972.

Lewis E B (1963) *Science*, pp. 1494, 1492.

Lewis H (1978) *Risk Assessmnent Review Group Report to the US Nuclear Regulation Commission.* NUREG/CR-0400. USNRC.

Lindell B (1996) 'A history of radiation protection', *Radiation Protection Dosimetry*, 68(1–2), pp. 83–95.

Lindell B, Dunster H J and Valentin J (no date) *International Commission on Radiological Protection: History, Policies, Procedures*. ICRP.

Lodge O (1896) 'The surviving hypothesis concerning the X-rays', *The Electrician*, 36, pp. 471–473.

Longworth T C (1965) 'The GEM Monte Carlo code', in *IAEA Conference Proceedings: Criticality Control of Fissile Materials*.

Longworth T C (1968) *The GEM4 code*. AHSB(S)R146. UKAEA.

Lucas A C (1993) 'The history and direction of TL Instrumentation', *Rad Prot Dosimetry*, 47(1–4), pp. 451–456.

Lucas H F (1957) 'Improved low-level scintillation counter for radon', *Rev Sci Instrum*, 28, p. 680.

Luckey T D (1980) *Hormesis with ionizing radiation*. Boca Raton,FL: CRC Press.

Luckey T D (1999) 'Radiation hormesis overview', *RSO Magazine*, 8(4), pp. 22–41.

Luria S E (1947) 'Reactivation of irradiated bacteriophage by transfer of self-reproducing units', *Proc Nat Acad Sci*, 33, pp. 253–264.

Mancuso T F, Stewart A M and Kneale G W (1977) 'Radiation exposures of Hanford workers dying from cancer and other causes', *Health Physics*, 33, pp. 369–385.

Mandeville C E and Scherb M V (1950) 'Photosensitive Geiger couters: Their applications', *Nucleonics*, 7(5), pp. 34–38.

March H C (1944) 'Leukemia in radiologists', *Radiology*, 43(3), pp. 275–278.

March H C (1950) 'Leukemia in radiologists in a 20 year period', *Am J Med Sci*, 220, pp. 282–286.

March H C (1961) 'Leukemia in radiologists ten years later. With a review of the pertinent evidence for radiation leukemia', *Am J Med Sci*, 242, pp. 135–149.

Marinelli L D, Miller C E, Gustafson P F and Rowlands R E (1955) 'The quantitative determination of gamma-ray emitting elements in living persons', *Am J Roentgenol Rad Ther Nucl Med*, 73, pp. 661–671.

Marinelli L D, Miller C E, Rowland R E and Rose J E (1955) 'The measurement in vivo of Ra gamma-ray activities lower than K-40 levels existing in the human body', *Radiology*, 64, p. 116.

Marley W G and Fry T M (1956) in *Proc UN Intnational Conf Peaceful Uses of Atomic Energy, August 1955*. Geneva: UN, p. 102.

Martin J H (1988) *History of the Society for Radiological Protection 1963-1988*.

Martland H S (1929) 'Occupational poisoning in manufacture of luminous watch dials', *J Am Med Assoc*, 92(6&7), pp. 466–473&552–559.

Martland H S (1931) 'The occurrences of malignancy in radioactive persons', *Am J Cancer*, 15(4), pp. 2435–2516.

Martland H S, Conlon P and Knef J P (1925) 'Some unrecognised dangers in the use and handling of radioactive substances with especial reference to the storage of insoluble products of radium and mesothorium in the reticula-endothelial system', *J Am Med Assoc*, 85(23), pp. 1769–1775.

Mavor J W (1924) 'The production of non-disjunction by x-rays', *J Exptl Zoology*, 39(2), pp. 381–432.

Mayneord W V (1934) 'The physical basis of the biological effects of high voltage radiations', *Proc Roy Soc A*, 146, pp. 867–879.

McCombs R S and McCombs R P (1930) 'A hypothesis on the causation of cancer', *Science*, 72, pp. 423–424.

McCullough C R (ed.) (1957) *Safety Aspects of Nuclear Reactors*. Princeton, NJ: Van Nostrand.

McGregor J H (1908) 'Abnormal development of frog embryos as a result of treatment of ova and sperm with Roentgen rays', *Science*, 27, p. 445.

McKay K G (1951) 'Electron-hole production in germanium by alpha particles', *Phys Rev*, 84, pp. 829–832.

McKenzie J M (1979) 'Development of the semiconductor radiation detector', *Nucl Instrum Meth*, 169, pp. 49–73.

McLaughlin P M, M. S. P. (2000) *A Review of Criticality Accidents 2000 Revision*. LANL / No: LA-13638.

Meggitt G C (2006) 'Fission, Critical Mass and Safety – a Historical Review', *J Radiol Prot*, 26, pp. 141–159.

Meggitt G C (2016) *Genes, Flies, Bombs and a Better Life - In the footsteps of Hermann Muller*. Pitchpole Books.

Metcalf G F and Thompson B J (1930) 'A low grid-current vacuum tube', *Phys Rev*, 36, p. 1489.

Metivier H (2007) *Fifty years of radiological protection*. Paris:OECD/NEA / No: No 6280.

Metropolis N (1987) 'The Beginning of the Monte Carlo Method', *Los Alamos Science*, Special Issue, pp. 125–130.

Miller C E (1958) 'Low intensity spectrometry of the gamma radiation emitted by human beings', *Proc 2nd Int Conf Peaceful Uses Atomic Energy*, pp. 113–122.

Misra K B (1992) *Reliability Analysis and Prediction: A Methodology Oriented Treatment*. Amsterdam: Elsevier.

Moloney W C and Kastenbaum M A (1955) 'Leukemogenic effects of ionizing radiation on atomic bomb survivors in Hiroshima city', *Science*, 121, pp. 308–309.

Moolgavkar S H (1983) 'Model for human carcinogenesis', *Environ Health Persps*, 50, pp. 285–291.

Moolgavkar S H (1986) 'Carcinogenesis modelling from molecular biology to epdemiology', *Ann Rev Public Health*, 7, pp. 151–169.

Moore J A (1972) *Heredity and development*. NY: OUP.

Moore J G (1974) *The general Monte Carlo code MONK*. ANL-75-2 / No: ANL-75-2 & NEA-CRP-L-118.

Moore M E, G. H. J. (2002) 'Laser illuminated etched track scattering (LITES)dosimetry system', *Rad Prot Dosimetry*, 101, pp. 1–4.

Morgan K Z (1995) *Oral history of Karl Z Morgan*. DOE/EH-0475. USDOE.

Morone J G and Woodhouse E J (1986) *Averting Catastrophe: Strategies for Regulating Risky Technologies*. Univ Cal Press. Available at: http://publishing.cdlib.org/ucpressebooks/view?docId=ft2k4004pp;brand=ucpress.

Moseley H G J (1913) 'The high frequency spectra of the elements', *Phil Mag*, 26, p. 1024.

Moss W and Eckhardt R (1995) 'The plutonium injection experiments', *Los Alamos Science*, 23, pp. 177–233.

Mould R F (1993) *A century of x-rays and radioactivity in medicine:...* Bristol: IOPP.

Mould R F (1995) 'The early history of x-ray diagnosis with emphasis on the contributions of physics 1895-1915', *Phys Med Biol*, 40(11), pp. 1741–1787.

Mould R F (1998) 'The discovery of radium in 1898 by', *B J Radiology*, 71, pp. 1229–1254.

MRC (1956) *The hazards to man of nuclear and allied radiations.* London:HMSO / No: Cmnd 9780.

MRC (1960) *The hazards to man of nuclear and allied radiations.* London:HMSO / No: Cmnd 1225.

Muirhead C R et al (2003) *Mortality and Cancer Incidence 1952-1998 in UK Participants in the UK atmospheric nuclear weapons tests and experimental programmes.* NRPB / No: NRPB-W27.

Muirhead C R, Goodill A A, Haylock R G et al (1999) 'Occupational radiation exposure and mortality: second analysis of the National Registry for Radiation Workers', *J Radiol Prot*, 19, pp. 3–26.

Muller H J (1927) 'Artificial transmutation of the gene', *Science*, 66, pp. 84–87.

Muller H J (1954) 'The relation of neutron dose to chromosome changes and point mutations in Drosophila I Translocations', *Am Nat*, 88, pp. 437–459.

Muller H J (1958) 'General survey of mutational effects of radiation', in *Radiation Biology and Medicine (ed W D Claus).* Reading MA: Addison-Wesley Pub Co.

Muller H J (1964) *The production of mutations in Nobel Lectures:Physiology or Medicine 1942-1962.* Amsterdam: Elsevier Publishing.

Mullner R (1999) *Deadly Glow. The Radium Dial Worker Tragedy.* Am Pub Health Assoc.

Mulvihill R J (1966) *A Probabilistic Methodology for the Safety Analysis of Nuclear Power Reactors*. PRC-R-657.

Mulvihill R J, Arnold D R, Bloomquist C E, Epstein B (1965) *Analysis of United States Power Reactor Accident Probability*. PRC R-695. Los Angeles: Planning Research Corp.

Mutscheller A (1925) 'Physical standards of protection against roentgen ray dangers', *Am J Roentgenol*, 13, pp. 65–70.

Mutscheller A (1928) 'Safety standards of protection against X-ray dangers', *Radiology*, 10, p. 466.

Mutscheller A (1934) 'More on X-ray protection standards', *Radiology*, 22, p. 739.

NBS (1931) *X-Ray Protection*. Wash:US Govt Printing Office / No: Handbook 15.

NBS (1936) *X-ray Protection*. Wash:US Govt Printing Office / No: Handbook 20.

NBS (1953) *Maximum permisssible amounts of radioisotopes in the human body and maximum permissible concentrations in air and water*. Handbook 52. Washington DC: US Govt Printing Office.

Neary G J (1952) 'The evaluation of tolerance levels', in *Biological Hazards of Atomic Energy (ed A Haddow)*. Oxfor: Clarenden Press.

Neel J V et al (1990) 'The children of parents exposed to atomic bombs: estimates of genetic doubling dose of radiation for humans', *Am J Hum Genet*, 46, pp. 1053–1072.

Neher H V and Harper W W (1936) 'A high speed Geiger-counter circuit', *Phys Rev*, 49, pp. 940–943.

Neher H V and Pickering W H (1938) 'Modified high speed Geiger ounter circuit', *Phys Rev*, 53, p. 316.

Neuzil M and Kovarik W (1996) *Mass Media and Environmental Conflict: America's Green Crusades*. Thousand Oaks, CA: Sage Publications, 1996: Sage Publications.

New York World (1903) 'Edison Fears Hidden Perils of the X-rays', 3 August.

Newton Hayes F, Hiebert R D and Schuch R L (1952) 'Low energy counting with a new scintillation solute', *Science*, 116, p. 140.

Nordling C O (1953) 'A new theory of the cancer inducing mechanism', *Br J Cancer*, 7, pp. 68–72.

Norris W P, Speckman T W and Gustafson P F (1955) 'Studies of the metabolism of radium in man', *Am J Roentgenology*, 73, pp. 785–802.

Nowell P C (1976) 'The clonal evolution of tumor cell populations', *Science*, 194, pp. 23–28.

NRC (1983) *Safety goals for nuclear power plant operations.* NUREG-0880 Rev1. USNRC.

NRC (1986) *Safety goals for the operations of nuclear power plants.* 10 CFR Part 50.

NRC (no date) *A short history of nuclear regulation 1946-1999.* NRC / No:

NRPB/RCR (1990) 'Patient dose reduction in diagnostic radiology', *Doc NRPB*, p. 0.

Omar R Z, Barber J A and Smith P G (1999) 'Cancer mortality and morbidity among plutonium workers at the Sellafield plant of British Nuclear Fuels', *Br J Cancer*, 79, pp. 1288–1301.

O'Riordan M (2007) *Radiation protection: A memoir of the National Radiological Protection Board.* Chilton,Oxon:HPA / No:

Orr J S and Laurie J (1975) *The pool and the initial slope of survival curves for high and low LET radiations*. London: IOP/Wiley.

Ortiz P, Oresegun M and Wheatley J (2000) 'Lessons from major radiation accidents', in. *IRPA10*, Hiroshima.

Pachoa A S, Nogueira de Oliviera C A et al (1993) 'History and development of whole body counting in Brazil', *Environ Int*, 19, pp. 519–526.

Packard C (1931) 'The biological effects of short radiations', *Quart Rev Biol*, 6, pp. 253–280.

Pardue, G. and W. (1944) *Photographic film as a pocket dosimeter*. Met Lab Report / No: CH-1553, April 1944.

Parker H (1950) 'Tentative dose units for mixed radiations', *Radiology*, 54, pp. 257–262.

Paxton H C and Pruvost (1986) *Critical dimensions of systems containing U-235, Pu-239 and U-233*. LANL / No: LA-10860.

Perrin F (1939) 'Calcul relatif aux conditions eventuelles de transmutation en chaine de l'uranium', *Comptes Rendus B 1 May 1939*, pp. 1394–1396.

Perry R T (2005) *Raymond Elliott Zirkle 1902-1988*. Biographical Memoirs Wash DC: National Academies Press.

Pfahler G E (1906) 'A roentgen filter and a universal diaphragm and protecting screen', *Trans Am Roentgen Ray Soc*, pp. 217–224.

Pfahler G E (1922) 'Protection in radiology', *Am J Roentgenology*, 9, p. 467.

Pierce D A (2002) 'Age-time patterns of radiogenic cancer risk: their nature and likely explanations', *J Radiol Prot*, 22, pp. A147-154.

Plan for developing a safety goal (1980). NUREG-0735. NRC.

Platt R (1955) 'Clonal ageing and cancer', *Lancet*, i, p. 867.

Plummer G (1952) 'Anomalies occurring in children exposed in utero to the atomic bomb at Hiroshima', *Pediatrics*, 10, pp. 692–687.

Pochin E E (1983) 'The first twenty years of UNSCEAR', *J Rad Prot*, 3, pp. 5–8.

Powers E L (1962) 'Considerations of survival curves and target theory', *Phys Med Biol*, 7, pp. 3–28.

Present R D (1947) 'On self-quenching halogen counters', *Phys Rev*, 72, pp. 243–244.

Preston D L, Shimizu Y, Pierce D A Suyama A and Mabuchi K (2003) 'Studies of mortality of atomic bomb survivors. Report 13: Solid cancer and noncancer disease mortality: 1950-1997', *Radiat Res*, 160, pp. 381–407.

Preston D L,Kato H, Kopecky K J and Fujita S (1987) 'Studies of the mortality of A-bomb survivors 8 Cancer Mortality 1950-1982', *Radiat Res*, 111, pp. 151–178.

Puck T T and Marcus P I (1956) 'Action of X-rays on mammalian cells', *J Exp Med*, 103, pp. 653–666.

Quimby E H (1926) 'A method for the study of scattered and secondary radiation in c-ray and radium laboratories', *Radiology*, 7, p. 211.

Raben M S and Bloembergen N (1951) 'Determination of radioactivity by solution in liquid scintillator', *Science*, 114, pp. 363– 364.

Radium, its production and therapeutic applications (1925). Radium Belge / No:

Rajchman J A and Snyder R L (1940) 'An electrically-focussed multiplier phototube', *Electronics*, 13, pp. 20–23.

Ramsden D and Foster P P (1984) 'Whole body monitors for internal dosimetry 1963-1984-2005', *J Soc Radiol Prot*, 34(3), pp. 153–155.

Randall J T and Wilkins M A F (1945) 'Phosphorescence and electron traps', *Proc Roy Soc*, A184, pp. 365–389 & 390–407.

Reactor Safety Study An Assessment of Accident Risks in U.S. Commercial Nuclear Power Plants (1975). WASH-1400. Available at: http://nuclearsafety.info/probabilistic-risk-assessment/.

Regaud C and Dubreuil G (1908) 'Perturbations dans le developpement des oeufs fecondes par des spermatozoides rontgenises chez le lapin', *C R Soc Biol(Paris)*, 64, pp. 1014–1016.

Regener V H (1946) 'Decade counting circuits', *Rev Sci Instrum*, 7, pp. 185–189.

Reider R (1971) *An Early History of Criticality Safety*. LANL / No: LA-4671.

Reines F, Schuch R L, Cowan C L, Harrison F B, Anderson E C and Hayes F N (1953) 'Determination of total body radioactivity using liquid scintillation detectors', *Nature*, 172, pp. 521–523.

Reverdy L et al (2003) 'CAAS - An Experts Working Group Revisits the Safety Files for the French Nuclear Safety Authority', in *Proc Int Conf Nucl Criticality*. Tokai Mura, Japan, Oct 2003.

Reynolds G T, Harrison F B and Salvini G (1950) 'Liquid scintillation counters', *Phys Rev*, 78, p. 488.

Robinette C D, Jablon S and Preston D L (1985) *Studies of Participants in Nuclear Tests*. Washington DC: National Academy Press.

Rogers D W O (1979) 'Why not trust a neutron rem-meter', *Health Physics*, 37, pp. 735–742.

Rontgen W C (1896) 'On a New Kind of Ray', *Nature*, 53, pp. 274–277.

Rowland R E (1994) *Radium in humans*. ANL/ER-3. Argonne Nat Lab.

Royal College of Radiology (2007) *Making the Best Use of Clinical Radiology Services: referral guidelines*. London: RCR.

Rundo J (1958) 'Body radioactivity measurement as an aid in assessing contamination by radionuclides', *Prog Nucl Energy I Ser XII*, pp. 283–306.

Rundo J (1993) 'History of the determination of radium in man since 1915', *Environ Int*, 19(5), pp. 425–438.

Russell C R (1962) *Reactor Safeguards*. Oxford: Pergamon Press. Available at: https://babel.hathitrust.org/cgi/pt?id=uc1.b4540521;view=1up;seq=11.

Russell W L (1951) 'X-ray- induced mutations in mice', *Cold Spring Harbor Symp Quant Biol*, 16, pp. 327–335.

Rutherford E and Geiger H (1908) 'An electrical method of counting the number of alpha-pareticles from radio-active substances', *Proc Roy Soc A*, 81, pp. 141–161.

Sanders B S (1970) 'Commentary on: Health of workers in the United Kingdom Atomic Energy Authority', *Health Physics*, 19, pp. 321–322.

Sankaranarayanan K (2002) 'Genetic risks of exposure to ionizing radiation', in *Health effects of low-level radiation, BNES, 2002 1. Health effects of low-level radiation, BNES, 2002 1*, pp. 1–6.

Sankaranarayanan K and Chakraborty R (2000) 'Ionizing radiation and genetic risks XI The doubling dose estimates from the mid-1950s to the present and the conceptual change to the use of human data for spontaneous mutation rates and mouse data for induced rates for doubling dose calculations', *Mutat Res*, 453, pp. 107–127.

Savage J R (1998) 'A brief survey of aberration origin theories', *Mutat Res*, 404, pp. 139–147.

Savage J R K (2000) 'Proximity matters', *Science*, pp. 63, 62.

Sax K (1938) 'Chromosome aberrations induced by X-rays', *Genetics*, 23, p. 494.

Schales F (1978) 'Brief history of Ra-224 usage in radiotherapy and radiobiology', *Health Physics*, 35, pp. 25–32.

Schall W E (1932) *X Rays: their origin, dosage, and practical application*. Bristol: John Wright and Sons Ltd.

Schlundt H and Barker H H et al (1929) 'The detection and estimation of radium and mesothorium in living persons I', *Am J Roentgenology Rad Ther*, 21, pp. 345–354.

Schlundt H, Nerancy J T and Morris J P, (1933) 'Detection and estimation of radium in living persons IV Retention of soluble radium salts administered intravenously', *Am J Roentgenology Rad Ther*, 30, pp. 515–522.

Schmitt O E (1938) 'A thermionic trigger', *J Sci Instrum*, 15, pp. 24–26.

Schorr M G and Torney F L (1950) 'Solid non-crystalline scintillation phosphors', *Phys Rev*, 80, p. 474.

Schraub A, Aurand K and Jacobi W, (1957) 'The importance of radon and its decay products in relation to the normal radiation dose in humans', *Br J Radiol*, Supp 7, pp. 114–119.

Schull W J (1991) 'Ionising radiation and the developing human brain', *Ann ICRP*, 22(1), pp. 95–118.

Schull W J (1995) *Effects of Atomic Radiation: A Half-Century of Studies from Hiroshima and Nagasaki*. New York: John Wiley.

Schultz W W (1968) 'Track density measurement in dielectric track detectors with scattered light', *Rev Sci Instrum*, 39, pp. 1893–1896.

Schuske C L and Paxton H C (1976) 'History of fissile array measurements in the United States', *Proc Am Nucl Soc*, 30, pp. 101–137.

Seil H A, Viol C H and Gordon M A (1915) 'Elimination of soluble radium salts taken intravenously and per os', *New York Med J*, 101, pp. 896–898.

Serber R (1992) *The Los Alamos Primer:the first lectures on how to build an atomic bomb*. Berkeley;Oxford: Univ Cal Press.

Shapiro J (1956) 'Radiation dosage from breathing radon and its daughter products', *Am Med Assoc Arch Ind Health*, 14, pp. 169–177.

Siddall E (1957a) *Reactor safety standards and their attainment*. AECL-498. AECL.

Siddall E (1957b) 'Reliable reactor protection', *Nucleonics*, 15(6), pp. 124–129.

Sievert R M (1951) 'Measurement of gamma radiation from the human body', *Arkiv für Fysik*, 3, pp. 317–346.

Silk E C H and Barnes R S (1959) 'Examination of fission fragment tracks with an electron microscope', *Phil Mag*, 4, pp. 970–971.

Silver L M (1995) *Mouse genetics*. Oxford: OUP.

Simister D N and Clemson P D (2003) 'A historical review of critical experiment research facilities in the United Kingdom', *Nucl Sci Eng*, 145, pp. 64–71.

Simpson J A (1948) 'Air proportional counters', *Rev Sci Instrum*, 19, pp. 733–743.

Smith E E (1975) *Radiation science at the National Physical Laboratory 1912-1955*. London: HMSO.

Smith P G and Doll R G (1981) 'Mortality from cancer and all causes among British radiologists', *Br J Radiol*, 54, pp. 187–194.

Smith PG and Douglas AJ (1986) 'Mortality of workers at the Sellafield plant of British Nuclear Fuels', *Br Med J*, 293, pp. 845–854.

Socolow R L, Hashizume E, Neriishi S and Niitani R (1963) 'Thyroid carcinoma in man after exposure to ionizing radaiation: a summary of the findings in Hiroshima and Nagasaki', *New Eng J Med*, 268, pp. 406–410.

Soddy F (1913) 'Intra-atomic charge', *Nature*, 92, pp. 399–400.

Solomon I (1926) *Precis de Radiotherapie profonde*. Paris: Masson.

Sommer A and Turk W E (1950) 'New multiplier photo-tubes of high sensitivity', *J Sci Instrum*, 27, pp. 113–117.

Sorahan T and Roberts P J (1993) 'Childhood cancer and parental exposure to ionizing radiation: preliminary findings from the Oxford Survey of Childhood Cancers', *Am J Ind Med*, 23, pp. 343–354.

Sorensen J N (1997) 'Historical notes on defense in depth - Memorandum'.

Sowby F D and Valentin J (2003) 'Forty years on: how radiological protection has evolved internationally', *J Radiol Prot*, 23, pp. 157–171.

Spiess H (2002) 'Peteosthor - a medical disaster due to radium-224', *Radiat Environ Biophys*, 41(3), pp. 163–172.

Spiess H (2005) 'Late effects of radium-224 injected in children and adults', in *Health Effects of Incorporated Radionuclides. Ninth Int Conference HEIR 2004*, GSF -Forschunszentrum, pp. 61–66.

Stadler L J (1928) 'Mutations in barley induced by X-rays and radium', *Science*, 68, pp. 186–187.

Stannard J N (1988) *Radioactivity and Health: A History*. Battelle Press.

Starling S G and Woodall A J (1953) *Electricity and Magnetism for Degree Students*. London: Longmans, Green and Co.

Steel G G (1996) 'From targets to genes: a brief history of radiosensitivity', *Phys Med Biol*, 41, pp. 205–222.

Stewart A, Webb J, Giles D and Hewitt D (1956) 'Malignant disease in childhood and diagnostic irradiation in utero', *Lancet*, 268(6940), pp. 421–472.

Strangeways T S P and Oakley H E H (1923) *Proc Roy Soc B*, 95, pp. 373–381.

Stratton W R (1960) 'A review of criticality accidents', *Prog Nuclear Energy Ser IV*, 3, p. 3.

Stratton W R (1967) *A review of criticality accidents*. Los Alamos Scientific Laboratory / No: LA-3611.

Sturtevant A H (1965) *A History of Genetics*. NY: Harper and Row. Available at: http://www.esp.org/books/sturt/history/readbook.html.

Sutton W S (1902) 'On the morphology of the chromosome group in Brachystola magna', *Biological Bulletin*, 4, pp. 24–39.

Tanguy P (1988) 'Three decades of nuclear safety', *IAEA Bull*, 2, pp. 51–57.

Taylor D (1950) 'Radioactive surveying and monitoring instruments', *J Sci Instrum*, 27, pp. 81–88.

16 References

Taylor L S (1958a) 'History of the ICRP', *Health Physics*, 1, pp. 97–104.

Taylor L S (1958b) 'History of the International Commission on Radiation Units and Measurement(ICRU)', *Health Physics*, 1(3), pp. 306–314.

Taylor L S (1967) 'An early portable radiation survey meter', *Health Physics*, 13, p. 1347.

Taylor L S (1984) *Tripartite conferences on radiation protection: Canada, United Kingdom, United States (1949-1953)*. Wash DC:DOE OSTI / No: NVO-271.

Taylor L S (1989) *80 years of quantities and units Personal reminiscences Part 1*. ICRU News: No 1.

Taylor L S (1990) *80 years of quantities and units Personal reminiscences Part 2*. ICRU News / No: June.

Taylor L S (2002) 'Brief history of NCRP', *Health Physics*, 82, pp. 776–781.

Teller E (1953) 'Reactor hazards predictable', *Nucleonics*, 11(11), p. 80.

Temin H M and Mizutani S (1970) 'RNA-dependent DNA polymerase in virions of Rous sarcoma virus', *Nature*, 226, pp. 1211–1213.

Thaul S, P. W. (2000) *The Five Series Study: Mortality of Military Participants in th US Nuclear Weapons Tests*. Washington, DC: National Academy Press.

The Chernobyl Forum (2006) *Chernobyl's legacy: health environmental ans socio-economic impacts*. Vienna: IAEA (IAEA/PI/A.87.Rev2/06-09181).

Theoretical Possibilities and Consequences of Major Accidents in Large Nuclear Power Plants (1957). WASH-740. USAEC.

Thomas A F and Abbey F (1973) *Calculational methods for interacting arrays of fissile material.* Oxford: Pergamon Press.

Thomas A M K, Isherwood I and Wells P N T(eds) (1995) *The Invisible Light. 100 years of Medical Radiology.* Oxford: Blackwell Science Ltd.

Thomas A M K and Banerjee A K (2013) *The History of Radiology.* Oxford: Oxford University Press.

Thomas J T (ed) (1978) *Nuclear safety guide.* TID-7016rev 2 NUREG/CR0095 / No:

Thomson J J (1904) 'On the structure of the atom', *Phil Mag S6*, 7, pp. 237–265.

Thorne P R, Bowden R L and Venner, J. (2003) '"Oh, No it isn't!" "Oh, Yes it is!" The Omission of Criticality Incident Detection Systems in the UK', in *Nuclear criticality safety; Challenge in the pursuit of global nuclear criticality safety. Nuclear criticality safety; Challenge in the pursuit of global nuclear criticality safety*, JAERI, pp. 758–763.

Tijo J H and Levan A (1956) 'The chromosome number of man', *Hereditas*, 42, pp. 1–6.

Timofeeff-Ressovsky N W and Zimmer K G (1947) *Biophysik I; Das Trefferprinzip in der Biologie.* Leipzig: Hirzel.

Tobias C A (1985) 'The repair-misrepair model in radiobiology: comparison to other models', *Radiat Res*, 104, pp. S77-95.

Tobias C A, Blakely E A, Ngo F Q H and Yang T C H (1980) *The repair-misrepair model of cell survival.* New York: Raven Press.

Townsend J S (1901) 'The conductivity produced in gases by the motion of negatively charged ions', *Phil Mag*, 1, pp. 198–227.

16 References

Trost A (1937) 'Uber zahlrohre mit dampfzusatz', *Zeits f Phys*, pp. 399–444.

Tsuruta T (no date) *Research and development of solid state track detectors for external dosimetry in Japan.* / No:

Tyzzer E E (1916) 'Tumor immunology', *J Cancer Res*, 1, pp. 125–155.

Ulrich H (1946) 'Incidence of leukemia in radiologists', *N Eng J Med*, 234, pp. 45–46.

Unknown (1997) 'The Reines-Cowan experiments', *Los Alamos Science*, (25).

UNSCEAR (1958) *Report of the United Nations Scientific Committee on the Effects of Atomic Radiation. Suppl. 17 (A/3838).* UNSCEAR.

UNSCEAR (1962) *Report of the United Nations Scientific Committee on the Effects of Atomic Radiation.* UNSCEAR.

UNSCEAR (1966) *Report of the United Nations Scientific Committee on the Effects of Atomic Radiation.* UNSCEAR.

UNSCEAR (1972) *Ionizing radiation :Levels and effects.* UNSCEAR.

UNSCEAR (1977) *Sources and effects of ionizing radiation.* UNSCEAR.

UNSCEAR (1982) *Ionizing radiations:sources and biological effects.* UNSCEAR.

UNSCEAR (1986) *Genetic and somatic effects of ionizing radiation.* UNSCEAR.

UNSCEAR (1988) *Sources, effects and risks of ionizing radiation.* UNSCEAR.

UNSCEAR (1993) *Sources and effects of ionizing radiation.* UNSCEAR.

UNSCEAR (2000) *Sources and effects of ionizing radiation.* UNSCEAR.

UNSCEAR (2001) *Hereditary effects of radiation.* UNSCEAR.

Upton A C (1991) 'Risk estimates for carcinogenic effects of radiation', *Ann ICRP*, 22(1), pp. 1–29.

USAEC (1973) *The Safety of Nuclear Power Reactors (Light Water Cooled) and Related Facilities.* WASH-1250. USAEC.

USDOE (2015) *Facility Safety: Order O 420.1C.* DOE O 420.1C Chg1. DOE.

Use of probabilistic safety assessment methods in nuclear activities: Final policy statement (1995). Fed Register 60 FR 42622.

USNRC (no date) *Three Mile Island Accident.* USNRC / No: Fact Sheet.

Van Hardeen P J (1945) *The Crystal Counter.* Amsterdam: N Holland.

Walker J S (1992) *Containing the Atom: Nuclear Regulation in a Changing Environment 1963-1971.* U California Press.

Walker J S and Wellock T R (2010) *A Short History of Nuclear Regulation 1946-2009.* NUREG/BR-0175 Rev2. USNRC.

Walker R M, Price P B and Fleischer R L (1963) 'A versatile disposable dosimeter for slow and fast neutrons', *App Phys Lett*, 3, pp. 28–29.

Warren S et al (1949) *Minutes of the permissible doses conference held at Chalk River, Canada, September 29-20th, 1949.* / No: RM10.

Watanabe K K, Kang H K and Dalager N A (1995) 'Cancer mortality risk among military participants of a 1958 atmospheric nuclear weapons test', *Am J Pub Health*, 8, pp. 523–527.

Weiss J (1944) 'Radiochemistry in aqueous solutions', *Nature*, 153, p. 748.

Whitesides G E (1971) 'A difficulty in computing the k-eff of the world', *Trans Am Nucl Soc*, 14, p. 68.

Whitesides G E (2004) *Private communication.* / No:

Whitesides G E, W. R. M. and H. C. (2003) 'Criticality Safety Methods', *Oak Ridge Mathematics and Computation Division Seminar, Gatlinburg, TE April 2003*, p. 0.

Wilson C T R (1927) *On the cloud track method of making visible ions and the tracks of ionizing particles.* Nobel Lecture.

Woodcock E R and Murphy T (1965) *Techniques used in the GEM code.* USAEC / No: ANL 7050.

Wright S (1934) 'On the genetics of subnormal development of the head (otocephaly) in the guinea pig', *Genetics*, 19, p. 471505.

Wright S and Eaton O N (1923) 'Factors which determine otocephaly in the guines pig', *J Agric Res*, 26, pp. 161–182.

Wynn-Williams C E (1932) 'The use of thyratrons for high speed automatic ounting of physical phenomena', *Proc Roy Soc A*, 132(819), pp. 295–310.

X-ray and Radium Protection Committee (1915) *X-ray and radium protection.* XRPC / No:

Young D A (1958) 'Etching of radiation damage in lithium fluoride', *Nature*, 182, p. 375.

Zirkle R E (1935) 'Biological effectiveness of alpha particles as a function of ion concentration produced in their paths', *Am J Cancer*, 23, pp. 558–567.

---

17 Index

absolute risk model, 193
Absorbed Dose Index, 125
accidents,
   criticality, 83
   Goiania, 98
   irradiation facilities, 96
   Kyshtym and northwest of, 82
   San Salvador, 96
   Three Mile Island, 77
Adler, Gustave , 34 Adult Health Study, 182
Advanced Gas-cooled Reactor, 62
Advisory Committee on X-ray and Radium Protection, 285
Ailinginae, 54
Aldermaston, 52, 198
American Board of Health Physics, 296
American Journal of Roentgenology, 284
American Roentgen Ray Society, 284
Ames, Bruce, 174
amplifiers, 225
Andersson and Braun, 230 Andrade, E N da C, 30
ankylosing spondylitis, 191

Anticipated Transient without Scram, 355
APM pilot plant, 64
Archives of Clinical Skiagraphy, 276
Archives of Clinical Skiagraphy , 3
Archives of Radiology, 276
Archives of Skiagraphy, 276
Archives of the Rontgen Ray, 3
Arco, lit by BORAX reactor, 61
Argonne, 47, 256
Armitage and Doll, 172
Arrays, criticality of, 323
atmospheric discharges, 69
atmospheric dispersion, 365
Atomic Bomb Casualty Commission (ABCC), 180
Atomic Energy Commission (AEC), 285
Atomic Weapons Research Establishment, 280
ausenium, 40
Authority Health and Safety Branch, 76
B204/B205 reprocessing plants, 64
Bacq and Alexander, 117 Baltimore, David, 174

17 Index

Barclay and Cox, 301
Bardeen, C R, 139
Barendsen, G W, 110
Barkla, C G, 32 Barnwell plant, 64 Baskerville, Charles, 35 Bateson, William, 136 Bauer, K H, 171
BEAR, 55, 146, 154, 285, 288, 294
BEAR21, 146
Becquerel, Henri, 26
 radiation burns, 33
Behnken, H, 121
Bell, P R, 232
Belot, J, 36, 121
Bender and Gooch30, 110
Benoist, L, 20
BEPO (British Experimental Pile Zero), 52
Bergonie and Tribondeau, 139
berzelium, 35
Bikini atoll, 54
Biological Effects of Ionizing Radiation (BEIR), Committee, 148, 151, 153, 154, 158-161, 195-197, 203, 207, 288, 294, 295
bismuth phosphate process, 63
Black, Sir Douglas, 198 Bocage, André, 93
Bohr, Niels,
 atomic stability, 32
 news of fission, 41
 U-235 as fissioning nuclide, 42
BORAX reactor, 61
Bovari, T, 137, 170
Bragg curves, 101
Bragg-Gray theory, 122
Bragg-Kleeman rule, 101
Bragg, W H, 30, 101, 123
Bridges, Calvin, 138
British Institute of Radiology, 276

British Journal of Radiology, 276
British Nuclear Fuels, 280
British Radiation Protection Association, 298
Bronk, Detlev, 55
Brookhaven, 359
Brookhaven Report, 359
Bruce, W Ironside, 18, 277
Brues, Austin, 305
bubble detector, 239
Bucky, Gustav, 6
Bulletin of the Atomic Scientists, 55
Burch, P R J, 123
Burghfield, 198
Bush, Vannevar, 46
Byers, Eben, 192
Calder Hall, 62
Cameron, John, 240
Campbell Swinton, A A, 2, 3, 7
cancer theories, 168
CANDU reactors, 63
Cantril, S T, 166
Cap de la Hague, 65, 68 Capenhurst, 53
carolinium, 35
Cart Poppy, 227
Carter, T C, 143
CAT scanning, 94
cell killing models,
 Blau and Altenburger 11, 105
 Condon and Terrill, 106
 Crowther, J A, 106
 Dessauer, F, 105
 Glocker, 107
 Holweck and Lacassagne, 107
 Lea, Douglas, 106
 LPL Unified Repair Model, 113
 Mayneord, 106
 Packard, Charles, 106
 Repair–Misrepair (RMR) model, 113

Central Electricity Generating Board, 281
CERRIE, 199
Chadwick and Leenhouts, 111, 115
Chadwick, James, 43
Chalk River, 51, 265
Chang and Eng, 228
Chelyabinsk, 65, 82
Chernobyl, 62, 63, 73, 79, 81, 82, 205, 270, 275, 292
Cherwell, Lord, 45
Chevaline, 55
China Syndrome, 79
Chinese testing,
    medical reviews, 58
    Partial Test Ban Treaty, 58
Chiroscope, 18
Christen, T, 121
Christmas Island, 54, 56
chromosome damage, 114
civil defence, 60
Clinton, 48
cloud chamber, 31
Cockcroft, John, 40, 51, 74
Collinson, A J L, 89 Colwell and Russ, 177 COMARE, 198 COMARE85, 199
Compton, Arthur, 47
Computerised Axial Tomography (CAT), 94
Conant, James B, 47
containment, 350
Coolidge, W D, 8
Cormack, Allen M, 94
Correns, Carl, 135
Cosmic rays, 87
Court Brown and Doll, 167, 186
Court Brown, W M, 178
CP-1, 47
CR-39, 238

critical mass, 319
criticality accidents, 338
criticality alarms, 341
criticality dosimetry, 344
criticality handbooks, 334
criticality measurements, 326
criticality, demonstrating safety, 337
Crookes, W, 31
CROSSROADS tests, 201
Crow Committee, 183 Cryptoscope, 8
Curie, Marie, 18, 27
    fluorescence of diamonds, 35
    hands, 36
    Radiumgraphs, 35
Curie, Pierre, 27 Curtis, Stanley B, 113 Dally, Clarence, 16 Daniels, F, 240
Danlos, Henri and radium therapy, 33
Darwin, Charles, 135
de Vathaire, Florent, 57
de Vries, Hugo, 135, 138 Debierne, A-L, 116
Defence in Depth, 352
Degrais, Paul, 33
Dekatron tube, 225
Density Analogue method, 324
Department of Energy (DOE), 293
Design Basis Accident, 352
Dessauer, F, 105
deterministic criticality codes, 331
Doll, R, 172, 279
Dorn F E and radon, 29
Dose and Dose Rate Reduction Factor (DDREF), 196
dose quantities,
    absorbed dose, 125
    Ambient dose equivalent:, 128
    CRU, 125

directional dose equivalent, 129
dose equivalent, 128
dose equivalent index, 128
effective dose, 130
effective dose equivalent, 130, 312
equivalent dose, 130
ICRU, 125
Individual dose equivalent, penetrating,, 128
Individual dose equivalent, superficial,, 129
Parker, Herb, 126
personal dose equivalent, 129
quality factor, 127
radiation weighting factor, 130
RBE dose, 127
tissue weighting factor, 130
doubling dose, 144
Dounreay, 64, 198 Dragon Experiment, 323
Drosophila melanogaster, 138, 140, 155
Duane, William, 215
Duncan and Howell97, 202
Dunning, J R, 225
Eastman, 7
Ebert, H, 86
EBR-1 and 2 reactors, 60
Eccles and Jordan49, 225
Eiffel Tower, 87
electrometers, 214
Elgin State Hospital, 259
Elkind and Sutton, 112
Elster, J, 86
Elugelab, 54
Emergency Core Cooling System, 78, 79, 290, 291, 354
Emu Fields, 56
emulsions for fast neutron dosimetry, 237

Energy Research and Development Administration (ERDA), 291
Enewatak, 54
Environment Agency, 282
Environment and Heritage Service, 282
Environmental Protection Agency (EPA), 288
Ernst Report, 361
EURATOM Directives, 275
European Atomic Energy Community (EURATOM), 274
Evans, Robley D, 189, 225, 254
Event Tree Analysis, 362 evolution of a criticality accident, 340
Ewers, P, 86
f-N curves, 368
Failla, G, 259, 302, 303
Fajans, K, 32
Farabee, W C, 144
Farmer curves, 376
Farmer, F R, 75, 371
Fat Man, 50
Fault Tree Analysis, 362
Federal Radiation Council, 286
Fermi, Enrico, 40
    moderation, 42
Fibiger, J, 169
film badges, 232
Finite Locus Threshold Model (FLTM), 161
FISH-painting, 115
Fisheries Research Laboratory, 282
Five Series study, put in place, 200
Fluoroscope, 8
fluoroscopy, 92
FMEA, 361
FMECA, 361
Focus tube, 5
Folley, Borges and Yamawaki, 181

Forsmark, 67
Four Factor Formula, 318
fractional crystallisation, 28
Francis Committee, 182
Franke, Heinrich, 233
Freund, Leopold, 11
Frisch-Peierls Memorandum, 43
Frisch, Otto,
   bomb unlikely, 43
   interpretation of fission, 41
   kicksorter, 226
   U-235 critical mass about 600 gm, 43
Frost, Edwin, 9
Fryer, Grace, 188
G2 Marcoule reactor, 62
Gardner, Martin, 198
Garrod, Archibald, 144
gas amplification, 218 gas tubes, 6, 8
Geiger-Muller(G-M) tubes, 219
Geiger, Hans(J H W),
   alpha ranges, 31 Geisel, F, 33
Geitel, Hans, 86
GEM criticality code, 330
German Röntgen Society, 300
Glasser and Fricke2, 215
GLEEP, 52
Goodhead, D T, 111, 113
Goodspeed, A W, 3
Gowing, Margaret, 50
Gray, L H, 109, 123
Greulich, W W, 181
Gross, L, 170
Grubbé, E H, 11
Guilleminot, H, 18
HABOG, 66
Haeckel, Ernst, 137
Hahn, Otto, 40
Hall-Edwards, J,
   first clinical x-ray, 3
   his ten rules, 22
Hanford, 48, 203 Haque, A K M M, 89 Hardtack I tests, 200 Harnack, Ernest, 12, 16 Hart and Boag, 117 Harvey, E B, 187 Harwell, 51
Hawks, Herbert, 14
Health and Safety Executive (HSE), 281
Health Physics, 296
Health Physics Society, 295
HeLa cells, 109
Hertwiga, Oskar, 137
hesperium, 40
Hess, Victor, 87, 255
Himsworth, Harold, 55
Hinton, Christopher, 52
Hiroshima, 50, 179, 184
hit theory, 108
HMS Plym, 53
Hofstadter, R, 222
Holzknecht, G, 19
hormesis, 206
Hounsfield, Godfrey, 94
House Joint Resolution 87, 66
Howard and Pelc, 109
Huebner and Todaro, 174 Hultqvist, B, 88
HUMCO, 256
Hurd Deep, 67
Ichiban project, 185
ICRP,
   1985 Statement41, 311
   ALARA, 310 committees, 266
   cost-benefit analysis, 310
   Justification, Optimisation and Dose Limitation, 310

17 Index

linear no-threshold hypothesis, 309
 origin, 264
 permissible public exposure, 308
 Publication 1, 308
 Publication 22, 310
 Publication 26, 89, 309, 312
 Publication 30, 312
 Publication 37, 311
 Publication 5539, 311
 Publication 60, 312
 Publication 77, 313
 Publication 8, 309
 Publication 9, 308, 309
 Quality Factor, 308
 structure, 266
Idaho Falls, 60, 64, 77
Imperial Chemical Industries (ICI), 46
implosion device, 49
In Ecker, 56
in utero effects, 197
intensifying screen, 7
Interaction Parameter method, 325
Intermediate Power Breeder Reactor, 350
internal dosimetry,
 air sampling, 249
 Annual Limits on Intake (ALIs), 133
 body burden, 130
 body content measurement, 253
 critical organ, 131, 132
 cross-fire doses, 133
 Derived Air Concentration, 133
 excretion monitoring, 258
 germanium detectors, 258 ICRP, 131
 ICRP Publication 10, 132, 257
 ICRP Publication 2, 131
 ICRP Publication 30, 132, 133

ICRP Publication 60, 133
Lucas cell2, 251
lung counting, 258
maximum permissible amounts, 131
maximum permissible concentrations (MPCs), 132
measurement, 249
microdosimetry, 133
radon measurement, 250
Reference Man, 132
Schlundt, H, 253
Standard Man, 132
Sugar Bowl, 257
Tripartite Conference, 130
whole body monitors, 255
Working Level, 251
International Atomic Energy Agency (IAEA), 272
International Commission on Radiation Units (ICRU),
 absorbed dose, 125
 ICRU sphere, 125
 Index Quantities, 125
International Congress of Radiology(ICR), 121
 absorbed dose, 125
International Congress of Radiology, 263
 the Rontgen, 121
International Labour Organisation (ILO), 270
International Radiation Protection Association (IRPA, 296 International X-ray and Radium Protection Committee, 165, 264, 278, 300
International X-ray Units and Measurement Committee, 263
ion chambers, 217
Jackson, Herbert, 5

Jacobi, Wolfgang, 312
Janssens, F A, 138
Joliot- Curie, F and I, 40
Joliot-Curie, F, 42
Journal of Radiological Protection, 298
Journal of Radiology, 284
Journal of Roentgenology, 284
Journal of the Röntgen Society, 276
Journal of the Society for Radiological Protection, 298
k effective, 319
Kalpakkam, 65
Kappos and Pohlit 40, 112
Kara Sea, 67
Kassabian, Mihran, 16
Kaye, G W C, 264, 277, 284
Kellerer and Rossi, 110
Kelvin, Lord, 31
Kemeny Commission, 378
kerma, 125
Kerr, George, 185
Kienböck, R, 19
Kinlen, L, 198
Kleeman, R D, 30
Kneale, George, 202
Knudson and Moolgavkar25, 174
Knudson, A, 174
Koernicke, M, 139
Kowarski, Lew, 42, 50 Krasnoyarsk, 65
Kronig and Friedrich, 121
Kunz, George F, 35
Kusnetz, H L, 251
Langham, Wright, 297, 305
Lanzhou Nuclear Fuel Complex , 65
Lapp, Ralph, 55
Laurence, G C, 123
Lauritsen electroscope, 216
Lauritsen, C C, 231
laverbread, 71

Lawrence, Ernest, 47
Lea, Douglas, 114, 116
Leake, J, 230
Lennard, C L, 121 Lewis Committee, 369 Lewis, E B, 178
liability and multifactorial inheritance, 157
Life Span Study, 182
Limerick Nuclear Power Station, 89
Limiting Surface Density method, 324
limits and standards,
   body burden, 306
   critical organ, 306 critical tissue, 307
   early, 300
   erythema doses, 301 ICRP, 306
   lead shielding, 302
   permissible intakes, 304
   plutonium, 305
   radium gamma-rays, 302
   Stone, Robert, 305
   tolerance dose, 301
   tolerance level, 303, 304
   US Advisory Committee, 303
linear-quadratic formula, 111
liquid scintillators , 223
liquid wastes, 68
Lisco, Herman, 132
Lister, Joseph , 14
lithium fluoride, 240
Little Boy, 50
Lodge, Oliver, 5
London Hospital, The, 12
Los Alamos Primer, 49
Loss of Coolant Accident, 353
Luckey, T D, 207
Lucky Dragon, 54
Lysenko, T D, 142

MacMahon, B, 187
Magnox, 62 Malden Island, 56
mammography examinations, 92
Mancuso, T F, 202
Manhattan Project, 180
   air sampling, 249
   and AEC, 285
   dose limit, 303
   film badge, 233
   film badges, 233
   Groves, Leslie R , 47
   Manhattan Engineer District, 47
   Mavor, J W, 140
   neutron measurement, 228
   rep, 124
   separation methods, 47
   the French, 50
   tolerance level, 305
   worker followup, 202
Maralinga, 56
March, H C, 178
Marinelli, L D, 256
Marsden, Ernest, 32
Marshall Islanders, 54
Martland, Harrison S, 188, 253
martyr memorial, 17
Mass chest x-ray screening, 92
MAUD,
   committee set up, 43
   handover of report to USA, 46
Maximum Credible Accident, 351
Mayak, 65, 205
McCombs and McCombs, 171
Medical Research Council,
   Hazards to Man of Nuclear and Allied Radiations, 55, 156, 178, 279
   Hazards to Man...2nd Report, 279
Medical Research Council (MRC), 278
Medical Research Council committee, 55
Meitner, Lise, 40
Mendel, Gregor, 135
Metallurgical Laboratory, 47
Milham, Samuel, 202
Ministry of Agriculture, Fisheries and Food (MAFF), 282 Minometer, 230
Moloney and Kastenbaum, 181
MONK criticality code, 330
Monte Bello Islands, 53
Monte Carlo methods, 328 Montreal project, 50, 51
Morgan, K Z, 228, 296
Morgan, T H, 138
Morris, IL reprocessing plant , 64
Moseley, Henry, 32
Muller, H J, 140, 155, 172
multichannel analyser, 226
multifactorial disease, 158
multifactorial inheritance, 157
multigroup diffusion codes, 327
mutation, 138
Mutation Component, 161
Mutation Component (MC), 151, 160
mutations, 140
Mutscheller, A, 166, 301
MVK model, 174
Nagaoka, Hantaro, 31
Nagasaki, 50, 179, 183, 184
National Council on Radiation Protection and Measurement (NCRP), 265
National Academy of Sciences, 294
National Bureau of Standards(NBS, 122

National Committee on Radiation Protection, 285
National Council on Radiation Protection and Measurements (NCRP), 185, 265, 266, 286-288, 307
National Physical Laboratory (NPL), 122, 233
National Radiological Protection Board, 76, 283
National Registry for Radiation Workers, 203
Neary, G J, 166
Neel, J V, 180
neutron detectors, 227
Noddack, Ida, 41
Noiré, H, 19
Nordling, C, 172
Novaya Zemlya, 57
NRX, 51
Nuclear Energy Agency, 269
Nuclear Installations Inspectorate, 282
nuclear model, 32
Nuclear Regulatory Commission (NRC, 291
Oak Ridge,
   Clinton, 48
   electromagnetic separation plant Y-12, 48
   K-25, the gaseous diffusion plant, 48
   PUREX process, 63
   S-50, the liquid thermal diffusion plant, 48
   site X, 48
Obninsk reactor, 62
occupational exposures, 72
Ohu, Gensaku, 182
oncogene, 175 Oppenheimer, J Robert, 47
optically-stimulated luminescence, 243
Orr, J S, 112
Oskarshamm, 66
Osteoscope, 18
Painter, T S, 155
pangenesis theory, 135
Parker, Herb, 124, 126, 127, 166, 228
Partial Test Ban Treaty, 56 Pasquill, Frank, 366
pastilles, 19
Pathology Study, 183
Pearl Harbour, 46
Peierls, Rudolf,
   critical mass, 43
   Memorandum, 43
   U-235 critical mass about 600 gm, 43
Penney, William, 52, 179
permissible dose, 166
Perrin, Francis and critical mass, 43
Personal Air Samplers, 252 Perthes, 139
Peteosthor, 191
Pfahler, G, 232
Pfahler, George E, 22
Phillips, C E S, 36, 121
photomultiplier tubes, 222
plastic scintillators, 223
Pliotron, 216
plutonium, 260
   air monitoring, 252
   and bomb, 46
   and the bomb, 46
   discovery, 45
   epidemiology, 205
   hormesis, 207
   implosion, 49
   July 1941 report, 45
   lung counting, 258

17 Index

tolerance level, 305
polonium, discovery, 28
Polynesia, weapons test effects, 57
Porphyra umbilicalis, 71
potential exposures, 312
Potential Recoverability Correction Factor, PRCF, 152
Potter, Hollis E, 7
Powers, E L, 103, 112
Price-Anderson Act, 290, 358
Pripyat, 81
Probabilistic Risk Assessment (PRA, 358
Project Y, 47
proportional counter, 218
Puck and Marcus, 109
Pupin, Michael, 7, 11, 14
PUREX, 63
Quebec Agreement, 51
quenching of G-M tubes, 219
Quimby, Edith, 232
Radiation Effects Research Foundation (RERF), 183
Radiation units,
   D-unit , 122
   dose equivalent indexes, 128
   e-unit, 121
   equivalent rontgen, 124 French R-unit , 121
   German Röntgen, 121
   gram rontgen, 124
   Gray's energy unit, 124
   lmc, 122
   Mallet and Proust, 122
   nominal rontgen, 124
   R-unit, 121
   reb , 127
   rem, 126
   rep, 124
   rhegma, 124
   Röntgen (r), 121
   rontgen equivalent, 124
   Sievert, 122
   Sievert , 128
   tissue rontgen, 124
radioactivity,
   of the earth, 86
Radiological Protection Service, 233
Radiological Society of North America, 284
radiologists, 178
radiophotoluminescence, 235, 243
Radithor, 192
radium,
   atomic weight, 29
   Debierne, A-L , 29
   discovery, 28
   improving supply, 33
   injuries from, 37
   leukaemia, 37
   milligram-hour, 36
   protective measures, 38
   unit of measurement, 35
Radon, 88
Rajchman and Snyder, 221
Randall and Wilkins, 240
Rasmussen, Norman, 291, 293, 363
ratemeter, 225
RBMK reactor series, 62
reactor capacity evolution, 63 reactor dumping by USSR, 67 reactor pressure vessel failure, 357
Reactor Safeguard Committee, 349
Reactor Safeguards Advisory Committee, 350
Reactor Safety Study, 363 Regener, V, 225
Reggan, 56
relative risk model, 193
reliability engineering, 361
Repair–Misrepair (RMR) model, 113

Resnick, M A, 115
retinoblastoma, 174
retroviruses, 174
Revell, S H, 114
Reynolds, Earl, 181
Ringhals, 70
Risley, 52
Rock Carling, E, 265
Rokkasho, 65 Rollins, William,
    film dosimeter, 232
    radium therapy, 34
    x-ray precautions, 20
Rongelap, 54
Rongerik, 54
Röntgen Society, 275, 276
Röntgen, Wilhelm Conrad, 1-6, 8, 12, 13, 15, 18, 20
Rossi, H H, 110, 185
Rous, Peyton, 169
Rowland, Sydney, 4
RT-1 plant, 65
RT-2, 65
Rundo, John, 256
Russ, Sydney, 276
Russell, W L (Bill), 142 Rutherford, Ernest,
    alpha and beta rays, 9, 29-33, 86, 218, 219
    alpha velocity and e/m, 29
    alphas are helium atoms, 30
    random transmutation, 9, 29-33, 86, 218, 219
    velocities of beta rays, 30
Sabouraud, R, 19
Salvioni, Enrico, 8
Sankaranarayanan and Chakraborty37, 152
Sankaranarayanan, K, 151, 152, 160
Sarahan and Roberts, 198 Savannah River, 63

Sax, K, 114
Schmidt, G and thorium, 28
Schmitt trigger, 226
Schorr and Torney, 223
Schull, W J, 180, 197
Schuster, A, 2
scintillation detector for body content., 256
scintillators, 222
Scott Russell, R, 309
Scottish Environment Protection Agency, 282
sea dumping of wastes, 66 Seascale, 198
Selection Coefficient, s, 151
Sellafield, 64, 66, 71 semiconductor detector, 224 Semipalatinsk, 57
Semiworks, 48
Sequeira, James, 34
Serber, Robert, 49, 179 Serebrovsky and Muller, 114 Serratia marcescens, 103
SETP, 68
Shippingport PWR, 61
Siddall, Ernie, 363
Sievert, Rolf, 255, 301
Simon, Francis and gaseous diffusion, 45
SIXEP, 68
Sizewell PWR, 62
Smoky5 test, 200
Society for Radiological Protection, 297
Soddy, F, 29, 32
sodium iodide scintillator, 222 Solid Angle method, 325
somatic mutation theory, 167
Sosnovy Bor reactor, 62
South of Scotland Electricity Board, 281

17 Index

Spencer and Attix, 123
Spiess, H, 191
Spinthariscope, 31, 221
Springfields, 52 Stadler, Lewis, 141 Staggs Field, 47
Stereoscopic Roentgen Pictures, 93
Sternglass, E J, 297
Stewart, Alice, 187, 202
Stone, Robert, 305
STR reactor, 61
Strassmann, Fritz, 40
Strath report, 55
Sturtevant, Arthur, 138
Surface Density method, 324 Sutton, Walter S, 137
Szilard, Leo,
  chain reaction, 42
  radiation unit, 121
Tarapur, 65
target theory, 107, 112
Taylor, Lauristen S, 215, 216, 264, 284, 285, 288, 296, 303
Teller, 351
Teller, Edward, 321, 349
The Effects of Atomic Weapons 50, 182
The Tolerability of Risk..., 380
Thermally Stimulated Electron Emission, 244 thermoluminescence, 185, 239-243
thermoluminescence dosimetry, 239, 240
Thompson, Silvanus, 4
Thomson, Elihu, 14
Thomson, George, 43
Thomson, J J, 31
Thoron, 88
Thorotrast, 190
THORP reprocessing plant, 64 Three Mile Island, 291

Three-country study, 203
Timoféeff-Ressovsky, N V, 109
Tobias, C A, 113
Tokai-mura, 65, 83
tolerance dose, 165, 166
Tomography, 93
track etch detectors, 237
Tri-State study, 187
Trident, 55
Trinity, 50
Tripartite Conference, 130-132, 166, 265, 305-307
Troch, P, 191
Trombay, 65
Tuamoto, 56
Tube Alloys, 46
Tuck, James, 49
Tyzzer, E E, 171
UKAEA, 76, 202, 203, 280, 282, 283
Ulrich, H, 178
Umbrathor, 189
UN Atomic Energy Commission (UNAEC), 271
UNSCEAR,
  chromosomes, 155
  collective dose from fuel cycle, 70
  creation, 267
  fallout model, 59
  genetic disease, 149
  genetics, 147
  multifactorial disease, 158
  multifactorial inheritance, 158
  natural doses, 87
  occupational doses, 72
  Patient dose review, 91
  Radon model, 88
  setting up, 55
UP1 PUREX plutonium recovery plan, 64

447

TAMING THE RAYS

UP3 plant, 65
Upton, A C, 194, 195
Upton69, 194
Urey, Harold, 47
US Radium Corporation, 188
USS Nautilus, 61
USS Seawolf, 61
Utirik, 54
Vallecitos prototype BWR, 61
van der Broek, A, 32
Varmus and Bishop, 175 Vezey, J J, 12
Victoreen, 215, 217, 219, 230, 231, 240
Villard, Paul,
    discovery of gamma rays, 30
    radiation unit, 19, 121
Virchow, R, 169
viruses as cause of cancer, 170
von Halban, Hans, 42, 50
von Tschermak, Erik, 135
VVER reactors, 62
VX-41, 217
Waldeyer, William, 136 Walkie Talkie, 220
Walkoff, F, 33
Walton, E T S, 40
WASH-1400, 363
WASH-1400 contents, 368
WASH-740, 359
Waste Isolation Pilot Plant, 67
waste management, 65
Watras, Stan, 89
Watson and Crick, 109 weapons testing,
    Antler series, 56
    Bravo shot, 54
    Buffalo series, 56
    Chinese, 57
    Comprehensive Test Ban Treaty, 53

Crossroads, 54
Desert Rock, 54
French, 56 Grapple series, 56 Harry shot, 54 Hurricane, 55 Indian, 58
Ivy Mike, 54
Limited Test Ban Treaty, 290
North Korean, 58
Pakistani, 58
Partial Test Ban Treaty, 53
participant exposures, 201
Russian, 57
Smoky shot, 54
Totem, 56
Webster, William, 12
West Valley NY, 64
Wickham, L, 33
Wigner energy, 74
Wilkinson, D H, 226
Williams, F H, 34
Willis, G S, 37
Wilson, C T R, 31, 86, 101
Windscale, 52, 280
Windscale Piles21, 74
Wollan, E, 234
World Health Organisation, 90
Wright, A E, 3
Wulf, Theodore, 87
Wynn-Williams, C E, 225
X-ray and Radium Protection Committee, 38, 277, 300
x-rays at war, 9
Yamigawa and Ichigawa, 170
Yankee Rowe PWR, 61
Yucca Mountain, 66
ZEEP, 51
Ziedes des Plantes, B G, 93
Zirkle, R E, 104

17 Index

TAMING THE RAYS

END **********H4

www.ingramcontent.com/pod-product-compliance
Lightning Source LLC
Chambersburg PA
CBHW070211240426
43671CB00007B/619